Vinay A.

Jan '2003

Baltimore

Joint Mathematics
Meeting

Undergraduate Texts in Mathematics

Readings in Mathematics

Editors

S. Axler
F.W. Gehring
K.A. Ribet

Springer

New York
Berlin
Heidelberg
Barcelona
Hong Kong
London
Milan
Paris
Singapore
Tokyo

Graduate Texts in Mathematics

Readings in Mathematics

Ebbinghaus/Hermes/Hirzebruch/Koecher/Mainzer/Neukirch/Prestel/Remmert: *Numbers*
Fulton/Harris: *Representation Theory: A First Course*
Murty: *Problems in Analytic Number Theory*
Remmert: *Theory of Complex Functions*
Walter: *Ordinary Differential Equations*

Undergraduate Texts in Mathematics

Readings in Mathematics

Anglin: *Mathematics: A Concise History and Philosophy*
Anglin/Lambek: *The Heritage of Thales*
Bressoud: *Second Year Calculus*
Hairer/Wanner: *Analysis by Its History*
Hämmerlin/Hoffmann: *Numerical Mathematics*
Isaac: *The Pleasures of Probability*
Laubenbacher/Pengelley: *Mathematical Expeditions: Chronicles by the Explorers*
Samuel: *Projective Geometry*
Stillwell: *Numbers and Geometry*
Toth: *Glimpses of Algebra and Geometry,* Second Edition

Gabor Toth

Glimpses of Algebra and Geometry

Second Edition

With 183 Illustrations, Including 18 in Full Color

 Springer

Gabor Toth
Department of Mathematical Sciences
Rutgers University
Camden, NJ 08102
USA
gtoth@camden.rutgers.edu

Front cover illustration: The regular compound of five tetrahedra given by the face-planes of a colored icosahedron. The circumscribed dodecahedron is also shown. Computer graphic made by the author using Geomview. *Back cover illustration*: The regular compound of five cubes inscribed in a dodecahedron. Computer graphic made by the author using *Mathematica*®.

Mathematics Subject Classification (2000): 15-01, 11-01, 51-01

Library of Congress Cataloging-in-Publication Data
Toth, Gabor, Ph.D.
    Glimpses of algebra and geometry/Gabor Toth.—2nd ed.
      p. cm. — (Undergraduate texts in mathematics. Readings in mathematics.)
    Includes bibliographical references and index.
    ISBN 0-387-95345-0 (hardcover: alk. paper)
      1. Algebra.   2. Geometry.   I. Title.   II. Series.
    QA154.3 .T68   2002
    512′.12—dc21   2001049269

Printed on acid-free paper.

Production managed by Francine McNeill; manufacturing supervised by Jeffrey Taub.
Typeset from the author's LaTeX2e files using Springer's UTM style macro by The Bartlett Press, Inc., Marietta, GA.
Printed and bound by Hamilton Printing Co., Rensselaer, NY.
Printed in the United States of America.

9 8 7 6 5 4 3 2 1

ISBN 0-387-95345-0   SPIN 10848701

Springer-Verlag   New York Berlin Heidelberg
A member of BertelsmannSpringer Science+Business Media GmbH

This book is dedicated to my students.

Preface to the
Second Edition

Since the publication of the Glimpses in 1998, I spent a considerable amount of time collecting "mathematical pearls" suitable to add to the original text. As my collection grew, it became clear that a major revision in a second edition needed to be considered. In addition, many readers of the Glimpses suggested changes, clarifications, and, above all, more examples and worked-out problems. This second edition, made possible by the ever-patient staff of Springer-Verlag New York, Inc., is the result of these efforts. Although the general plan of the book is unchanged, the abundance of topics rich in subtle connections between algebra and geometry compelled me to extend the text of the first edition considerably. Throughout the revision, I tried to do my best to avoid the inclusion of topics that involve very difficult ideas.

The major changes in the second edition are as follows:

1. An in-depth treatment of root formulas solving quadratic, cubic, and quartic equations à la van der Waerden has been given in a new section. This can be read independently or as preparation for the more advanced new material encountered toward the later parts of the text. In addition to the Bridge card symbols, the dagger † has been introduced to indicate more technical material than the average text.

2. As a natural continuation of the section on the Platonic solids, a detailed and complete classification of finite Möbius groups à la Klein has been given with the necessary background material, such as Cayley's theorem and the Riemann–Hurwitz relation.

3. One of the most spectacular developments in algebra and geometry during the late nineteenth century was Felix Klein's theory of the icosahedron and his solution of the irreducible quintic in terms of hypergeometric functions. A quick, direct, and modern approach of Klein's main result, the so-called *Normalformsatz*, has been given in a single large section. This treatment is independent of the material in the rest of the book, and is suitable for enrichment and undergraduate/graduate research projects. All known approaches to the solution of the irreducible quintic are technical; I have chosen a geometric approach based on the construction of canonical quintic resolvents of the equation of the icosahedron, since it meshes well with the treatment of the Platonic solids given in the earlier part of the text. An algebraic approach based on the reduction of the equation of the icosahedron to the Brioschi quintic by Tschirnhaus transformations is well documented in other textbooks. Another section on polynomial invariants of finite Möbius groups, and two new appendices, containing preparatory material on the hypergeometric differential equation and Galois theory, facilitate the understanding of this advanced material.

4. The text has been upgraded in many places; for example, there is more material on the congruent number problem, the stereographic projection, the Weierstrass $\wp$-function, projective spaces, and isometries in space.

5. The new Web site at http://mathsgi01.rutgers.edu/~gtoth/ Glimpses/ containing various text files (in PostScript and HTML formats) and over 70 pictures in full color (in gif format) has been created.

6. The historical background at many places of the text has been made more detailed (such as the ancient Greek approximations of π), and the historical references have been made more precise.

7. An extended solutions manual has been created containing the solutions of 100 problems.

I would like to thank the many readers who suggested improvements to the text of the first edition. These changes have all been incorporated into this second edition. I am especially indebted to Hillel Gauchman and Martin Karel, good friends and colleagues, who suggested many worthwhile changes. I would also like to express my gratitude to Yukihiro Kanie for his careful reading of the text and for his excellent translation of the first edition of the Glimpses into Japanese, published in early 2000 by Springer-Verlag, Tokyo. I am also indebted to April De Vera, who upgraded the list of Web sites in the first edition. Finally, I would like to thank Ina Lindemann, Executive Editor, Mathematics, at Springer-Verlag New York, Inc., for her enthusiasm and encouragement throughout the entire project, and for her support for this early second edition.

Camden, New Jersey   Gabor Toth

Preface to the First Edition

Glimpse: 1. a very brief passing look, sight or view. 2. a momentary or slight appearance. 3. a vague idea or inkling.

—*Random House College Dictionary*

At the beginning of fall 1995, during a conversation with my respected friend and colleague Howard Jacobowitz in the Octagon Dining Room (Rutgers University, Camden Campus), the idea emerged of a "bridge course" that would facilitate the transition between undergraduate and graduate studies. It was clear that a course like this could not concentrate on a single topic, but should browse through a number of mathematical disciplines. The selection of topics for the Glimpses thus proved to be of utmost importance. At this level, the most prominent interplay is manifested in some easily explainable, but eventually subtle, connections between number theory, classical geometries, and modern algebra. The rich, fascinating, and sometimes puzzling interactions of these mathematical disciplines are seldom contained in a medium-size undergraduate textbook. The Glimpses that follow make a humble effort to fill this gap.

The connections among the disciplines occur at various levels in the text. They are sometimes the main topics, such as Rationality and Elliptic Curves (Section 3), and are sometimes hidden in problems, such as the spherical geometric proof of diagonalization of Euclidean isometries (Problems 1 to 2, Section 16), or the proof of Euler's theorem on convex polyhedra using linear algebra (Problem 9, Section 20). Despite numerous opportunities throughout the text, the experienced reader will no doubt notice that analysis had to be left out or reduced to a minimum. In fact, a major source of difficulties in the intense 8-week period during which I produced the first version of the text was the continuous cutting down of the size of sections and the shortening of arguments. Furthermore, when one is comparing geometric and algebraic proofs, the geometric argument, though often more lengthy, is almost always more revealing and thereby preferable. To strive for some originality, I occasionally supplied proofs out of the ordinary, even at the "expense" of going into calculus a bit. To me, "bridge course" also meant trying to shed light on some of the links between the first recorded intellectual attempts to solve ancient problems of number theory, geometry, and twentieth-century mathematics. Ignoring detours and sidetracks, the careful reader will see the continuity of the lines of arguments, some of which have a time span of 3000 years. In keeping this continuity, I eventually decided not to break up the Glimpses into chapters as one usually does with a text of this size. The text is, nevertheless, broken up into subtexts corresponding to various levels of knowledge the reader possesses. I have chosen the card symbols ♣, ◇, ♡, ♠ of Bridge to indicate four levels that roughly correspond to the following:

♣ College Algebra;
◇ Calculus, Linear Algebra;
♡ Number Theory, Modern Algebra (elementary level), Geometry;
♠ Modern Algebra (advanced level), Topology, Complex Variables.

Although much of ♡ and ♠ can be skipped at first reading, I encourage the reader to challenge him/herself to venture occasionally into these territories. The book is intended for (1) students (♣ and ◇) who wish to learn that mathematics is more than a set of tools (the way sometimes calculus is taught), (2) students (♡ and ♠) who love mathematics, and (3) high-school teachers ($\subset \{\clubsuit, \diamondsuit, \heartsuit, \spadesuit\}$) who always had keen interest in mathematics but seldom time to pursue the technicalities.

Reading what I have written so far, I realize that I have to make one point clear: Skipping and reducing the size of subtle arguments have the inherent danger of putting more weight on intuition at the expense of precision. I have spent a considerable amount of time polishing intuitive arguments to the extent that the more experienced reader can make them withstand the ultimate test of mathematical rigor.

Speaking (or rather writing) of danger, another haunted me for the duration of writing the text. One of my favorite authors, Iris Murdoch, writes about this in *The Book and the Brotherhood*, in which Gerard Hernshaw is badgered by his formidable scholar Levquist about whether he wanted to write mediocre books out of great ones for the rest of his life. (To learn what Gerard's answer was, you need to read the novel.) Indeed, a number of textbooks influenced me when writing the text. Here is a sample:

1. M. Artin, *Algebra*, Prentice-Hall, 1991;
2. A. Beardon, *The Geometry of Discrete Groups*, Springer-Verlag, 1983;
3. M. Berger, *Geometry I–II*, Springer-Verlag, 1980;
4. H.S.M. Coxeter, *Introduction to Geometry*, Wiley, 1969;
5. H.S.M. Coxeter, *Regular Polytopes*, Pitman, 1947;
6. D. Hilbert and S. Cohn-Vossen, *Geometry and Imagination*, Chelsea, 1952.
7. J. Milnor, *Topology from the Differentiable Viewpoint*, The University Press of Virginia, 1990;
8. I. Niven, H. Zuckerman, and H. Montgomery, *An Introduction to the Theory of Numbers*, Wiley, 1991;
9. J. Silverman and J. Tate, *Rational Points on Elliptic Curves*, Springer-Verlag, 1992.

Although I (unavoidably) use a number of by now classical arguments from these, originality was one of my primary aims. This book was never intended for comparison; my hope is that the Glimpses may trigger enough motivation to tackle these more advanced textbooks.

Despite the intertwining nature of the text, the Glimpses contain enough material for a variety of courses. For example, a shorter version can be created by taking Sections 1 to 10 and Sections 17 and 19 to 23, with additional material from Sections 15 to 16 (treating Fuchsian groups and Riemann surfaces marginally via the examples) when needed. A nonaxiomatic treatment of an undergraduate course on geometry is contained in Sections 5 to 7, Sections 9 to 13, and Section 17.

The Glimpses contain a lot of computer graphics. The material can be taught in the traditional way using slides, or interactively in a computer lab or teaching facility equipped with a PC or a workstation connected to an LCD-panel. Alternatively, one can create a graphic library for the illustrations and make it accessible to the students. Since I have no preference for any software packages (although some of them are better than others for *particular* purposes), I used both *Maple*®[1] and *Mathematica*®[2] to create the illustrations. In a classroom setting, the link of either of these to Geomview[3] is especially useful, since it allows one to manipulate three-dimensional graphic objects. Section 17 is highly graphic, and I recommend showing the students a variety of slides or three-dimensional computer-generated images. Animated graphics can also be used, in particular, for the action of the stereographic projection in Section 7, for the symmetry group of the pyramid and the prism in Section 17, and for the cutting-and-pasting technique in Sections 16 and 19. These Maple® text files are downloadable from my Web sites

http://carp.rutgers.edu/math-undergrad/science-vision.html
and
http://mathsgi01.rutgers.edu/~gtoth/.

Alternatively, to obtain a copy, write an e-mail message to

gtoth@camden.rutgers.edu

---

[1] Maple is a registered trademark of Waterloo Maple, Inc.

[2] *Mathematica* is a registered trademark of Wolfram Research, Inc.

[3] A software package downloadable from the Web site: http://www.geom.umn.edu.

or send a formatted disk to Gabor Toth, Department of Mathematical Sciences, Rutgers University, Camden, NJ 08102, USA.

A great deal of information, interactive graphics, animations, etc., are available on the World Wide Web. I highly recommend scheduling at least one visit to a computer or workstation lab and explaining to the students how to use the Web. In fact, at the first implementation of the Glimpses at Rutgers, I noticed that my students started spending more and more time at various Web sites related to the text. For this reason, I have included a list of recommended Web sites and films at the end of some sections. Although hundreds of Web sites are created, upgraded, and terminated daily, every effort has been made to list the latest Web sites currently available through the Interent.

Camden, New Jersey   Gabor Toth

Acknowledgments

The second half of Section 20 on the four color theorem was written by Joseph Gerver, a colleague at Rutgers. I am greatly indebted to him for his contribution and for sharing his insight into graph theory. The first trial run of the Glimpses at Rutgers was during the first six weeks of summer 1996, with an equal number of undergraduate and graduate students in the audience. In fall 1996, I also taught undergraduate geometry from the Glimpses, covering Sections 1 to 10 and Sections 17 and 19 to 23. As a result of the students' dedicated work, the original manuscript has been revised and corrected, some of the arguments have been polished, and some extra topics have been added. It is my pleasure to thank all of them for their participation, enthusiasm, and hard work. I am particularly indebted to Jack Fistori, a mathematics education senior at Rutgers, who carefully revised the final version of the manuscript, making numerous worthwhile changes. I am also indebted to Susan Carter, a graduate student at Rutgers, who spent innumerable hours at the workstation to locate suitable Web sites related to the Glimpses. In summer 1996, I visited the Geometry Center at the University of Minnesota. I lectured about the Glimpses to an audience consisting of undergraduate and graduate students and high-school teachers. I wish to thank them for their valuable comments, which I took into account in the final version of the manuscript. I am especially indebted to Harvey Keynes, Education Director of the Geometry Center, for his enthusiastic support of the Glimpses. During my stay, I produced a 10-minute film *Glimpses of the Five Platonic Solids* with Stuart Levy, whose dedication to the project surpassed all my expectations. The typesetting of the manuscript started when I gave the first 20 pages to Betty Zubert as material with which to practice LaTeX. As the manuscript grew beyond any reasonable size, it is my pleasure to record my thanks to her for providing inexhaustible energy that turned 300 pages of chicken scratch into a fine document.

Camden, New Jersey   Gabor Toth

Contents

"A Number Is a Multitude Composed of Units" — Euclid

SECTION 1

♣ We adopt Kronecker's phrase: "God created the natural numbers, and all the rest is the work of man," and start with the set

$$\mathbf{N} = \{1, 2, 3, 4, 5, 6, \ldots\}$$

of all *natural numbers*. Since the sum of two natural numbers is again a natural number, $\mathbf{N}$ carries the operation[1] of addition $+ : \mathbf{N} \times \mathbf{N} \to \mathbf{N}$.

Remark.
Depicting natural numbers by arabic numerals is purely traditional. Romans might prefer

$$\mathbf{N} = \{\text{I, II, III, IV, V, VI}, \ldots\},$$

and computers work with

$$\mathbf{N} = \{1, 10, 11, 100, 101, 110, \ldots\}.$$

Notice that converting a notation into another is nothing but an *isomorphism* between the respective systems. Isomorphism respects

---

[1]If needed, please review "Sets" and "Groups" in Appendices A and B.

addition; for example, $29 + 33 = 62$ is the same as XXIX + XXXIII = LXII or $11101 + 100001 = 111110$.

From the point of view of group theory, $\mathbf{N}$ is a failure; it does not have an identity element (that we would like to call zero) and no element has an inverse. We remedy this by extending $\mathbf{N}$ to the (additive) *group of integers*

$$\mathbf{Z} = \{0, \pm 1, \pm 2, \pm 3, \pm 4, \pm 5, \pm 6, \ldots\}.$$

$\mathbf{Z}$ also carries the operation of multiplication $\times : \mathbf{Z} \times \mathbf{Z} \to \mathbf{Z}$. Since distributivity holds, $\mathbf{Z}$ forms a *ring* with respect to addition and multiplication.

Although we have 1 as the identity element with respect to $\times$, we have no hope for $\mathbf{Z}$ to be a multiplicative group; remember the saying: "Thou shalt not divide by zero!" To remedy this, we delete the ominous zero and consider

$$\mathbf{Z}^{\#} = \mathbf{Z} - \{0\} = \{\pm 1, \pm 2, \pm 3, \pm 4, \pm 5, \pm 6, \ldots\}.$$

The requirement that integers have inverses gives rise to fractions or, more appropriately, *rational numbers*:

$$\mathbf{Q} = \mathbf{Q}^{\#} \cup \{0\} = \{a/b \mid a, b \in \mathbf{Z}^{\#}\} \cup \{0\},$$

where we put the zero back to save the additive group structure. All that we learned in dealing with fractions can be rephrased elegantly by saying that $\mathbf{Q}$ is a *field*: $\mathbf{Q}$ is an additive group, $\mathbf{Q}^{\#}$ is an abelian (i.e., commutative) multiplicative group, and addition and multiplication are connected through distributivity.

After having created $\mathbf{Z}$ and $\mathbf{Q}$, the direction we take depends largely on what we wish to study. In elementary number theory, when studying divisibility properties of integers, we consider, for a given $n \in \mathbf{N}$, the (additive) group $\mathbf{Z}_n$ of *integers modulo n*. The simplest way to understand

$$\mathbf{Z}_n = \mathbf{Z}/n\mathbf{Z} = \{[0], [1], \ldots, [n-1]\}$$

is to start with $\mathbf{Z}$ and to *identify* two integers a and b if they differ by a multiple of n. This identification is indicated by the square bracket; $[a]$ means a plus all multiples of n. Clearly, no numbers are identified among $0, 1, \ldots, n-1$, and any integer is identified

Figure 1.1

with exactly one of these. The (additive) group structure is given by the usual addition in $\mathbf{Z}$. More explicitly, $[a]+[b] = [a+b]$, $a, b \in \mathbf{Z}$. Clearly, $[0]$ is the zero element in $\mathbf{Z}_n$, and $-[a] = [-a]$ is the additive inverse of $[a] \in \mathbf{Z}_n$. Arithmetically, we use the division algorithm to find the quotient q and the remainder $0 \leq r < n$, when $a \in \mathbf{Z}$ is divided by n:

$$a = qn + r,$$

and set $[a] = [r]$. The geometry behind this equality is clear. Consider the multiples of n, $n\mathbf{Z} \subset \mathbf{Z}$, as a *one-dimensional lattice* (i.e., an infinite string of equidistantly spaced points) in $\mathbf{R}$ as in Figure 1.1. Now locate a and its closest left neighbor qn in $n\mathbf{Z}$ (Figure 1.2). The distance between qn and a is r, the latter between 0 and $n-1$. Since a and r are to be identified, the following geometric picture emerges for $\mathbf{Z}_n$: Wrap $\mathbf{Z}$ around a circle infinitely many times so that the points that overlap with 0 are exactly the lattice points in $n\mathbf{Z}$; this can be achieved easily by choosing the radius of the circle to be $n/2\pi$. Thus, $\mathbf{Z}_n$ can be visualized as n equidistant points on the perimeter of a circle (Figure 1.3). Setting the center of the circle at the origin of a coordinate system on the Cartesian plane $\mathbf{R}^2$ such that $[0]$ is the intersection point of the circle and the positive first axis, we see that addition in $\mathbf{Z}_n$ corresponds to *addition of angles of the corresponding vectors*. A common convention is to choose the positive orientation as the way $[0], [1], [2], \ldots$ increase. This picture of $\mathbf{Z}_n$ as the vertices of a *regular n-sided polygon* (with angular addition) will recur later on in several different contexts.

Figure 1.2

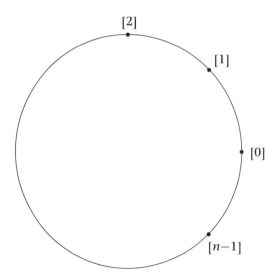

Figure 1.3

Remark.

In case you've ever wondered why it was so hard to learn the clock in childhood, consider $\mathbf{Z}_{60}$. Why the Babylonian choice[2] of 60? Consider natural numbers between 1 and 100 that have the largest possible number of small divisors.

♡ The infinite $\mathbf{Z}$ and its finite offsprings $\mathbf{Z}_n$, $n \in \mathbf{N}$, share the basic property that they are generated by a single element, a property that we express by saying that $\mathbf{Z}$ and $\mathbf{Z}_n$ are *cyclic*. In case of $\mathbf{Z}$, this element is 1 or -1; in case of $\mathbf{Z}_n$, a generator is [1].

♣ You might be wondering whether it is a good idea to reconsider multiplication in $\mathbf{Z}_n$ induced from that of $\mathbf{Z}$. The answer is yes; multiplication in $\mathbf{Z}$ gives rise to a well-defined multiplication in $\mathbf{Z}_n$ by setting $[a] \cdot [b] = [ab]$, $a, b \in \mathbf{Z}$. Clearly, [1] is the multiplicative identity element. Consider now multiplication restricted to $\mathbf{Z}_n^{\#} = \mathbf{Z}_n - \{[0]\}$. There is a serious problem here. If n is composite, that is, $n = ab$, $a, b \in \mathbf{N}$, $a, b \geq 2$, then $[a], [b] \in \mathbf{Z}_n^{\#}$, but $[a] \cdot [b] = [0]$! Thus, multiplication restricted to $\mathbf{Z}_n^{\#}$ is not even an operation.

---

[2]Actually, a number system using 60 as a base was developed by the Sumerians about 500 years before it was passed on to the Babylonians around 2000 B.C.

We now pin our hopes on $\mathbf{Z}_p$, where p a prime. Elementary number theory says that if p divides ab, then p divides either a or b. This directly translates into the fact that $\mathbf{Z}_p^{\#}$ is closed under multiplication. Encouraged by this, we now go a step further and claim that $\mathbf{Z}_p^{\#}$ is a multiplicative group! Since associativity follows from associativity of multiplication in $\mathbf{Z}$, it remains to show that each element $a \in \mathbf{Z}_p^{\#}$ has a multiplicative inverse. To prove this, multiply the complete list $[1], [2], \ldots, [p-1]$ by $[a]$ to obtain

$$[a], [2a], \ldots, [(p-1)a].$$

By the above, these all belong to $\mathbf{Z}_p^{\#}$. They are mutually disjoint. Indeed, assume that $[ka] = [la], k, l = 1, 2, \ldots, p-1$. We then have $[(k-l)a] = [k-l] \cdot [a] = [0]$, so that $k = l$ follows. Thus, the list above gives $p-1$ elements of $\mathbf{Z}_p^{\#}$. But the latter consists of exactly $p-1$ elements, so we got them all! In particular, $[1]$ is somewhere in this list, say, $[\bar{a}a] = [1], \bar{a} = 1, \ldots, p-1$. Hence, $[\bar{a}]$ is the multiplicative inverse of $[a]$. Finally, since distributivity in $\mathbf{Z}_p$ follows from distributivity in $\mathbf{Z}$, we obtain that $\mathbf{Z}_p$ is a field for p prime.

We give two applications of these ideas: one for $\mathbf{Z}_3$ and another for $\mathbf{Z}_4$. First, we claim that if 3 divides $a^2 + b^2$, $a, b \in \mathbf{Z}$, then 3 divides both a and b. Since divisibility means zero remainder, all we have to count is the sum of the remainders when a^2 and b^2 are divided by 3. In much the same way as we divided all integers to even $(2k)$ and odd $(2k+1)$ numbers $(k \in \mathbf{Z})$, we now write $a = 3k, 3k+1, 3k+2$ accordingly. Squaring, we obtain $a^2 = 9k^2, 9k^2 + 6k+1, 9k^2 + 12k+4$. Divided by 3, these give remainders 0 or 1, with 0 corresponding to a being a multiple of 3. The situation is the same for b^2. We see that when dividing $a^2 + b^2$ by 3, the possible remainders are $0+0, 0+1, 1+0, 1+1$, and the first corresponds to a and b both being multiples of 3. The first claim follows.

Second, we show the important number theoretical fact that no number of the form $4m + 3$ is a sum of two squares of integers. (Notice that, for $m = 0$, this follows from the first claim or by inspection.) This time we study the remainder when $a^2 + b^2$, $a, b \in \mathbf{Z}$, is divided by 4. Setting $a = 4k, 4k+1, 4k+2, 4k+3$, a^2 gives remainders 0 or 1. As before, the possible remainders for $a^2 + b^2$ are $0+0, 0+1, 1+0, 1+1$. The second claim also follows.

♡ The obvious common generalization of the two claims above is also true and is a standard fact in number theory. It asserts that if a prime p of the form $4m + 3$ divides $a^2 + b^2$, $a, b \in \mathbf{Z}$, then p divides both a and b. Aside from the obvious decomposition $2 = 1^2 + 1^2$, the primes that are left out from our considerations are of the form $4m + 1$. A deeper result in number theory states that any prime of the form $4m + 1$ is always representable as a sum of squares of two integers. Fermat, in a letter to Mersenne in 1640, claimed to have a proof of this result, which was first stated by Albert Girard in 1632. The first published verification, due to Euler, appeared in 1754. We postpone the proof of this result till the end of Section 5.

Problems

1. Use the division algorithm to show that (a) the square of an integer is of the form $3a$ or $3a + 1$, $a \in \mathbf{Z}$; (b) the cube of an integer is of the form $7a$, $7a + 1$ or $7a - 1$, $a \in \mathbf{Z}$.

2. Prove that if an integer is simultaneously a square and a cube, then it must be of the form $7a$ or $7a + 1$. (Example: $8^2 = 4^3$.)

3. (a) Show that $[2]$ has no inverse in $\mathbf{Z}_4$. (b) Find all $n \in \mathbf{N}$ such that $[2]$ has an inverse in $\mathbf{Z}_n$.

4. Write $a \in \mathbf{N}$ in decimal digits as $a = a_1 a_2 \cdots a_n$, $a_1, a_2, \ldots, a_n \in \{0, 1, \ldots, 9\}$, $a_1 \neq 0$. Prove that $[a] = [a_1 + a_2 + \cdots + a_n]$ in $\mathbf{Z}_9$.

5. Let $p > 3$ be a prime, and write

$$\frac{1}{1^2} + \frac{1}{2^2} + \cdots + \frac{1}{(p-1)^2}$$

as a rational number a/b, where a, b are relatively prime. Show that $p | a$.

Web Sites

1. www.utm.edu/research/primes

2. daisy.uwaterloo.ca/~alopez-o/math-faq/node10.html

2

SECTION

"... There Are No Irrational Numbers at All"—Kronecker

♣ Although frequently quoted, the epigraph to this chapter and some other statements of Kronecker on irrational numbers have been shown to be distortions.[1] (See H.M. Edwards's articles on Kronecker in *History and Philosophy of Modern Mathematics* (Minneapolis, MN, 1985) 139–144, Minnesota Stud. Philos. Sci., XI, Univ. Minnesota Press, Minneapolis, MN, 1988.)

In calculus, the field of rational numbers **Q** is insufficient for several reasons, including convergence. Thus, **Q** is extended to the *field of real numbers* **R**. It is visualized by passing the Cartesian Bridge[2] connecting algebra and geometry; each real number (in infinite decimal representation) corresponds to a single point on the real line. Do we actually get new numbers? Here is a simple answer known to Pythagoras (c. 570–490 B.C.):

---

[1] I am indebted to Victor Pambuccian for pointing this out.

[2] Cartesius is the Latinized name of René Descartes, who, contrary to widespread belief, did not invent the coordinate axes, much less analytic geometry. Here we push this gossip a little further. Note that analytic geometry was born in Fermat's *Introduction to Plane and Solid Loci* in 1629; although circulated from 1637 on, it was not published in Fermat's lifetime. The notion of perpendicular coordinate axes can be traced back to Archimedes and Apollonius. Both Descartes and Fermat used coordinates but only with nonnegative values; the idea that coordinates can also take negative values is due to Newton.

Proposition 1.

$\sqrt{2}$ (*the unique positive number whose square is* 2) *is not rational.*

Proof.

¬[3] Assume that $\sqrt{2}$ is rational; i.e., $\sqrt{2} = a/b$ for some $a, b \in \mathbf{Z}$. We may assume that a and b are relatively prime, since otherwise we cancel the common factors in a and b. Squaring, we get $a^2 = 2b^2$. A glimpse of the right-hand side shows that a^2 is even. Thus a must be even, say, $a = 2c$. Then $a^2 = 4c^2 = 2b^2$. Hence b^2 and b must be even. Thus 2 is a common factor of a and b. ¬   ∎

Remark.

Replacing 2 by any prime p, we see that $\sqrt{p}$ is not rational. Instead of repeating the argument above, we now describe a more powerful result due to Gauss. (If you study the following proof carefully, you will see that it is a generalization of the Pythagorean argument above.) The idea is very simple (and will reoccur later) and is based on the fact that $\sqrt{p}$ is a solution of the quadratic equation $x^2 - p = 0$. To show irrationality of $\sqrt{p}$, we study the rational solutions of this equation. More generally, assume that the polynomial equation

$$P(x) = c_0 + c_1 x + \cdots + c_n x^n = 0, \quad c_n \neq 0,$$

with integer coefficients $c_0, c_1, \ldots, c_n \in \mathbf{Z}$, has a rational root $x = a/b$, $a, b \in \mathbf{Z}$. As usual, we may assume that a and b are relatively prime. Substituting, we have

$$c_0 + c_1(a/b) + \cdots + c_n(a/b)^n = 0.$$

Multiplying through by b^{n-1}, we obtain

$$c_0 b^{n-1} + c_1 a b^{n-2} + \cdots + c_{n-1} a^{n-1} + c_n a^n/b = 0.$$

This says that $c_n a^n/b$ must be an integer, or equivalently, b divides $c_n a^n$. Since a and b are relatively prime, we conclude that b *divides*

---

[3] ¬ indicates indirect argument; that is, we assume that the statement is false and get (eventually) a contradiction (indicated by another ¬).

c_n. In a similar vein, if we multiply through by b^n/a, we get

$$c_0 b^n/a + c_1 b^{n-1} + \cdots + c_n a^{n-1} = 0,$$

and it follows that *a divides c_0*.

Specializing, we see that if $x = a/b$ is a solution of $x^n = c$, then $b = \pm 1$ and so $c = (\pm a)^n$. Thus, if c is not the nth power of an integer, then $x^n = c$ does not have any rational solutions. This is indeed a vast generalization of the Pythagorean argument above!

Algebraically, we think of a real number as an infinite decimal. This decimal representation is unique (and thereby serves as a definition), assuming that we exclude representations terminating in a string of infinitely many 9's; for example, instead of $1.2999\ldots$ we write 1.3. How can we recognize the rational numbers in this representation? Writing $1/3$ as $0.333\ldots$ gives a clue to the following:

Proposition 2.

An infinite decimal represents a rational number iff it terminates or repeats.

Proof.

We first need to evaluate the finite geometric sum

$$1 + x + x^2 + \cdots + x^n.$$

For $x = 1$, this is $n+1$, hence we may assume that $x \neq 1$. This sum is telescopic; in fact, after multiplying through by $1 - x$, everything cancels except the first and last term. We obtain

$$(1 - x)(1 + x + x^2 + \cdots + x^n) = 1 - x^{n+1},$$

or equivalently

$$1 + x + x^2 + \cdots + x^n = \frac{1 - x^{n+1}}{1 - x}, \qquad x \neq 1.$$

Letting $n \to \infty$ and assuming that $|x| < 1$ to assure convergence, we arrive at the geometric series formula[4]

$$1 + x + x^2 \cdots = \frac{1}{1-x}, \quad |x| < 1.$$

We now turn to the proof. Since terminating decimals are clearly rationals, and after multiplying through by a power of 10 if necessary, we may assume that our repeating decimal representation looks like

$$0.a_1 a_2 \cdots a_k a_1 a_2 \cdots a_k a_1 a_2 \cdots a_k \cdots,$$

where the decimal digits a_i are between 0 and 9. We rewrite this as

$$a_1 a_2 \cdots a_k (10^{-k} + 10^{-2k} + 10^{-3k} + \cdots)$$
$$= a_1 \cdots a_k 10^{-k} (1 + 10^{-k} + (10^{-k})^2 + \cdots)$$
$$= \frac{a_1 \cdots a_k 10^{-k}}{1 - 10^{-k}} = \frac{a_1 \cdots a_k}{10^k - 1}$$

where we used the geometric series formula. The number we arrive at is clearly rational, and we are done. The converse statement follows from the division algorithm. Indeed, if a, b are integers and a is divided by b, then each decimal in the decimal representation of a/b is obtained by multiplying the remainder of the previous step by 10 and dividing it by b to get the new remainder. All remainders are between 0 and $b - 1$, so the process necessarily repeats itself. ∎

Irrational numbers emerge quite naturally. The two most prominent examples are

$$\pi = \text{half of the perimeter of the unit circle}$$

and

$$e = 1 + \frac{1}{1!} + \frac{1}{2!} + \frac{1}{3!} + \cdots.$$

---

[4]A special case, known as one of Zeno's paradoxes, can be explained to a first grader as follows: Stay 2 yards away from the wall. The goal is to reach the wall in infinite steps, in each step traversing half of the distance made in the previous step. Thus, in the first step you cover 1 yard, in the second 1/2, etc. You see that $1 + 1/2 + 1/2^2 + \cdots = 2 = 1/(1 - 1/2)$.

e is also the principal and interest of $1 in continuous compounding after 1 year with 100% interest rate:

$$e = \lim_{n \to \infty} \left(1 + \frac{1}{n}\right)^n.$$

The equivalence of the two definitions of e is usually proved[5] in calculus.

◇ To prove that π and e are irrational, we follow Hermite's argument, which dates back to 1873. Consider, for fixed $n \in \mathbf{N}$, the function $f : \mathbf{R} \to \mathbf{R}$ defined by

$$f(x) = \frac{x^n(1-x)^n}{n!} = \frac{1}{n!} \sum_{k=n}^{2n} c_k x^k.$$

Expanding $(1-x)^n$, we see that $c_k \in \mathbf{Z}$, $k = n, n+1, \ldots, 2n$. In fact, $c_k = (-1)^k \binom{n}{n-k} = (-1)^k \binom{n}{k}$ (by the binomial formula), but we will not need this. For $0 < x < 1$, we have

$$0 < f(x) < \frac{1}{n!}$$

and $f(0) = 0$. Differentiating, we obtain

$$f^{(m)}(0) = \begin{cases} 0, & \text{if } m < n \text{ or } m > 2n \\ \dfrac{c_m m!}{n!}, & \text{if } n \leq m \leq 2n. \end{cases}$$

In any case $f^{(m)}(0) \in \mathbf{Z}$ (since $n!$ divides $m!$ for $n \leq m$). Thus, f and all its derivatives have integral values at $x = 0$. Since $f(x) = f(1-x)$, the same is true for f at $x = 1$.

Theorem 1.
  e^k is irrational for all $k \in \mathbf{N}$.

Proof.
¬ Assume that $e^k = a/b$ for some $a, b \in \mathbf{Z}$. Define $F : \mathbf{R} \to \mathbf{R}$ by

$$F(x) = k^{2n}f(x) - k^{2n-1}f'(x) + k^{2n-2}f''(x) + \cdots - kf^{(2n-1)}(x) + f^{(2n)}(x).$$

[5]For a direct proof based on the binomial formula (with integral coefficients), see G.H. Hardy, *A Course of Pure Mathematics*, Cambridge University Press, 1960.

The sum $F'(x) + kF(x)$ is telescopic; in fact, we have

$$F' + kF(x) = k^{2n+1}f(x).$$

We now use a standard trick in differential equations (due to Bernoulli) and multiply both sides by the integrating factor e^{kx} to obtain

$$[e^{kx}F(x)]' = e^{kx}[F'(x) + kF(x)] = k^{2n+1}e^{kx}f(x).$$

Integrating both sides between 0 and 1:

$$k^{2n+1}\int_0^1 e^{kx}f(x)\,dx = [e^{kx}F(x)]_0^1 = e^k F(1) - F(0)$$

$$= (1/b)(aF(1) - bF(0)).$$

In other words,

$$bk^{2n+1}\int_0^1 e^{kx}f(x)\,dx = aF(1) - bF(0)$$

is an integer. On the other hand, since $f(x) < 1/n!$, $0 < x < 1$, we have

$$0 < bk^{2n+1}\int_0^1 e^{kx}f(x)\,dx < \frac{bk^{2n+1}e^k}{n!} = bke^k\frac{(k^2)^n}{n!}.$$

For fixed k, $(k^2)^n/n!$ tends to zero as $n \to \infty$. Thus, for n sufficiently large, the entire right-hand side can then be made < 1. We found an integer

$$0 < aF(1) - bF(0) < 1$$

strictly between 0 and 1! ¬ ∎

Remark.
Another proof of irrationality of e can be given using the geometric series formula as follows: Let $[\cdot] : \mathbf{R} \to \mathbf{Z}$ be the *greatest integer function* defined by $[r] =$ the greatest integer $\leq r$. Let $k \in \mathbf{N}$. We claim that

$$[k!e] = k!\sum_{j=0}^k \frac{1}{j!}.$$

Indeed, since the left-hand side is an integer, all we have to show is

$$k! \sum_{j=k+1}^{\infty} \frac{1}{j!} < 1.$$

For this, we estimate

$$k! \sum_{j=k+1}^{\infty} \frac{1}{j!}$$

$$= \frac{k!}{(k+1)!} + \frac{k!}{(k+2)!} + \frac{k!}{(k+3)!} + \cdots$$

$$= \frac{1}{k+1} + \frac{1}{(k+1)(k+2)} + \frac{1}{(k+1)(k+2)(k+3)} + \cdots$$

$$< \frac{1}{2} + \frac{1}{2^2} + \frac{1}{2^3} + \cdots = \frac{1}{1-(1/2)} - 1 = 1.$$

Clearly, $k! \sum_{j=0}^{k} 1/j! < k!e$, so that, for $k \in \mathbf{N}$, we have

$$[k!e] < k!e.$$

But this implies that e is irrational. ¬ Indeed, assume that $e = a/b$, $a, b \in \mathbf{N}$. If we choose $k = b$, then the number $b!e = b!(a/b)$ is an integer, so that $[b!e] = b!e$. ¬

Corollary.

e^q is irrational for all $0 \neq q \in \mathbf{Q}$.

Proof.

If e^q is rational, then so is any power $(e^q)^m = e^{qm}$. Now choose m to be the denominator of q (written as a fraction) to get a contradiction to Theorem 1. ∎

Remark.

$\mathbf{Q}$ is *dense* in $\mathbf{R}$ in the sense that, given any real number r, we can find a rational number arbitrarily close to r. This is clear if we write r in a decimal representation and chop off the tail arbitrarily far away from the decimal point. For example, $\sqrt{2} = 1.414213562\ldots$ means that the rationals $1, 1.4 = 14/10, 1.41 = 141/100, 1.414 =$

1414/1000 ... get closer and closer to $\sqrt{2}$. Indeed, we have $1.4^2 = 1.96$, $1.41^2 = 1.9981$, $1.414^2 = 1.999396$.

Theorem 2.

π is irrational.[6]

Proof.

It is enough to prove that π^2 is irrational. ¬ Assume that $\pi^2 = a/b$ for some $a, b \in \mathbf{Z}$. Using the idea of the previous proof, we define $F : \mathbf{R} \to \mathbf{R}$ by

$$F(x) = b^n[\pi^{2n}f(x) - \pi^{2n-2}f''(x) + \pi^{2n-4}f^{(4)}(x) - \cdots + (-1)^n f^{(2n)}(x)],$$

where f is given above. Since $\pi^{2n} = a^n/b^n$, $F(0)$ and $F(1)$ are again integers. This time, $F''(x) + \pi^2 F(x)$ is telescopic, so that we have

$$[F'(x)\sin(\pi x) - \pi F(x)\cos(\pi x)]' = [F''(x) + \pi^2 F(x)]\sin(\pi x)$$

$$= b^n \pi^{2n+2} f(x)\sin(\pi x)$$

$$= a^n \pi^2 f(x)\sin(\pi x).$$

Integrating, we obtain

$$\int_0^1 a^n \pi f(x)\sin(\pi x)\,dx = \frac{1}{\pi}[F'(x)\sin(\pi x) - \pi F(x)\cos(\pi x)]_0^1$$

$$= F(1) + F(0) \in \mathbf{Z}.$$

On the other hand,

$$0 < \int_0^1 a^n \pi f(x)\sin(\pi x)\,dx < \frac{a^n \pi}{n!} < 1$$

if n is large enough. ¬　　　　　　　　　　　　　　　　　　■

♣ Although π is irrational, it is not only tempting but very important to find good approximations of π by rational numbers[7] such as the Babylonian 25/8 (a clay tablet found near Susa) or the Egyptian

---

[6] Irrationality of π was first proved by Lambert in 1766.

[7] For an interesting account on π, see P. Beckmann, *A History of π*, The Golem Press, 1971. For a recent comprehensive treatment of π, see L. Berggren, J. Borwein, and P. Borwein, *π: A Source Book*, Springer, 1997.

$2^8/3^4$ (Rhind papyrus, before 1650 B.C.). The Babylonians obtained the first value by stating that the ratio of the perimeter of a regular hexagon to the circumference of the circumscribed circle was "equal" to $57/60 + 36/60^2$. (Recall that the Babylonians used 60 as a base.) The Egyptians obtained the second value by inscribing a circle into a square of side length 9 units, putting a 3×3 grid on the square by trisecting each edge, and approximating the circle by the octagon obtained by cutting off the corner triangles (Figure 2.1). The area of the octagon, 63 units, was rounded up to 8^2, and the approximate value of π turned out to be $8^2/(9/2)^2 = 2^8/3^4$.

A well-known recorded approximation of π was made by Archimedes (c. 287–212 B.C.), who applied Eudoxus's *method of exhaustion*, the approximation of the circumference of the unit circle by the perimeters of inscribed and circumscribed regular polygons. By working out the perimeters of regular 96-sided polygons inscribed in and circumscribed about a given circle, he established the estimate

$$3\frac{10}{71} < \pi < 3\frac{1}{7}.$$

The difference between the upper and lower bounds is $1/497 \approx 0.002$, a remarkable accomplishment! To arrive at this estimate, Archimedes first worked out the perimeters of the inscribed and

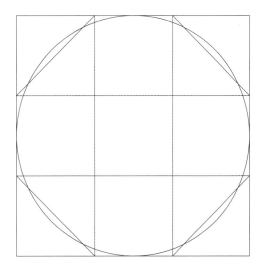

Figure 2.1

circumscribed hexagons and obtained

$$3 < \pi < \frac{6}{\sqrt{3}}.$$

He then estimated $\sqrt{3}$ from below by 265/153 (4-decimal precision!) and obtained 918/265 as an upper bound. Finally, to arrive at a polygon with $96 = 6 \times 2^4$ sides, he progressively doubled the sides of the polygons and made more and more refined estimates of certain radicals (such as $\sqrt{349450} = 591 \; 1/8$). The algorithm Archimedes used is treated in Problem 12.

Other rational approximations of π are 22/7 (3-decimal precision) and 355/113 (6-decimal precision). $\sqrt{10}$ was used by Brahmagupta (c. A.D. 600). Numerous ingenious approximations of π were given by Ramanujan, such as the 9-decimal precision

$$\frac{63}{25}\left(\frac{17 + 15\sqrt{5}}{7 + 15\sqrt{5}}\right).$$

and the infinite series expansion

$$\frac{1}{\pi} = \frac{2\sqrt{2}}{9801}\sum_{j=0}^{\infty}\frac{(4j)!(1103 + 26390j)}{(j!)^4 396^{4j}},$$

where each term of the series produces an additional 8 correct digits in the decimal representation of π.

◇ Another way to approximate π is to use the integral formula

$$\tan^{-1} y = \int_0^y \frac{dt}{1 + t^2}$$

from calculus, expand the integrand into a power series and integrate. In fact, replacing x by $-t^2$ in the geometric series formula, we obtain

$$\frac{1}{1 + t^2} = 1 - t^2 + t^4 - t^6 + t^8 - t^{10} + \cdots,$$

so that

$$\tan^{-1} y = \int_0^y \frac{dt}{1 + t^2} = y - \frac{y^3}{3} + \frac{y^5}{5} - \frac{y^7}{7} + \frac{y^9}{9} - \frac{y^{11}}{11} + \cdots,$$

and, setting $y = 1$, we arrive at the Gregory–Leibniz series

$$\frac{\pi}{4} = 1 - \frac{1}{3} + \frac{1}{5} - \frac{1}{7} + \frac{1}{9} - \frac{1}{11} + \cdots .$$

(The Scottish mathematician David Gregory discovered the infinite series expansion of $\tan^{-1}$ in 1671, three years before Leibniz derived the alternating series for $\pi/4$. Published by Leibniz in 1682, the expansion is sometimes named after him. For Størmer's approach to the Gregory numbers $t_{a/b} = \tan^{-1}(b/a)$, $a, b \in \mathbf{N}$, see Problem 14.) Since the series on the right-hand side converges very slowly (several hundred terms are needed for even 2-decimal precision!), numerous modifications of this procedure have been made, using, for example,

$$\pi/4 = 4\tan^{-1}(1/5) - \tan^{-1}(1/239)$$

(see Problem 14) and

$$\pi/4 = 6\tan^{-1}(1/8) + 2\tan^{-1}(1/57) + \tan^{-1}(1/239).$$

To close this section, we give a brief account on a geometric method devised by Gauss that produces rational approximations of π with arbitrary decimal precision. Consider the *two-dimensional square lattice* $\mathbf{Z}^2 \subset \mathbf{R}^2$ of points with integral coordinates. For $r > 0$, let $N(r)$ denote the total number of lattice points contained in the (closed) disk $\bar{D}_r$ with center at the origin and radius r:

$$N(r) = |\bar{D}_r \cap \mathbf{Z}^2|$$

(see Figure 2.2). Each lattice point $(a, b) \in \bar{D}_r$, contributing to $N(r)$, can be considered as the lower left corner of a unit square with sides parallel to the coordinate axes. Thus, $N(r)$ can be thought of as the total area of the squares whose lower left corner is contained in $\bar{D}_r$. To see how far $N(r)$ is from the area πr^2 of $\bar{D}_r$, let $B(r)$ denote the total area of those squares that intersect the boundary circle of $\bar{D}_r$. We have $|N(r) - \pi r^2| \leq B(r)$. Moreover, since the diagonal of a unit square has length $\sqrt{2}$,

$$B(r) < \pi((r + \sqrt{2})^2 - (r - \sqrt{2})^2) = 4\sqrt{2}\pi r.$$

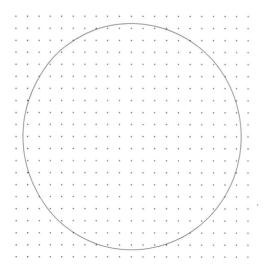

Figure 2.2

Dividing by r^2, we obtain the estimate

$$\left| \frac{N(r)}{r^2} - \pi \right| < \frac{4\sqrt{2\pi}}{r}.$$

Choosing r to be an integer, we see that $N(r)/r^2$ gives a rational approximation of π with precision $4\sqrt{2\pi}/r$. For example, $r = 300$ gives $N(300)/300^2 = 3.14107$ (3-decimal precision).

♡ To see an interesting connection with number theory, notice that $N(r)$ counts the number of ways that nonnegative integers $n \leq r^2$ can be written as sums of two squares of integers. As shown in number theory,[8] the number of representations of a positive integer n as a sum of two squares is four times the excess in the number of divisors of n of the form $4m + 1$ over those of the form $4m + 3$ (cf. also the discussion at the end of Section 1). In terms of the greatest integer function, the total number of divisors of the form $4m + 1$ of all positive integers $n \leq r^2$ is

$$[r^2] + \left[\frac{r^2}{5} \right] + \left[\frac{r^2}{9} \right] + \cdots,$$

---

[8]See I. Niven, H. Zuckerman, and H. Montgomery, *An Introduction to the Theory of Numbers*, Wiley, 1991.

and those of the form $4m + 3$ is

$$\left[\frac{r^2}{3}\right] + \left[\frac{r^2}{7}\right] + \left[\frac{r^2}{11}\right] + \cdots.$$

Taking into account the trivial case ($0 = 0^2 + 0^2$), we obtain

$$\frac{N(r) - 1}{4} = [r^2] - \left[\frac{r^2}{3}\right] + \left[\frac{r^2}{5}\right] - \left[\frac{r^2}{7}\right] + \left[\frac{r^2}{9}\right] - \left[\frac{r^2}{11}\right] + \cdots.$$

The right-hand side contains only finitely many nonzero terms. Assuming from now on that r is an odd integer, we find that the number of nonzero terms is $(r^2 + 1)/2$. To estimate this alternating series, we remove the greatest integer function from the first r terms and observe that we make an error less than r. The remaining terms are dominated by the leading term $[r^2/(r + 1)]$ since the series is alternating. We thus have

$$\frac{N(r) - 1}{4} = r^2 - \frac{r^2}{3} + \frac{r^2}{5} - \frac{r^2}{7} + \frac{r^2}{9} - \frac{r^2}{11} + \ldots \pm \frac{r^2}{r} \pm cr,$$

where $0 \leq c \leq 2$. Dividing by r^2 and letting $r \to \infty$ through odd integers, we recover the Gregory–Leibniz series

$$\frac{\pi}{4} = 1 - \frac{1}{3} + \frac{1}{5} - \frac{1}{7} + \frac{1}{9} - \frac{1}{11} + \cdots.$$

$\diamond$ For the approximation of π above we used unit squares with vertices in $\mathbf{Z}^2$ to more or less cover the closed disk $\bar{D}_r$ for r large. Instead of squares we can use parallelograms $F_{a,b} = F + (a, b)$, $a, b \in \mathbf{Z}$ ($F_{a,b}$ is F translated by the vector (a, b)), where the *fundamental parallelogram* $F = F_{0,0}$ (and hence $F_{a,b}$) has vertices in $\mathbf{Z}^2$ but no other points in $\mathbf{Z}^2$ (not even on the boundary). (See Figure 2.3.) The parallelograms $F_{a,b}$ *tesselate* the whole plane in the sense that they cover $\mathbf{R}^2$ with no overlapping interiors. To make F unique we can assume that a specified vertex of F is the origin. The two vertices of F adjacent to the origin are then given by vectors v and w (with integral components), and the vertex of F opposite to the origin is $v + w$. The argument of Gauss goes through with obvious modifications. We obtain

$$|N_F(r) - \pi r^2| < B_F(r) < \pi((r + \delta)^2 - (r - \delta)^2) = 4\delta\pi r,$$

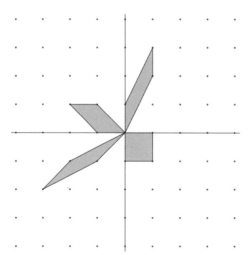

Figure 2.3

where δ is the diameter, the length of the longer diagonal, of F; $N_F(r)$ is the total area of the parallelograms whose specified vertex lies in $\bar{D}_r$; and $B_F(r)$ is the total area of the parallelograms that intersect the boundary circle of $\bar{D}_r$. Since $N_F(r) = \text{area}(F)N(r)$, letting $r \to \infty$, we obtain

$$\lim_{r \to \infty} \frac{N_F(r)}{r^2} = \text{area}(F) \lim_{r \to \infty} \frac{N(r)}{r^2}.$$

Since both limits are equal to π, we conclude that the area of a fundamental parallelogram in a square lattice is always unity.

Since F is fundamental, the integral linear combinations of v and w exhaust $\mathbf{Z}^2$; i.e., we have

$$\mathbf{Z}^2 = \{kv + lw \mid k, l \in \mathbf{Z}\}.$$

The passage from the unit square to the fundamental parallelogram F is best described by the transfer matrix A between the bases $\{(1, 0), (0, 1)\}$ and $\{v, w\}$. The column vectors of A are v and w. In particular, A has integral entries. Viewing v and w as vectors in $\mathbf{R}^3$ with zero third coordinates, we have

$$\text{area}(F) = |v \times w| = |\det(A)|.$$

This is unity, since F is fundamental. We thus have $\det(A) \pm 1$. Switching v and w changes the sign of the determinant. A 2×2 matrix with integral entries and determinant 1 (and the associated linear transformation) is called *unimodular*. The inverse of a unimodular matrix is unimodular, and the product of unimodular matrices is also unimodular. ♡ Thus, under multiplication, the unimodular matrices form a group. This is called the *modular group*

$$SL(2, \mathbf{Z}) = \left\{ A = \begin{bmatrix} a & b \\ c & d \end{bmatrix} \mid \det(A) = ad - bc = 1, \ a, b, c, d \in \mathbf{Z} \right\}.$$

The unimodular transformations preserve the square lattice $\mathbf{Z}^2$ (in fact, this property can be used to define them). Finally, the unimodular transformations are area-preserving. The proof of this is an application of the argument of Gauss above to planar sets more general than disks. The only technical difficulty is to define what area means in this general context.

Problems

1. Let a be a positive integer such that $\sqrt[n]{a}$ is rational. Show that a must be the nth power of an integer.

2. ◇ Show that cos 1 is irrational.

3. Show that the cubic polynomial $P(x) = 8x^3 - 6x - 1$ has no rational roots. Use the trigonometric identity $\cos(\theta) = 4\cos^3(\theta/3) - 3\cos(\theta/3)$ to conclude that $\cos 20°$ is a root of P, and thereby irrational.

4. ◇ Work out the integral $\int_1^a x^n \, dx$, $a > 1$, $n \neq -1$, using the geometric series formula in the following way: Let $m \in \mathbf{N}$ and consider the Riemann sum (approximating the integral) on the partition $\{\alpha^k\}_{k=0}^m$, where $\alpha = \sqrt[m]{a}$.

5. ◇ Show that, for $n \in \mathbf{N}$, we have
$$1 + 2 + \cdots + n = \frac{n(n+1)}{2}.$$

6. Solve Problem 5 in the following geometric way: Consider the function $f : [0, n] \to \mathbf{R}$, defined by $f(x) = 1 + [x]$, $0 \leq x \leq n$, and realize that the area of the region S under the graph of f gives $1 + 2 + \cdots + n$. (S looks like a "staircase.") Cut S into triangles and work out the area of the pieces. Generalize this to obtain the sum of an arithmetic progression.

7. ◇ Show that, for $n \in \mathbf{N}$, we have
$$1^2 + 2^2 + \cdots + n^2 = \frac{n(n+1)(2n+1)}{6}.$$

8. Solve Problem 7 along the lines of Problem 6 by building a 3-dimensional pyramid staircase whose volume represents $1^2 + 2^2 + \cdots + n^2$. Cut the staircase into square pyramids and triangular prisms.

9. Show that

$$1^3 + 2^3 + \cdots + n^3 = \left(\frac{n(n+1)}{2}\right)^2$$

in the following geometric way (this was known to the Arabs about 1000 years ago): For $k = 1, \ldots, n$, consider the square $S_k \subset \mathbf{R}^2$ with vertices $(0, 0)$, $(k(k+1)/2, 0)$, $(k(k+1)/2, k(k+1)/2)$, and $(0, k(k+1)/2)$. For $n \geq 2$, view S_n as the union of S_1 and the L-shaped regions $S_k - S_{k-1}$, $k = 2, \ldots, n$. By splitting $S_k - S_{k-1}$ into two rectangles, show that the area of $S_k - S_{k-1}$ is k^3.

10. Given $n \in \mathbf{N}$, define a finite sequence of positive integers: $a_0 = a$, $a_1 = [a_0/2]$, $a_2 = [a_1/2]$, $\ldots$. Show that for $b \in \mathbf{N}$ we have

$$ab = \sum_{a_n \text{ odd}} 2^n b.$$

(This algorithm of multiplication, consisting of systematic halving and doubling, was used by the ancient Greeks, and was essentially known to the ancient Egyptians.)

11. A triangular array of dots consists of $1, 2, \ldots, n$ dots, $n \in \mathbf{N}$, stacked up to form a triangle. The total number of dots in a triangular array is the *nth triangular number* $Tri(n) = 1 + 2 + 3 + \cdots + n = n(n+1)/2$ (cf. Problem 5).

(a) Define the *nth square number* $Squ(n) = n^2$, $n \in \mathbf{N}$. By fitting two triangular arrays into a square, show that $Squ(n) = Tri(n) + Tri(n-1)$. (This is due to Nicomachus, c. A.D. 100.) Define the *nth pentagonal number* $Pen(n) = Squ(n) + Tri(n-1)$ and arrange the corresponding dots in a pentagonal array. Generalize this to define all *polygonal numbers*.

(b) By fitting 8 triangular arrays into a square, show that $a \in \mathbf{N}$ is a triangular number iff $8a + 1$ is a perfect square. (This is attributed to Plutarch, c. A.D. 100.)

(c) Prove that

$$\sum_{n=1}^{\infty} \frac{1}{Tri(n)} = 2.$$

(d) Define the *nth k-gonal number* inductively[9] as the sum of the nth $(k-1)$-gonal number and $Tri(n-1)$. Show that the nth k-gonal number is equal to $((k-2)n^2 - (k-4)n)/2$ by observing that this number is the sum of the first n terms of an arithmetic progression with first term 1 and common difference $k - 2$. Conclude that a hexagonal number is triangular, and a pentagonal

[9]See L.E. Dickson, *History of the Theory of Numbers*, II, Chelsea Publishing, 1971.

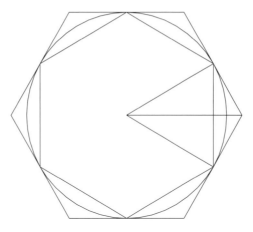

Figure 2.4

number is 1/3 of a triangular number. What is the geometry behind these conclusions?

12. Let s_n and S_n denote the side lengths of the inscribed and circumscribed regular n-sided polygons to a circle.

(a) Derive the formulas

$$\frac{S_{2n}}{S_n - S_{2n}} = \frac{s_n}{S_n} \quad \text{and} \quad 2s_{2n}^2 = s_n S_{2n}$$

by looking for similar triangles in Figure 2.4.

(b) Let $p_n = ns_n$ and $P_n = nS_n$ be the corresponding perimeters. Use (a) to conclude that

$$P_{2n} = \frac{2P_n p_n}{P_n + p_n} \quad \text{and} \quad p_{2n} = \sqrt{p_n P_{2n}}.$$

(c) Assume that the circle has unit radius and show that

$$p_6 = 6, \quad P_6 = 4\sqrt{3}, \quad p_{12} = 12\sqrt{2 - \sqrt{3}}, \quad P_{12} = 24\left(2 - \sqrt{3}\right).$$

How well do p_{12} and P_{12} approximate 2π?

(d) Use the inequality

$$\frac{2ab}{a + b} \leq \sqrt{ab}, \quad a, b > 0,$$

to show[10] that

$$P_{2n} p_{2n}^2 < P_n p_n^2$$

and

$$\frac{1}{P_{2n}} + \frac{2}{p_{2n}} < \frac{1}{P_n} + \frac{2}{p_n}.$$

---

[10]This interpolation technique is due to Heinrich Dörrie, 1940.

Define

$$A_n = \sqrt[3]{P_n p_n^2} \quad \text{and} \quad B_n = \frac{3 P_n p_n}{2 P_n + p_n}.$$

Translate the inequalities above to prove monotonicity of the sequences $\{A_n\}$ and $\{B_n\}$, and conclude that

$$B_n < 2\pi < A_n.$$

Finally, use (c) to show that

$$A_{12} = 12\sqrt[3]{2\left(2 - \sqrt{3}\right)^2} \quad \text{and} \quad B_{12} = \frac{72(2 - \sqrt{3})}{4\sqrt{2 - \sqrt{3}} + 1}.$$

How well do A_{12} and B_{12} approximate 2π? Does this method surpass Archimedes' approximation of π using 96-sided polygons?

13. Evaluate the expansion of $\tan^{-1}$ at $1/\sqrt{3}$, and obtain

$$\frac{\pi}{6} = \frac{1}{\sqrt{3}} \left(1 - \frac{1}{3 \cdot 3} + \frac{1}{3^2 \cdot 5} - \frac{1}{3^3 \cdot 7} + \cdots \right).$$

(This expansion was used by Sharp (1651–1742) to calculate π with 72 decimal precision.)

14. Define the rth *Gregory number* t_r, $r \in \mathbf{R}$, as the angle (in radians) in an uphill road that has slope $1/r$. Equivalently, let $t_r = \tan^{-1}(1/r)$. (As indicated in the text, Gregory found the infinite series expansion

$$t_r = \frac{1}{r} - \frac{1}{3r^3} + \frac{1}{5r^5} - \cdots,$$

an equivalent form of the Taylor series for $\tan^{-1}$.) ◇ Use Størmer's observation[11] that the argument of the complex number $a + bi$ is $t_{a/b}$ (and additivity of the arguments in complex multiplication) to derive Euler's formulas

$$t_1 = t_2 + t_3,$$
$$t_1 = 2t_3 + t_7,$$
$$t_1 = 5t_7 + 2t_{18} - 2t_{57}.$$

Generalize the first equation to prove Lewis Caroll's identity

$$t_n = t_{n+s} + t_{n+t} \quad \text{iff} \quad st = n^2 + 1,$$

and Machin's formula

$$t_1 = 4t_5 - t_{239}.$$

The latter translates into

$$\pi/4 = 4\tan^{-1}(1/5) - \tan^{-1}(1/239).$$

---

[11] For more details, see J. Conway and R. Guy, *The Book of Numbers*, Springer, 1996.

(This was used by Machin, who calculated π with 100-decimal precision in 1706.) Derive the last formula using trigonometry. (Let $\tan \theta = 1/5$ and use the double angle formula for the tangent to work out $\tan(2\theta) = 5/12$ and $\tan(4\theta) = 120/119$. Notice that this differs from $\tan(\pi/4) = 1$ by $1/119$. Work out the error $\tan(4\theta - \pi/4) = 1/239$ using a trigonometric formula for the tangent.)

15. ◇ Given a basis $\{v, w\}$ in $\mathbf{R}^2$, the set $L = \{kv + lw \mid k, l \in \mathbf{Z}\}$ is called a *lattice* in $\mathbf{R}^2$ with generators v and w. Generalize the concept of fundamental parallelogram (from $\mathbf{Z}^2$) to L, and show that every possible fundamental parallelogram of a lattice L has the same area. Use this to prove that the transfer matrix, with respect to the basis $\{v, w\}$, between two fundamental parallelograms is unimodular.

Web Sites

1. www.maa.org

2. www.math.psu.edu/dna/graphics.html

3. www.dgp.utoronto.ca/people/mooncake/thesis

4. www.mathsoft.com/asolve/constant/pi/pi.html

3

SECTION

Rationality, Elliptic Curves, and Fermat's Last Theorem

♣ A point in $\mathbf{R}^2$ with rational coordinates is called *rational*. The rational points $\mathbf{Q}^2$ form a dense set in $\mathbf{R}^2$ (that is, any open disk in $\mathbf{R}^2$ contains a point in $\mathbf{Q}^2$). Theorem 1 of Section 2 can be reformulated by saying that the only rational point on the graph

$$\{(x, e^x) \mid x \in \mathbf{R}\} \subset \mathbf{R}^2$$

of the exponential function is $(0, 1)$. (It is quite amazing that this smooth curve misses all points of the dense subset $\mathbf{Q}^2 - \{(0, 1)\}$ of $\mathbf{R}^2$!) An easier example (to be generalized later to Fermat's famous Last Theorem) is the *unit circle*

$$S^1 = \{(x, y) \in \mathbf{R}^2 \mid x^2 + y^2 = 1\},$$

and we ask the same question of rationality: What are the rational points on the unit circle? Apart from the trivial $(\pm 1, 0)$, $(0, \pm 1)$, and the less trivial examples $(3/5, 4/5)$, $(5/13, 12/13)$, and $(8/17, 15/17)$, we can actually give a complete description of these points. Indeed, assume that x and y are rational and satisfy $x^2 + y^2 = 1$. Write x and y as fractions of relatively prime integers $x = a/c$ and $y = b/d$. Substituting and multiplying out both sides

26

of the equation by $c^2 d^2$, we obtain

$$a^2 d^2 + b^2 c^2 = c^2 d^2.$$

This immediately tells us that c^2 divides $a^2 d^2$, and d^2 divides $b^2 c^2$. Since a, c and b, d have no common divisors, c^2 and d^2 divide each other, and so they must be equal. The equation above reduces to

$$a^2 + b^2 = c^2.$$

We see that the question of rationality is equivalent to finding positive integer solutions of the Pythagorean equation above. A solution (a, b, c) with $a, b, c \in \mathbf{N}$ is called a *Pythagorean triple*. The obvious reason for this name is that each Pythagorean triple gives a right triangle with integral side lengths a, b, and c (with c corresponding to the hypotenuse). By the way, in case you have not seen a proof of the Pythagorean Theorem,[1] Figure 3.1 depicts one due to Bhaskara (1114–c. 1185). The rotated 4-sided polygon on the right is a square, which you can show from symmetry or by calculating angles.

Remark.

Another cutting-and-pasting gem of Greek geometry is due to Hippocrates of Chios (c. 430 B.C.), and it states that the crescent lune in Figure 3.2 has the same area as the inscribed triangle. To show this, verify that $A = 2a$ in Figure 3.3. The significance of this should not be underrated. In contrast to the futile attempts to

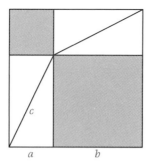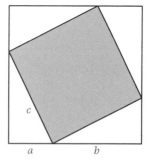

Figure 3.1

---

[1] Have you noticed that the Pythagorean Theorem, the Euclidean distance formula, the trigonometric identity $\sin^2 \theta + \cos^2 \theta = 1$, and the equation of the circle are all equivalent to each other?

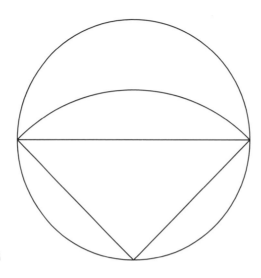

Figure 3.2

square the circle, this is the first recorded successful measurement of the exact area of a plane figure bounded by curves!

All Pythagorean triples have been known since the time of Euclid (see Book X of Euclid's *Elements*[2]) and can be found in the third century work *Arithmetica* by Diophantus. An ancient Babylonian tablet (c. 1600 B.C.) contains a list of Pythagorean triples including (4961, 6480, 8161)! It indicates that the Babylonians in the time of

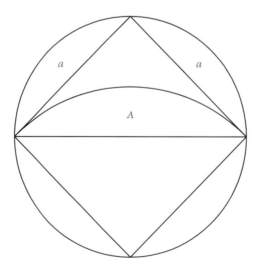

Figure 3.3

[2]T.L. Heath, *The Thirteen Books of Euclid's Elements*, Dover, New York, 1956.

Hammurabi were well aware of the significance of these numbers, including the fact that they are integral side lengths of a right triangle. The time scale is quite stunning; this is about 1000 years before Pythagoras! The first infinite sequence of all Pythagorean triples (a, b, c) *with b,c consecutive* was found by the Pythagoreans themselves:

$$(a, b, c) = (n, (n^2 - 1)/2, (n^2 + 1)/2),$$

where $n > 1$ is odd. Plato found Pythagorean triples (a, b, c) with $b + 2 = c$, namely, $(4n, 4n^2 - 1, 4n^2 + 1)$, $n \in \mathbf{N}$. Finally, note also that Euclid, although he obtained all Pythagorean triples, provided no proof that his method gave them all. We give now a brief account of the solution to the problem of Pythagorean triples.

We may assume that a, b, and c are relatively prime (i.e., a, b, and c have no common divisor). This is because (a, b, c) is a Pythagorean triple iff (ka, kb, kc) is a Pythagorean triple for any $k \in \mathbf{N}$. (Notice also that (ka, kb, kc) gives the same rational point on S^1 for all $k \in \mathbf{N}$.) It follows that a and b are also relatively prime. ($\neg$ If a prime p divides both a and b, then it also divides $c^2 = a^2 + b^2$. Being a prime, p also divides c, so that p is a common divisor of a, b, and c, contrary to our choice. $\neg$)

Similarly, b, c and a, c are also relatively prime. We now claim that a and b must have different parity. Since they are relatively prime, they cannot be both even. $\neg$ Assume that they are both odd, say, $a = 2k + 1$ and $b = 2l + 1$. Substituting, we have $c^2 = a^2 + b^2 = 4(k^2 + k + l^2 + l) + 2 = 2(2(k^2 + k + l^2 + l) + 1)$, but this is impossible since the square of a number cannot be the double of an odd number! $\neg$ (Compare this with the proof of Proposition 1 which shows the irrationality of $\sqrt{2}$.)

Without loss of generality, we may assume that a is even and b is odd. Hence, c must be odd. Since the difference and sum of odd numbers is even, $c - b$ and $c + b$ are both even. We can thus write

$$c - b = 2u \quad \text{and} \quad c + b = 2v,$$

where $u, v \in \mathbf{N}$. Equivalently, $c = v + u$ and $b = v - u$. These equations show that u and v are relatively prime (since b and c are). An argument similar to the one above shows that u and v have different parity. Recall now that a is even, so that $a/2$ is an integer. In fact, in terms of u and v, we have

$$\left(\frac{a}{2}\right)^2 = \left(\frac{c-b}{2}\right)\left(\frac{c+b}{2}\right) = uv.$$

Let p be a prime divisor of $a/2$. Then p^2 divides uv; and since u and v are relatively prime, p^2 must divide either u or v. It follows that u and v are pure square numbers: $u = s^2$ and $v = t^2$. In addition, s and t are also relatively prime and of different parity. With these, we have $a = 2\sqrt{uv} = 2st$. Putting everything together, we arrive at

$$a = 2st, \quad b = t^2 - s^2, \quad c = t^2 + s^2.$$

We claim that this is the general form of a Pythagorean triple (with relatively prime components) parametrized by two integers s and t, where $t > s$, and s, t are relatively prime and of different parity. To show this we just have to work backward:

$$a^2 + b^2 = 4s^2t^2 + (t^2 - s^2)^2 = (t^2 + s^2)^2 = c^2,$$

and we are done.

The following table shows a few values:

t	s	$a = 2st$	$b = t^2 - s^2$	$c = t^2 + s^2$
2	1	4	3	5
3	2	12	5	13
4	1	8	15	17
4	3	24	7	25
5	2	20	21	29
5	4	40	9	41
6	1	12	35	37

There is a lot of geometry behind the Pythagorean triples. For example, one can easily show that the radius of the inscribed circle of a right triangle with integral sides is always an integer (see Problem 2). On the number theoretical side, the components of a

Pythagorean triple (a, b, c) contain many divisors; for example, the Babylonian 60 always divides the product abc! (See Problem 3.)

Remark.
A much harder ancient problem is to find a simple test to determine whether a given positive rational number r is a *congruent number*, the area of a right triangle with rational side lengths a, b, and c. In general, any nonzero rational number r can be made a *square free* integer by multiplying r by the square of another rational number s. As the name suggests, square free means that the integer (in this case $s^2 r$) has no nontrivial square integral divisors. Now, if r is the area of a triangle with side lengths a, b, c, then $s^2 r$ is the area of a similar triangle with side lengths sa, sb, sc. Thus, we may (and will) assume that a congruent number is a square free natural number. As an example, we immediately know that 6 is a congruent number, since it is the area of a right triangle with side lengths 3, 4, 5. Euler showed that 7 is a congruent number, and Fermat that 1 is not. (The latter is a somewhat stronger form of Fermat's last theorem in the exponent 4, to be treated later in this section; see Problem 20.) As a further example, 5 is a congruent number, as shown in Figure 3.4. Eventually, it became clear that 1, 2, 3, 4 are not congruent numbers, while 5, 6, 7 are. A major breakthrough in the "congruent number problem" was provided by Tunnell in 1983. His deep result, in its simplest form, states that if an odd square free natural number n is congruent, then the number of triples of integers (x, y, z) satisfying $2x^2 + y^2 + 8z^2 = n$ is equal to twice the number of triples of integers (x, y, z) satisfying $2x^2 + y^2 + 32z^2 = n$. For n an even square free natural number, the same is true with $2x^2$ replaced by $4x^2$, and n replaced by $n/2$. The converse statements are also true, provided that the so-called

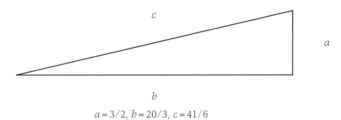

$a = 3/2,\ b = 20/3,\ c = 41/6$

Figure 3.4

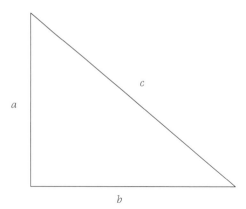

$$a = \frac{6803298487826435051217540}{411340519227716149383203}$$

$$b = \frac{411340519227716149383203}{21666555693714761309610}$$

Figure 3.5 $\quad c = \dfrac{22440351770433696992455751309067486316094847 2041}{891233226892885958802553517896716357001648 0830}$

Birch–Swinnerton–Dyer conjecture is true. To appreciate the subtlety of the problem, take a look at Figure 3.5, showing that 157 is a congruent number. (This is due to Zagier.)[3]

Returning to Pythagorean triples, we see that the rational points on the unit circle are of the form

$$\left(\frac{2st}{t^2 + s^2}, \frac{t^2 - s^2}{t^2 + s^2} \right)$$

where s, t are relatively prime integers.

There is a beautiful geometric way to obtain these points. Consider a line l through the rational point $(0, 1) \in S^1$ given by the equation

$$y = mx + 1.$$

For $m \neq 0$, l intersects S^1 in a point other than $(0, 1)$. This intersection point is obtained by coupling this equation with that of the

---

[3]Cf. N. Koblitz, *Introduction to Elliptic Curves and Modular Forms*, Springer, 1993.

unit circle. Substituting, we obtain

$$x^2 + (mx + 1)^2 = 1.$$

The left-hand side factors as

$$x((1 + m^2)x + 2m) = 0,$$

so that we obtain the coordinates of the intersection:

$$x = -\frac{2m}{1 + m^2} \quad \text{and} \quad y = mx + 1 = \frac{1 - m^2}{1 + m^2}.$$

This shows that (x, y) is a rational point on S^1 iff m is rational. (Indeed, $m \in \mathbf{Q}$ clearly implies $x, y \in \mathbf{Q}$. The converse follows from the equation $m = (y - 1)/x$.) Varying m in $\mathbf{Q}$, we obtain all rational points on S^1 (except $(0, -1)$, which corresponds to a vertical line intersection point). Setting $m = -s/t$, $t \neq 0$, $s, t \in \mathbf{Z}$, we recover the solution set above.

This "method of rational slopes" works for all quadratic curves described by $f(x, y) = 0$, where f is a quadratic polynomial with rational coefficients in the variables x, y, provided that there is at least one rational point on the curve. There may not be any; for example, there are no rational points on the circle (with radius $\sqrt{3}$) given by the equation

$$x^2 + y^2 = 3.$$

This follows because we showed in Section 1 that 3 does not divide any sums of squares $a^2 + b^2$, $a, b \in \mathbf{Z}$, with a and b relatively prime.

We are now tempted to generalize this approach to find rational points on all *algebraic curves*. An algebraic curve is, by definition, the locus of points on $\mathbf{R}^2$ whose coordinates satisfy a polynomial equation

$$f(x, y) = 0.$$

The locus as a point-set is denoted by

$$C_f = \{(x, y) \in \mathbf{R}^2 \mid f(x, y) = 0\}.$$

We say that C_f is *rational* if f has rational coefficients. Among all the polynomials whose zero sets give the algebraic curve, there is one with minimum degree. This degree is called the degree of the algebraic curve. We will consider cubic (degree-three) algebraic curves below.

⬦ As in calculus, it is convenient to consider the graph

$$\{(x, y, f(x, y)) \in \mathbf{R}^3 \mid (x, y) \in \mathbf{R}^2\}$$

of the polynomial f. This is a smooth surface in $\mathbf{R}^3$. The algebraic curve C_f is obtained by intersecting this graph with the horizontal plane spanned by the first two coordinate axes. (This is usually called the *level curve* of f corresponding to height 0.) We say that (x_0, y_0) is a *critical point* of f if the plane tangent to the graph at (x_0, y_0) is horizontal. In terms of partial derivatives, this means that

$$\frac{\partial f}{\partial x}(x_0, y_0) = \frac{\partial f}{\partial y}(x_0, y_0) = 0.$$

If $(x_0, y_0) \in C_f$ is not a critical point of f, then near (x_0, y_0), C_f is given by a smooth curve across (x_0, y_0). If C_f is cubic and $(x_0, y_0) \in C_f$ is a critical point of f, then near (x_0, y_0) the curve C_f displays essentially two kinds of singular behavior. At the assumed singular point, C_f either forms a *cusp* or it self-intersects in a *double point* (cf. Figure 3.6). The two typical nonsingular cubic curves are depicted in Figure 3.7. At present, we are most interested in the second type of singular point. This is the case when the singular point (x_0, y_0) of C_f is a saddle point for f. For example, compare the simple saddle point in Color Plate 1a depicting the graph of a quadratic polynomial, with Color Plate 1b, where the saddle is given by a cubic polynomial.

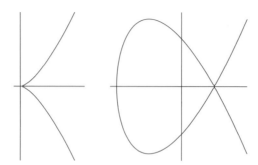

Figure 3.6

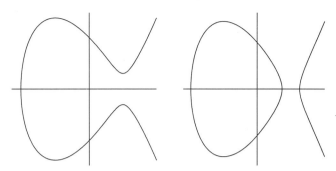

Figure 3.7

Again by calculus, this configuration is guaranteed by the "saddle condition"

$$\left(\frac{\partial^2 f}{\partial x^2}\right)\left(\frac{\partial^2 f}{\partial y^2}\right) - \left(\frac{\partial^2 f}{\partial x \partial y}\right)^2 < 0$$

near (x_0, y_0). Assuming this, if (x_0, y_0) is on C_f, then intersecting the surface with the horizontal coordinate plane, we obtain that, near (x_0, y_0), C_f consists of two smooth curves crossing over at (x_0, y_0). In this case, we say that the algebraic curve C_f has a *double point* at (x_0, y_0). The significance of this concept follows from the fact that any double point on a cubic rational curve is rational! We will not prove this in general. ♣ Instead, we restrict ourselves to the most prominent class of cubic rational curves given by the so-called *Weierstrass form*

$$y^2 = P(x),$$

where P is a cubic polynomial with rational coefficients. (♠ As far as rationality is concerned, restricting ourselves to Weierstrass forms results in no loss of generality. In fact, any cubic curve is "birationally equivalent" to a cubic in Weierstrass form. This means that for any cubic curve a new coordinate system can be introduced; the new coordinates depend rationally on the old ones; and in terms of the new coordinates, the cubic is given by a Weierstrass form. Because of the rational dependence of the old and new coordinates, the rational points on the original curve correspond to rational points on the new curve. (For a special case, see Problem 19.)

Remark.

♣ The origins of this equation go back to ancient times. This is easily understood if we consider the special case $P(x) = x^3 + c$, $c \in \mathbf{N}$, and realize that integer solutions of this equation are nothing but the possible ways to write c as the difference of a square and a cube:

$$y^2 - x^3 = c.$$

This special case is called the *Bachet equation*. The Bachet curve corresponding to the Bachet equation is nonsingular unless $c = 0$, in which case it reduces to a cusp. In 1917, Thue proved that for any c, there are only finitely many integral solutions to this equation. There may be none; for example, Bachet's equation has no solutions for $c = 7$. Although the proof is elementary, we will not show this, except to mention that the proof depends on writing the equation as

$$y^2 + 1 = x^3 + 8 = (x + 2)(x^2 - 2x + 4)$$

and studying the remainders of x and y under division by 4. However, Bachet's equation most often has infinitely many rational solutions. A little more about this later.

We now return to cubics in Weierstrass form. A good example on which to elaborate is given by the equation

$$y^2 = x^3 - 3x + 2.$$

The cubic curve is depicted in Figure 3.8. Here, $(1, 0)$ is a double point. (This will follow shortly.) Color Plate 1b shows the intersection of the graph of the polynomial $f(x, y) = x^3 - 3x + 2 - y^2$ in $\mathbf{R}^3$ with equidistant (horizontal) planes. Projecting the intersection curves to the coordinate plane spanned by the first two axes, we obtain the level curves of f (Figure 3.9). This indeed shows the three basic types discussed earlier!

◇ Let us try our machinery for the Weierstrass form. Since $f(x, y) = P(x) - y^2$, we have

$$\frac{\partial f}{\partial x} = P'(x) \quad \text{and} \quad \frac{\partial f}{\partial y} = -2y.$$

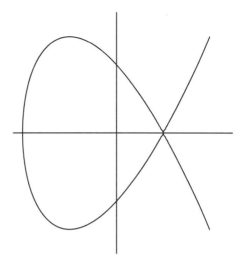

Figure 3.8

Let (x_0, y_0) be a critical point of f. Then $y_0 = 0$ and $P'(x_0) = 0$. If, furthermore, $(x_0, y_0) \in C_f$, then $y_0^2 = P(x_0) = 0$. Thus, a critical point of f on C_f is of the form $(r, 0)$, where $P(r) = P'(r) = 0$. Differentiating once more, we see that $(r, 0)$ is a saddle point if $P''(r) > 0$.

We now look at these conditions on P from an algebraic point of view. $P(r) = 0$ means that r is a root of P. By the factor theorem, the root factor $(x - r)$ divides P. The second condition $P'(r) = 0$ means that r is a root of P of multiplicity at least 2; that is, $(x - r)^2$

Figure 3.9

divides P. Finally, $P''(r) > 0$ means that r is a root of multiplicity exactly 2, or equivalently, $(x - r)^2$ exactly divides P. Since P is of degree 3, we obtain the complete factorization:

$$P(x) = a(x - r)^2(x - s), \quad r \neq s.$$

Since P has rational coefficients, multiplying out, it follows by easy computation that r, and hence s, is rational!

♣ Summarizing, we find that a double point on a cubic curve $y^2 = P(x)$ occurs at $(r, 0)$, where $r \in \mathbf{Q}$ is a root of multiplicity 2. In the example above, we see that $y^2 = x^3 - 3x + 2 = (x-1)^2(x+2)$, so that $(1, 0)$ is a double point. (Moreover, it also follows that the cusp corresponds to a triple root of P; in fact, the cusp in Figure 3.6 is given by the singular Bachet equation $y^2 = x^3$.)

Returning to the general situation, we now apply the method of rational slopes to describe all rational points on the cubic curve $y^2 = P(x)$ by considering lines through a rational double point $(r, 0)$ whose existence we assume. The general form of a line through $(r, 0)$ is given by

$$y = m(x - r).$$

To find the intersection point, we put this and the defining equation together and solve for x and y:

$$m^2(x - r)^2 = P(x).$$

Since r is a root of P with multiplicity two,

$$P(x) = a(x - r)^2(x - s), \quad r \neq s,$$

where a, r, and hence s, are rational. Thus,

$$x = s + m^2/a \quad \text{and} \quad y = m(s - r + m^2/a),$$

and these are rational iff m is. As before, varying m on $\mathbf{Q}$, we obtain all rational points on our cubic curve.

In the elusive case where P has three distinct roots, the method of rational slopes does not work. A cubic rational curve in Weierstrass form $y^2 = P(x)$, where the cubic polynomial P has no double or triple roots, is called *elliptic*. (◇ The name "elliptic" comes from the fact that when trying to determine the circumference of an ellipse one encounters elliptic integrals of the form $\int R(x, y)dx$, where R is a rational function of the variables x and y, and y is the square root of a cubic or quartic polynomial in x.) By the discussion above, an elliptic curve is everywhere nonsingular.

Remark.
The "congruent number problem" noted above is equivalent to a problem of rationality for a specific elliptic curve. As usual, we discuss only the beginning of this deep subject.[4] Let $n \in \mathbf{N}$ be a congruent number. By definition, there exists a right triangle with rational side lengths a, b, c, and area n. We have $a^2 + b^2 = c^2$ and $ab = 2n$. Adding or subtracting twice the second equation from the first, we get $(a \pm b)^2 = c^2 \pm 4n$. Setting $x = (c/2)^2$, we see that x, $x + n$, and $x - n$ are squares of rational numbers. We obtain that if a natural number n is congruent, then there exists a nonzero rational number x such that x, $x + n$, and $x - n$ are squares of rational numbers. The converse of this statement is also true, since $a = \sqrt{x+n} - \sqrt{x-n}$, $b = \sqrt{x+n} + \sqrt{x-n}$, and $c = 2\sqrt{x}$ are the rational side lengths of a right triangle with area n, provided that x is a nonzero rational number with x, $x+n$, and $x-n$ squares of rational numbers. If a, b, c are the rational side lengths of a right triangle with congruent area n, then multiplying the two equations $((a \pm b)/2)^2 = (c/2)^2 \pm n$ together, we get $((a^2 - b^2)/4)^2 = (c/2)^4 - n^2$. We see that the equation $u^4 - n^2 = v^2$ has a rational solution $u = c/2$ and $v = (a^2 - b^2)/4$. Setting $x = u^2 = (c/2)^2$ and $y = uv = (a^2 - b^2)c/8$, we obtain that (x, y) is a rational point on the elliptic curve in the Weierstrass form

$$y^2 = x^3 - n^2 x.$$

♣ The theory of elliptic curves displays one of the most beautiful interplays between number theory, algebra, and geometry. In what follows, all the examples of elliptic curves will be in Weierstrass form $y^2 = P(x)$, where we assume that P has rational coefficients and no multiple roots. (For an elliptic curve in Weierstrass form, P has either one or three real roots as in Figure 3.7.)

---

[4]See N. Koblitz, *Introduction to Elliptic Curves and Modular Forms*, Springer, 1993.

As a variation on the theme, we try the following "*chord-method*": Given an elliptic curve, we consider the line l though two of its rational points, say, A and B. The slope of l is rational since A and B have rational coordinates. Since the elliptic curve is cubic, there must be a third intersection point with l. We denote this by $A * B$. This point is rational. (This follows from the fact that if a cubic polynomial has rational coefficients and two rational roots, then the third root is also rational. Indeed, the sum $r_1 + r_2 + r_3$ of the three roots of a cubic rational polynomial $c_0 + c_1 x + c_2 x^2 + c_3 x^3 = c_3 (x - r_1)(x - r_2)(x - r_3)$, $c_3 \neq 0$, is equal to $-c_2/c_3 \in \mathbf{Q}$.)

The operation $(A, B) \mapsto A * B$ on the set of rational points of an elliptic curve does not define a group structure, since there is no point O playing the role of the identity element. Instead, we fix a rational point O (assuming it exists) and *define* $A + B$ as the third intersection of the line through O and $A * B$. Before going further, take a look at Figure 3.10.

Our aim is to show that the operation $(A, B) \mapsto A + B$ defines a group structure[5] on the set of rational points on the elliptic curve. Ignoring the finer point of what happens if the initial points co-

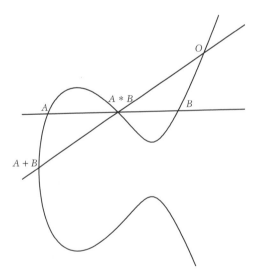

Figure 3.10

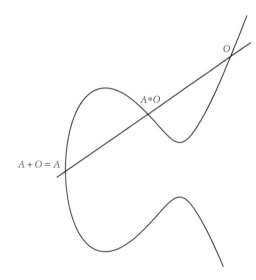

$A * O$

$A + O = A$

Figure 3.11

incide, it is easy to give visual proofs that the group axioms are satisfied.

We begin with the identity $A + O = A$ (Figure 3.11).

For the existence of the negative of a point, the identity $A + (-A) = O$ requires that the line through $A * (-A)$ and O should have no further intersection with the elliptic curve. It follows that this line must be tangent to the curve at O. (As will be shown below, a nonvertical line tangent to the elliptic curve meets the curve at exactly one other point.) The construction of $-A$ is depicted in Figure 3.12.

Associativity $A + (B + C) = (A + B) + C$ is slightly more complex[6] (Figure 3.13).

Since commutativity $(A + B = B + A)$ is clear, we are done!

Now, the finer point of coincidences. As we saw above, all is well and the chord construction applies if $A \neq B$; but what happens if $A = B$? At this point we have to rely on our geometric intuition. As usual, let the distinct A and B define $A * B$ as the third intersection point of the line l through A and B. Now pretend to be Newton and let B approach A. At the limit point $A = B$, the chord l becomes tangent to the elliptic curve at $A = B$! Thus, we define $A * A$ to be the intersection of the tangent at A with the elliptic curve. The big

---

[6] For details, see J. Silverman and J. Tate, *Rational Points on Elliptic Curves*, Springer, 1992.

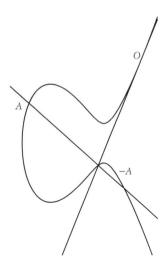

Figure 3.12

question, of course, is whether or not this intersection is a single point (including the case when it is empty)!

Remark.

For Bachet's equation $y^2 = x^3 + c$, $c \in \mathbf{Z}$, this so-called *tangent method* gives an interesting *duplication formula*. Given a rational solution $A \in \mathbf{Q}^2$, the line tangent to the algebraic curve at A

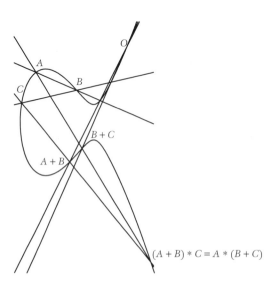

Figure 3.13

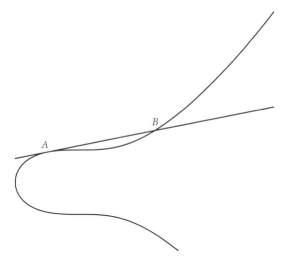

Figure 3.14

meets the curve at another point B that is easily seen to be rational (Figure 3.14).

Letting $A = (x, y)$, $y \neq 0$, computation shows that

$$B = \left(\frac{x^4 - 8cx}{4y^2}, \frac{-x^6 - 20cx^3 + 8c^2}{8y^3} \right).$$

By using ingenious algebraic manipulations, Bachet discovered this duplication formula in 1621, before calculus was developed (see Problem 9). Before going any further, let us play around with this formula in two special cases. Since $3^2 = 2^3 + 1$, we see that $A = (2, 3)$ is an integer point on the Bachet curve given by $y^2 = x^3 + 1$. The duplication formula gives $A * A = A * A * A * A = (0, -1)$. (For more on this, see Problem 18.) For a less trivial example, start with $3 = 2^2 - 1$, multiply both sides of this equation by $2^4 \cdot 3^3$, and obtain $(2^2 \cdot 3^2)^2 = (2^2 \cdot 3)^3 - 2^4 \cdot 3^3$. This shows that $A = (2^2 \cdot 3, 2^2 \cdot 3^2) = (12, 36)$ is an integer point on the Bachet curve given by $y^2 = x^3 - 432$. (If this is too ad hoc, take a look at Problem 19.) The duplication formula gives $A * A = A$! These two examples seem to crush our hope that by applying the duplication formula iteratively to an initial rational point we will obtain infinitely many distinct rational points on our Bachet curve. Fortunately, these are the only exceptional cases, and it can be proved[7] that starting with $A = (x, y)$, with $xy \neq 0$ and $c \neq 1, -432$, we do get infinitely many distinct rational points on the Bachet curve.

$\diamondsuit$ Back to $A * A$! Overflowing with confidence, we now believe that we can handle this technically and actually compute $A * A$. This needs a little calculus. Let $A = (x_0, y_0)$ be a rational point on the elliptic curve $y^2 = P(x)$ in question; $y_0^2 = P(x_0)$. Assuming $y_0 \neq 0$, the slope $m = y' = dy/dx$ of the tangent at x_0 can be obtained by implicit differentiation. We have

$$2yy' = P'(x),$$

which at x_0 gives

$$m = \frac{P'(x_0)}{2y_0} \in \mathbf{Q}.$$

(Note that the numerator is rational because P is.) Substituting this into the equation $y - y_0 = m(x - x_0)$, we obtain the equation of the tangent line

$$y = y_0 + \left(\frac{P'(x_0)}{2y_0} \right) (x - x_0).$$

To look for intersections, we combine this with the defining equation of the elliptic curve and obtain

$$\left(y_0 + \left(\frac{P'(x_0)}{2y_0} \right) (x - x_0) \right)^2 = P(x).$$

Equivalently,

$$(2P(x_0) + P'(x_0)(x - x_0))^2 = 4P(x_0)P(x).$$

Several terms look as if they jumped out of Taylor's formula. In fact, expanding P into Taylor series, we have

$$P(x) = P(x_0) + P'(x_0)(x - x_0) + \left(\frac{P''(x_0)}{2} \right) (x - x_0)^2$$

$$+ \left(\frac{P'''(x_0)}{6} \right) (x - x_0)^3.$$

---

[7]See N. Koblitz, *Introduction to Elliptic Curves and Modular Forms*, Springer, 1993.

(Notice that since P is cubic this is all we have.) After substituting, and allowing some mutual self-destruction of the opposite terms, we end up with

$$x = x_0 + 3\frac{P'(x_0)^2 - 2P''(x_0)P(x_0)}{2P'''(x_0)P(x_0)}.$$

This gives the first rational coordinate of the intersection $A * A$. (Do not worry about the denominator; it is twelve times the leading coefficient of P, and thereby nonzero.) Notice that, in the particular case $P(x) = x^3 + c$, our result reduces to Bachet's duplication formula! Since the point (x_0, y_0) is on a line with rational slope, the second coordinate is also rational. We thus obtain a unique rational point $A * A$. The remaining case $y_0 = 0$ sets the tangent line vertical (perpendicular to the first axis) at x_0 which, by assumption, is a rational root of P (since $y_0^2 = P(x_0) = 0$). Clearly, there is no further intersection of this line with the elliptic curve.

The usual way to circumvent this difficulty is to attach "ideal points" to the Euclidean plane to each "direction" or, more precisely, to each pencil of parallel lines. Attaching these ideal points to the Euclidean plane, we obtain the classical model of the *projective plane*. More about this in Section 16. ♣ Here we just say that the vertical tangent line intersects the extended elliptic curve at the ideal point given by the pencil of vertical lines. Since the vertical ideal point has to be attached to our elliptic curve, we might as well choose it to be the identity O! Addition then takes a simple form depicted in Figure 3.15.

The negative of a point is illustrated in Figure 3.16. (In fact, keep A fixed in Figure 3.15 and let B tend to $-A$. Then both $A * B$ and $A + B$ tend to the ideal point.)

Summarizing, we find that the rational points on a rational elliptic curve in the projective plane form a group under the addition rule $(A, B) \mapsto A + B$ defined above.

♠ A deep theorem of Mordell (1923) asserts that this group is finitely generated. More plainly, this means (in the spirit of Bachet's duplication formula) that on an elliptic curve with at least one rational point, there exist finitely many rational points such that all other rational points can be "generated" from these by the

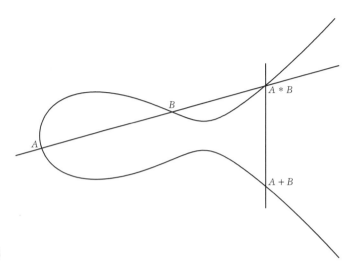

Figure 3.15

"chord-and-tangent" method. This gives a very transparent description of the rational points on any elliptic curve. As an example, the rational points of the elliptic curve[8] $y^2 + y = x^3 - x$ (y is shifted!) form an infinite cyclic group (thereby isomorphic with $\mathbf{Z}$). In Figure 3.17, the zero O is set at the vertical ideal point and the rational points are labelled with the integers (isomorphically).

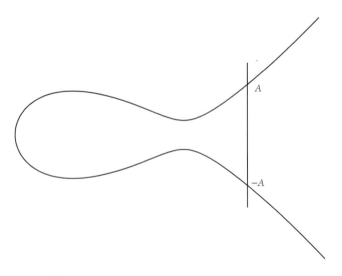

Figure 3.16

[8]See R. Hartshorne, *Algebraic Geometry*, Springer, 1977.

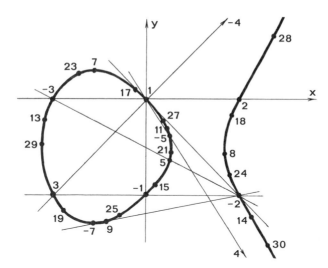

Figure 3.17.
R. Hartshorne,
Algebraic Geometry,
1977, 336. Reprinted
by permission of
Springer-Verlag New
York, Inc.

Remark.

We saw above that if n is a congruent number, then each right triangle with rational side lengths and area n can be made to correspond
to a specific rational point on the elliptic curve E_n given in Weierstrass form by $y^2 = x^3 - n^2 x$. It turns out[9] that the points A on E_n,
for n congruent, that correspond to "rational triangles" are exactly
the doubles of rational points ($A = B + B = 2B$ for B rational) on
E_n. On the other hand, in the group of rational points of E_n, the only
rational points of *finite* order are the four points of order two: the
identity O (the vertical ideal point), $(0, 0)$, and $(\pm n, 0)$ (cf. Problem
11). In this statement we need only to assume that n is a square free
natural number. If n is a congruent number, then the rational point
that corresponds to the right triangle with side lengths a, b, c, and
area n, has x-coordinate $x = (c/2)^2$ (see the computations above).
Since this is different from the x-coordinates of the four points of
order two above, we see that this rational point must be of infinite
order. The converse of this statement is also true. We obtain that
n is a congruent number iff the group of rational points of E_n is
infinite! This is the first (and most elementary) step in proving
Tunnell's characterization of congruent numbers. All that we said
here can be put in an elegant algebraic framework. If we denote

---

[9]See N. Koblitz, *Introduction to Elliptic Curves and Modular Forms*, Springer, 1993.

by $E_n(\mathbf{Q})$ the group of rational points on E_n, then, by Mordell's theorem, $E_n(\mathbf{Q})$ is finitely generated. Since this is an abelian group, its elements of finite order form a (finite) subgroup, the so-called torsion subgroup: $E_n(\mathbf{Q})_{\text{tor}}$. By what we said above, the order of this torsion subgroup is 4. By the fundamental theorem on finitely generated abelian groups,

$$E_n(\mathbf{Q}) \cong E_n(\mathbf{Q})_{\text{tor}} \times \mathbf{Z}^r,$$

where r is called the rank of $E_n(\mathbf{Q})$. Again by the above, $E_n(\mathbf{Q})$ has nonzero rank iff n is a congruent number. It turns out that determining r is much more difficult than locating the rational points of finite order on E_n.

♣ Another direction in which to generalize the Pythagorean problem is to find rational points on the algebraic curve

$$\{(x, y) \in \mathbf{R}^2 \mid x^n + y^n = 1\}$$

for $n \geq 3$. The graph of $x^4 + y^4 = 1$ is depicted in Figure 3.18.

The method of rational slopes breaks down, despite the fact that $(1, 0)$ and $(0, 1)$ are rational points. (Try to pursue this for $n = 3$.) As in the Pythagorean case, we can reformulate this problem to finding all positive integer solutions of the equation

$$a^n + b^n = c^n, \quad n \geq 3.$$

This problem goes back to Fermat[10] (1601–1665), who wrote the following marginal note in his copy of Diophantus's *Arithmetica*:

> It is impossible to write a cube as a sum of two cubes, a fourth power as a sum of fourth powers, and, in general, any power beyond the second as a sum of two similar powers. For this, I have discovered a truly wonderful proof, but the margin is too small to contain it.

Fermat thus claimed that there is no all-positive solution of this equation for any $n \geq 3$. His "truly wonderful proof" went with him to the grave, and despite intense efforts of many great minds, it was

---

[10] By profession, Fermat was a lawyer and a member of the supreme court in Toulouse. He did mathematics only as a hobby.

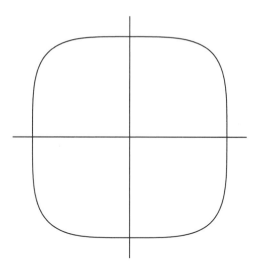

Figure 3.18

not until the year 1994 that this so-called Fermat's Last Theorem was finally proved by Andrew Wiles. On the modest side, as always, we will here reduce the problem to n an odd prime, and give some fragmentary historical notes. We set $n = 4$ and follow Fermat, who actually jotted down a sketch proof for this case, and Euler, who gave the first complete proof for this in 1747.

More generally, we claim that there is no all-positive integral solution of the equation

$$a^4 + b^4 = c^2, \quad a, b, c \in \mathbf{N}.$$

We employ here what is known as "Fermat's method of infinite descent." It starts with an all-positive solution a_0, b_0, c_0, that we assume to exist and creates another all-positive solution a, b, c with $c < c_0$. Since it cannot go on forever, we get a contradiction. (By the way, I almost forgot! ¬) We may assume that a_0, b_0, c_0 are relatively prime, as in the Pythagorean case. Substituting, we find

$$(a_0^2)^2 + (b_0^2)^2 = c_0^2,$$

so that (a_0^2, b_0^2, c_0) is a Pythagorean triple. By what we derived earlier, we have

$$a_0^2 = 2st, \quad b_0^2 = t^2 - s^2, \quad c_0 = t^2 + s^2,$$

where $t > s$, and s, t are relatively prime and of different parity. We claim that t is odd and s is even. ¬ If t is even and s is odd, then

$t^2 - s^2$ is a multiple of 4 minus 1. On the other hand, b_0 is odd, so that b_0^2 is a multiple of 4 plus 1. These cannot happen simultaneously! ¬

We can now write $s = 2r$. Substituting this into the expression for a_0^2, we obtain

$$\left(\frac{a_0}{2} \right)^2 = rt.$$

This reminds us of an analogous equation for Pythagorean triples! Since r and t are relatively prime, it follows in exactly the same way that r and t are pure squares: $t = c^2$ and $r = d^2$. Next we write the expression for b_0^2 in the Pythagorean form $s^2 + b_0^2 = t^2$ and apply the description of Pythagorean triples again to get

$$s = 2uv, \quad b_0 = u^2 - v^2, \quad t = u^2 + v^2,$$

where $u > v$, and u, v are relatively prime and of different parity. The equation

$$uv = s/2 = r = d^2$$

tells us that u and v are pure squares: $u = a^2$ and $v = b^2$. We finally have

$$c^2 = t = u^2 + v^2 = a^4 + b^4,$$

so that (a, b, c) is another solution! Comparing the values of c and c_0, we get

$$c \leq c^2 = t \leq t^2 < t^2 + s^2 = c_0,$$

and we are done. ¬

In particular, $a^4 + b^4 = c^4$ does not have any all-positive solutions. Now look at the general case $a^n + b^n = c^n$. If n is divisible by 4, say, $n = 4k$, then we can rewrite this as $(a^k)^4 + (b^k)^4 = (c^k)^4$, and, by what we have just proved, there is no positive solution in this case either. Assume now that n is not divisible by 4. Since $n \geq 3$, this implies that n is divisible by an odd prime, say p, and we have $n = kp$. We can then write the original equation as $(a^k)^p + (b^k)^p = (c^k)^p$ and Fermat's Last Theorem will be proved if we show that, for p an odd prime, no all-positive solutions exist for

$$a^p + b^p = c^p,$$

or equivalently, there are no rational points (with positive coordinates) on the *Fermat curve* defined by the equation

$$x^p + y^p = 1.$$

Now, some history. The first case, $p = 3$, although it seems to have attracted attention even before A.D. 1000, was settled by Euler in 1770 with a gap filled by Legendre. Around 1825, Legendre and Dirichlet independently settled the next case, $p = 5$. The next date is 1839, when Lamé succesfully completed a proof for $p = 7$. With the proofs getting more and more complex, it became clearer and clearer that a good way to attack the problem was to consider numbers that are more general than integers.

A good class of numbers turned out to be those that are roots of polynomial equations with integral or rational coefficients. We will investigate these in the next section. Kummer went further and, introducing the so-called "ideal numbers," managed to prove Fermat's Last Theorem for a large class of "regular" primes. Before Wiles' recent proof, one has to mention a result of Faltings in 1983 that implies that there may be only finitely many solutions (a, b, c) for a given odd prime p. A brief account on the final phase in proving Fermat's Last Theorem is as follows. ♠ The work of Hellegouarch between 1970 and 1975 revealed intricate connections between the Fermat curve and elliptic curves. This led to a suggestion made by Serre that the well-developed theory of elliptic curves should be exploited to prove results on Fermat's Last Theorem. In 1985, Frey pointed out that the elliptic curve

$$y^2 = x(x + a^p)(x - b^p),$$

where $a^p + b^p = c^p$, $a, b, c \in \mathbf{N}$, is very unlikely to exist due to its strange properties. Working on the so-called Taniyama–Shimura conjecture, in 1986, Ribet proved that the Frey curve above is not modular,[11] that is, it cannot be parametrized by "modular functions" (in a similar way as the unit circle given by the Pythagorean equation $x^2 + y^2 = 1$ can be parametrized by sine and cosine). Finally,

---

[11] For a good expository article, see R. Rubin and A. Silverberg, "A Report on Wiles' Cambridge Lectures," *Bulletin of the AMS*, *31*, 1 (1994) 15–38.

in a technical paper, Wiles showed that elliptic curves of the form

$$y^2 = x(x - r)(x - s)$$

can be parametrized by modular functions provided that r and s are relatively prime integers such that $rs(r - s)$ is divisible by 16. This, applied to the Frey curve with $r = -a^p$ and $s = b^p$ finally gives a contradiction since $rs(r - s) = a^p b^p c^p$ is certainly divisible by 16 for $p \geq 5$ (as one of the numbers a, b, c must be even).

Problems

1. Following Euclid, prove the Pythagorean theorem by working out the areas of triangles in Figure 3.19.

2. Show that the radius of the inscribed circle of a right triangle with integral side lengths is an integer.

3. Use the general form of Pythagorean triples to prove that $12|ab$ and $60|abc$ for any Pythagorean triple (a, b, c).

4. Show that the only Pythagorean triple that involves consecutive numbers is $(3, 4, 5)$.

5. Find all integral solutions $a, b \in \mathbf{Z}$ of the equation

$$a^{2b} + (a + 1)^{2b} = (a + 2)^{2b}.$$

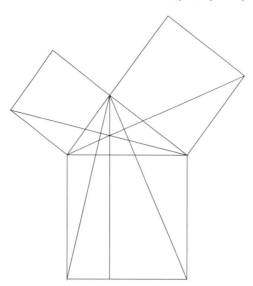

Figure 3.19

6. Find all right triangles with integral side lengths such that the area is equal to the perimeter.

7. Show that $a^2 + b^2 = c^3$ has infinitely many solutions.

8. Use the chord-method to find all rational points on the curves: (a) $y^2 = x^3 + 2x^2$; (b) $y^2 = x^3 - 3x - 2$.

9. Consider Bachet's curve $y^2 = x^3 + c$, $c \in \mathbf{Z}$. Show that if (x, y), $y \neq 0$, is a rational point, then
$$\left(\frac{x^4 - 8cx}{4y^2}, \frac{-x^6 - 20cx^3 + 8c^2}{8y^3} \right)$$
is also a rational point on this curve. $\heartsuit$ Use calculus to verify that this is the second intersection of the tangent line to the curve at (x, y).

10. Find all rational points on the circle with center at the origin and radius $\sqrt{2}$.

11. Prove that on an elliptic curve in Weierstrass form $y^2 = P(x)$, the points ($\neq O$) of order 2 are the intersection points of the curve with the first axis.

12. (a) Devise a proof of irrationality of $\sqrt{2}$ using Fermat's method of infinite descent. (b) Demonstrate (a) by paper folding: Assume that $\sqrt{2} = a/b$, $a, b \in \mathbf{N}$, consider a square paper of side length b and diagonal length a, and fold a corner of the square along the angular bisector of a side and an adjacent diagonal.

13. Use a calculator to work out the first two iterates of the duplication formula for the Bachet equation $y^2 = x^3 - 2$, starting from $(3, 5)$.

14. Does there exist a right triangle with integral side lengths whose area is 78?

15. Use the result of Gauss in Section 2 on the rational roots of a polynomial with integer coefficients to describe the rational points on the graph of the polynomial.

16. ♠ Let C be the cubic cusp given by $y^2 = x^3$. (a) Show that the set $C_{ns}(\mathbf{Q})$ of all nonsingular rational points ("ns" stands for nonsingular) forms a group under addition given by the chord method (the identity is placed at the vertical ideal point). (b) Verify that the map $\phi : C_{ns}(\mathbf{Q}) \to \mathbf{Q}$ defined by $\phi(x, y) = x/y$ and $\phi(O) = 0$ is an isomorphism. (Since the additive group of $\mathbf{Q}$ is not finitely generated, this shows that Mordell's theorem does not extend to singular curves!)

17. Derive the analytical formulas for adding points on an elliptic curve in Weierstrass form in the case where the identity is placed at the vertical ideal point.

18. Placing the identity O at the vertical ideal point, show that the integer point $A = (2, 3)$ has order 6 in the group of rational points of the Bachet curve given by $y^2 = x^3 + 1$. (Hint: Rewrite $A * A = A * A * A * A$. Note that this is the largest finite order a rational point can have on a Bachet curve.)

19. Show that the cubic curve given by $x^3 + y^3 = c$ can be transformed into a Weierstrass form by the rational transformation

$$(x, y) \mapsto \left(\frac{a+y}{bx}, \frac{a-y}{bx} \right).$$

Choose $a = 36c$ and $b = 6$ to obtain the Bachet equation $y^2 = x^3 - 432c^2$. (Notice that $c = 1$ gives the birational equivalence of the Fermat curve in degree 3 and the Bachet curve given by $y^2 = x^3 - 432$. Since the former has $(1, 0)$ and $(0, 1)$ as its only rational points, it follows that (away from O, the vertical ideal point) the Bachet curve also has only two rational points. Can you find them? With O, these form a cyclic group of order 3.)

20. Show that if 1 were a congruent number, then the equation $a^4 - c^4 = b^2$ would have an all-positive integral solution with b odd, and a, c relatively prime. (Hint: If 1 were congruent, then $u^4 - 1 = v^2$ would have a rational solution. Substitute $u = a/c$ and $v = b/d$, where a, c and b, d are relatively prime. Note that a Fermat's method of infinite descent, resembling the one in the text, can be devised to show that $a^4 - c^4 = b^2$ has no all-positive integral solutions. It thus follows that 1 is not a congruent number.)

21. Use Tunnell's theorem to show that 5, 6, 7 are congruent numbers.

Web Sites

1. www-groups.dcs.st-and.ac.uk/~history/HistTopics/ Fermat's_last_theorem.html

2. www.math.niu.edu/~rusin/papers/known-math/elliptic.crv/

4

SECTION

Algebraic or Transcendental?

♣ We managed to split the real numbers into two disjoint subsets: the rationals and irrationals. Is there a further split of the irrationals? For example, which is more subtle: $\sqrt{2}$ or e? For the answer, we go back to $\mathbf{Q}$ and make the following observation: If $x \in \mathbf{Q}$, then, writing $x = a/b$, $a, b \in \mathbf{Z}$, we see that x is the root of the linear equation

$$a - bx = 0$$

with integral coefficients. Raising the degree by one, we see that $\sqrt{2}$ has the same property; i.e., it is a root of the quadratic equation

$$x^2 - 2 = 0,$$

again with integral coefficients. We are now motivated to introduce the following definition: A real number r is *algebraic* if it is a root of a polynomial equation

$$c_0 + c_1 x + \ldots + c_n x^n = 0, \quad c_n \neq 0$$

with integral (or what is the same, rational; see Problem 1) coefficients. The least degree n is called the *degree* of r. A number is called *transcendental* if it is not algebraic.

We see immediately that the degree 1 algebraic numbers are the rationals and that $\sqrt{2}$ is an algebraic number of degree 2. What about e and π? It turns out that they are both transcendental, but the proof (especially for π) is not easy. A proof of transcendentality for e was first given by Hermite in 1873 and simplified considerably by Hilbert in 1902. Transcendentality of π was first proved by Lindemann.[1] To appreciate these revolutionary results, one may note the scepticism that surrounded transcendentality in those days (especially coming from the constructivists). As Kronecker, the leading contemporary to Lindemann, noted: "Of what use is your beautiful research on π? Why study such problems, since there are no irrational numbers at all?"

♡ From the point of view of abstract algebra, **Q** is a subfield of **R**. Are there any fields between **Q** and **R**? The answer is certainly yes; just pick an irrational number r and consider the smallest subfield of **R** that contains both **Q** and r. This subfield is denoted by **Q**(r). The structure of **Q**(r) depends on whether r is algebraic or transcendental.

♣ To see what happens when r is algebraic, consider $r = \sqrt{2}$. Since all even powers of $\sqrt{2}$ are in **N** $\subset$ **Q**, an element of **Q**$(\sqrt{2})$ is of the form

$$\frac{a_1 + b_1\sqrt{2}}{a_2 + b_2\sqrt{2}}, \quad a_1, a_2, b_1, b_2 \in \mathbf{Q},$$

with the assumption that a_2 and b_2 do not vanish simultaneously. Rationalizing the denominator now gives

$$\frac{a_1 + b_1\sqrt{2}}{a_2 + b_2\sqrt{2}} \cdot \frac{a_2 - b_2\sqrt{2}}{a_2 - b_2\sqrt{2}} = \frac{a_1a_2 - 2b_1b_2 + (a_2b_1 - a_1b_2)\sqrt{2}}{a_2^2 - 2b_2^2}$$

$$= \frac{a_1a_2 - 2b_1b_2}{a_2^2 - 2b_2^2} + \frac{a_2b_1 - a_1b_2}{a_2^2 - 2b_2^2}\sqrt{2}.$$

---

[1] For an interesting account, see F. Klein, *Famous Problems of Elementary Geometry*, Chelsea, New York, 1955. For a recent treatment, see A. Jones, S. Morris, and K. Pearson, *Abstract Algebra and Famous Impossibilities*, Springer, 1991.

The fractions are rational, so we conclude that $\mathbf{Q}(\sqrt{2})$ consists of numbers of the form

$$a + b\sqrt{2}, \quad a, b \in \mathbf{Q}.$$

In particular, $\mathbf{Q}(\sqrt{2})$ is a vector space of dimension 2 over $\mathbf{Q}$ (with respect to ordinary addition in $\mathbf{Q}(\sqrt{2}) \subset \mathbf{R}$ and multiplication of elements in $\mathbf{Q}(\sqrt{2})$ by $\mathbf{Q}$). The generalization is clear. Given an algebraic number r of degree n over $\mathbf{Q}$, the field $\mathbf{Q}(r)$ is a vector space over $\mathbf{Q}$ of dimension n.

EXAMPLE

For $n \in \mathbf{N}$, define $T_n, U_n : [-1, 1] \to \mathbf{R}$ by $T_n(x) = \cos(n \cos^{-1}(x))$ and $U_n(x) = \sin((n + 1) \cos^{-1}(x))/\sin(\cos^{-1}(x))$, $x \in [-1, 1]$. By the addition formulas, we have

$$\cos((n + 1)\alpha) = \cos(n\alpha)\cos(\alpha) - \sin(n\alpha)\sin(\alpha),$$

$$\sin((n + 2)\alpha) = \sin((n + 1)\alpha)\cos(\alpha) + \cos((n + 1)\alpha)\sin(\alpha).$$

For $\alpha = \cos^{-1}(x)$, these can be rewritten as

$$T_{n+1}(x) = xT_n(x) - (1 - x^2)U_{n-1}(x),$$

$$U_{n+1}(x) = xU_n(x) + T_{n+1}(x).$$

Since $T_1(x) = x$ and $U_0(x) = 1$, these recurrence relations imply that T_n and U_n are polynomials. We obtain that $\cos(\pi/n)$ is algebraic, since it is a root of the polynomial $T_n + 1$. (What about $\sin(\pi/n)$?) □

If r is transcendental, then $\mathbf{Q}(r)$ is isomorphic to the *field of all rational functions*

$$\frac{a_0 + a_1 r + \cdots + a_n r^n}{b_0 + b_1 r + \cdots + b_m r^m}, \quad a_n \neq 0 \neq b_m,$$

with integral coefficients. In this case, we have every reason to call r a *variable* over $\mathbf{Q}$.

◇ We finish this section by exhibiting infinitely many transcendental numbers. Let $a \geq 2$ be an integer. We claim that

$$r = \sum_{j=1}^{\infty} \frac{1}{a^{j!}}$$

is transcendental. To show convergence of the infinite series, we first note that

$$\frac{1}{a^{j!}} \le \frac{1}{2^j},$$

with sharp inequality for $j \ge 2$. Thus

$$\sum_{j=1}^{\infty} \frac{1}{a^{j!}} < \sum_{j=1}^{\infty} \frac{1}{2^j} = \frac{1}{1-(1/2)} - 1 = 1,$$

and so the series defining r converges, and the sum gives a real number $r \in (0, 1)$. Now for transcendentality: ¬ Assume that r is the root of a polynomial

$$P(x) = c_0 + c_1 x + \cdots + c_n x^n, \quad c_0, \ldots, c_n \in \mathbf{Z}, \quad c_n \ne 0.$$

For $k \in \mathbf{N}$, let

$$r_k = \sum_{j=1}^{k} \frac{1}{a^{j!}}$$

be the kth partial sum. We now apply the Mean Value Theorem of calculus for P on $[r_k, r]$ and obtain $\theta_k \in [r_k, r]$ such that the slope of the line through $(r_k, P(r_k))$ and $(r, P(r)) = (r, 0)$ is equal to the slope of the tangent to the graph of P at θ_k:

$$\frac{P(r) - P(r_k)}{r - r_k} = P'(\theta_k)$$

(Figure 4.1). Taking absolute values, we rewrite this as

$$|P(r_k)| = |P(r) - P(r_k)| = |r - r_k| \cdot |P'(\theta_k)|.$$

We now estimate each term as follows: For $l = 0, \ldots, n$, $c_l r_k^l$ is a rational number with denominator $a^{l \cdot k!}$. Thus,

$$|P(r_k)| = |c_0 + c_1 r_k + \cdots + c_n r_k^n| \ge 1/a^{n \cdot k!}.$$

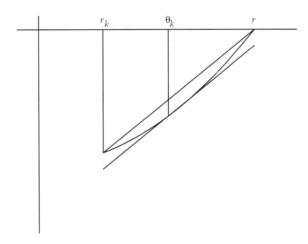

r_k θ_k r

Figure 4.1

Second, we have

$$|r - r_k| = \sum_{j=k+1}^{\infty} \frac{1}{a^{j!}}$$

$$= \frac{1}{a^{(k+1)!}} \left(1 + \frac{1}{a^{(k+2)!-(k+1)!}} + \frac{1}{a^{(k+3)!-(k+1)!}} + \cdots \right)$$

$$< \frac{1}{a^{(k+1)!}} \frac{1}{1 - (1/a)},$$

where we applied the geometric series formula (after comparison). Finally, taking the derivative of P:

$$|P'(\theta_k)| = |c_1 + 2c_2\theta_k + \cdots + nc_n\theta_k^{n-1}|$$

$$\leq |c_1| + 2|c_2| + \cdots + n|c_n| = c, \quad c \in \mathbf{N},$$

where we used $0 \leq \theta_k \leq 1$. Putting all these together, we arrive at

$$\frac{1}{a^{n \cdot k!}} \leq \frac{1}{a^{(k+1)!}} \frac{c}{1 - (1/a)}.$$

For k large, this is impossible since $a^{(k+1)!}$ grows faster than $a^{n \cdot k!}$. ¬
For $a = 10$, we obtain transcendentality of the number

$$\sum_{j=1}^{\infty} \frac{1}{10^{j!}} = 0.110001000000000000000001000\ldots.$$

This goes back to Liouville in 1844.

Remark 1.

A deep theorem of Gelfond and Schneider, proved in 1935, asserts that if $r \neq 0, 1$ is algebraic and s is not rational, then r^s is transcendental. (r, s can both be complex; see Section 5.) For example, $2^{\sqrt{2}}$ and e^π ($= i^{-2i}$; see Section 15) are transcendental. In 1966, Alan Baker[2] found effective close approximations for sums of natural logarithms, and proved transcendence results for sums of numbers of the form $a \ln b$, where a, b are algebraic. In particular, he reproved transcendence of $\sqrt{2}^{\sqrt{2}}$ and transcendence of the Gregory numbers $t_{a/b} = \tan^{-1}(b/a)$, $a, b \in \mathbf{N}$. ($t_{a/b}$ is the imaginary part of the complex logarithm of the Gaussian integer $a + bi \in \mathbf{Z}[i]$.) Finally, going back to geometry, we note that according to a result of Schneider in 1949, the perimeter of an ellipse with algebraic semimajor axes is transcendental.

Remark 2.

We saw in Section 2 that the only rational point on the graph of the exponential function is $(0, 1)$. As Lindemann proved, with the exception of $(0, 1)$, the graph actually avoids all points with algebraic coordinates.

Problems

1. Show that if a real number is the root of a polynomial with rational coefficients, then it is also the root of a polynomial with integer coefficients.

2. Show that the numbers $\sqrt{a} - \sqrt{b}$, $a, b \in \mathbf{N}$, are algebraic. What is the degree?

3. Let c be a positive real number. Prove that if c is algebraic, then so is $\sqrt{c}$.

4. Verify that if c_1 and c_2 are algebraic, then so is $c_1 + c_2$.

5. ♠ Show that the fields $\mathbf{Q}(\sqrt{2})$ and $\mathbf{Q}(\sqrt{3})$ are not isomorphic.

6. Prove the divison algorithm for polynomials: Given polynomials P, S there exist polynomials Q, R such that $P = QS + R$, where R is zero or has degree less than the degree of S. Moreover, Q and R are unique. In addition, show that if P and S have rational coefficients, then so do Q and R.

---

[2] See "Linear Forms in the Logarithms of Algebraic Numbers I–II–III–IV," *Mathematica*, 13 (1966) 204–216; 14 (1967) 102–107; 14 (1967) 220–228; 15 (1968) 204–216.

7. Let p be a prime. Assume that a polynomial P with rational coefficients has a root of the form $a + b\sqrt{p}$, where $a, b \in \mathbf{Q}$. Show that $a - b\sqrt{p}$ is also a root. (Hint: Divide P by $(x - (a + b\sqrt{p}))(x - (a - b\sqrt{p}))$, and consider the remainder.)

8. ♠ Generalize Problem 7 to polynomials over a field and a quadratic extension.

9. Use a calculator to show that the polynomial $x^6 - 7.5x^3 - 19x + 2.1$ evaluated on the Liouville's number $\sum_{j=1}^{\infty} 1/10^{j!}$ vanishes to 8-decimal precision.

10. ◇ Let $m \neq 1$ be positive. Show that the equation

$$e^x = mx + 1$$

has a unique nonzero solution x_0. In addition, show that x_0 is irrational if m is rational, and x_0 is transcendental if m is algebraic.

5

SECTION

Complex Arithmetic

♣ We now turn the setting around and ask the following: Given a polynomial equation

$$c_0 + c_1 x + \cdots + c_n x^n = 0, \qquad c_n \neq 0, \qquad c_0, c_1, \ldots, c_n \in \mathbf{R},$$

what can be said about the roots?

This is one of the oldest problems of algebra. The factor theorem says that there are at most n roots. It also says that it is enough to prove the existence of one root r, since after dividing by the root factor $(x - r)$, the quotient has degree $n - 1$, to which the existence result can again be applied. If we pretend that we do not know the FTA (fundamental theorem of algebra), then the existence of roots seems to carry some inherent difficulty. For example, the quadratic equation

$$x^2 + 1 = 0$$

is unsolvable in $\mathbf{R}$ since the left-hand side is always ≥ 1 for x real. This, of course, we knew all along. In general, the solutions of the quadratic polynomial equation

$$ax^2 + bx + c = 0$$

are given by the quadratic formula

$$r_1, r_2 = \frac{-b \pm \sqrt{b^2 - 4ac}}{2a},$$

with r_1 corresponding to the positive and r_2 to the negative sign in front of the square root. This is an easy exercise in completing the square. (To show this in the Greek way, we first divide through by a and obtain

$$x^2 + px + q = 0,$$

where $p = b/a$ and $q = c/a$. Consider the expression $x^2 + px$ as the area of a square with side length x plus the areas of two rectangles of side lengths $p/2$ and x. When the rectangles are joined with the square along two adjacent sides of the square, they form a larger square of side length $x + p/2$ with a small square of side length $p/2$ missing. Inserting the missing small square completes the big square, and we obtain $x^2 + px + (p/2)^2 = (x + p/2)^2$. The quadratic formula follows, since

$$x^2 + px + q = (x + p/2)^2 - (p/2)^2 + q,$$

so that the roots are $-p/2 \pm \sqrt{(p/2)^2 - q}$.) Now, depending on whether $b^2 - 4ac$ is positive, zero, or negative, we have two, one, or no roots among the reals. Thus, it is clear that real numbers are insufficient and that we have to introduce new numbers. These should form a field if we want to do 'decent' mathematics. Among the new numbers, there must be one that is a root of $x^2 + 1 = 0$, as above. Thus, we introduce the *complex unit i* with the property

$$i^2 + 1 = 0.$$

The complex unit i is sometimes denoted by $\sqrt{(-1)}$ for obvious reasons. We will use only the former to prevent silly errors such as $1 = \sqrt{1} = \sqrt{(-1)(-1)} = \sqrt{-1}\sqrt{-1} = -1$. For degree 2 polynomials, the problem is now solved. Indeed, for $b^2 - 4ac < 0$, the quadratic formula can be written as

$$r_1, r_2 = -\frac{b}{2a} \pm \frac{\sqrt{-(4ac - b^2)}}{2a} = -\frac{b}{2a} \pm \frac{\sqrt{4ac - b^2}}{2a} i.$$

(Notice that the first equality is somewhat heuristic; nevertheless the final expression does satisfy the quadratic equation.)

We now consider the field $\mathbf{R}(i)$ (an algebraic extension of $\mathbf{R}$ of degree 2) consisting of all elements of the form

$$\frac{a_1 + b_1 i}{a_2 + b_2 i}, \quad a_1, a_2, b_1, b_2 \in \mathbf{R},$$

with the assumption that a_2 and b_2 do not vanish simultaneously. (The reason for considering only linear fractions is clear; all even powers of i are real.) We now "rationalize the denominator":

$$\frac{a_1 + b_1 i}{a_2 + b_2 i} = \frac{a_1 + b_1 i}{a_2 + b_2 i} \cdot \frac{a_2 - b_2 i}{a_2 - b_2 i} = \frac{(a_1 a_2 + b_1 b_2)}{a_2^2 + b_2^2} + \frac{(a_2 b_1 - a_1 b_2)}{a_2^2 + b_2^2} i.$$

The fractions are real, so we conclude that $\mathbf{R}(i)$ consists of numbers of the form

$$z = a + bi, \quad a, b \in \mathbf{R}.$$

For notational convenience, we write $\mathbf{C} = \mathbf{R}(i)$ and call it the *complex number field*. The typical element z above is called a *complex number*. $a = \Re(z)$ is the *real part* of z and $b = \Im(z)$ is the *imaginary part* of z. The complex number z is real iff $\Im(z) = 0$ and *purely imaginary* iff $\Re(z) = 0$. Notice finally that rationalizing the denominator as above gives a formula for complex division!

In view of the analogy between the algebraic field extensions $\mathbf{Q}(\sqrt{2})$ in Section 4 and $\mathbf{R}(i)$ above, it is perhaps appropriate to remember Titchmarsh's words:

> There are certainly many people who regard $\sqrt{2}$ as something perfectly obvious, but jib at $\sqrt{(-1)}$. This is because they think they can visualize the former as something in physical space, but not the latter. Actually $\sqrt{(-1)}$ is a much simpler concept.

Summarizing, every quadratic polynomial equation has exactly two roots among the complex numbers. If $b^2 - 4ac = 0$, we say that we also have two roots that happen to coincide. (The elegant phrase "root with multiplicity 2" is not without reason; think again of the corresponding root factors!)

We now pin our hopes on $\mathbf{C}$ for solving cubic, quartic, quintic, etc. polynomial equations. Our doubts are not completely unfounded if we go back to Roman times when only rational numbers were accepted, and pretend that we want to invent reals by considering

$\mathbf{Q}(\sqrt{2})$ (and alike), which is definitely not the whole of $\mathbf{R}$. Fortunately this is not the case, and we can stop pretending that we did not know the

Fundamental Theorem of Algebra.

Every polynomial equation

$$c_0 + c_1 z + \cdots + c_n z^n = 0, \quad c_n \neq 0,$$

with complex coefficients $c_0, c_1, \ldots, c_n \in \mathbf{C}$, has a complex root.

Having been stated as early as 1629 by Albert Girard, the FTA was first proved by Gauss in 1799. Earlier attempts were made by D'Alembert (1746), Euler (1749), and Lagrange (1772). Many elementary proofs exist. A typical proof uses Liouville's theorem in complex analysis which, in turn, is based on a trivial estimate of the Cauchy formula. (By the way, have you noticed that almost everything in one-variable complex calculus goes back to the Cauchy formula?) To do something out of the ordinary, we prove the FTA by a "differential topological" argument.[1] Note that four additional proofs are outlined in Problems at the end of Section 8.

Before proving the FTA, we explore some geometric features of complex arithmetic. To define a complex number $z = a + bi$, we need to specify its real and imaginary parts a and b. Put together, they form a point (a, b) on the Cartesian plane or, similarly, they form a plane vector. This representation of a complex number as a plane vector is called the *Argand diagram*. This geometric representation of complex numbers is named after Jean Robert Argand, 1806, although nine years earlier it had been announced by the Norwegian cartographer Caspar Wessell before the Danish Academy of Sciences.

You may say, "Well, in Euclidean plane geometry we talked about points on the plane. In calculus, we were told that a vector drawn from the origin is essentially the same thing as a point. Now we are saying that a planar point can also be thought of as a complex

---

[1] See J. Milnor, *Topology from the Differentiable Viewpoint*, The University Press of Virginia, 1990. For a recent comprehensive study of the FTA and its relation to various mathematical disciplines, see B. Fine and G. Rosenberger, *The Fundamental Theorem of Algebra*, Springer, 1997.

number, an element of an abstract field. Why can we not decide on just one of these?" The answer is that all three are conceptually different, but are represented by the same mathematical model. As you go on, these kinds of identifications become more and more common.

Thus, there is a one-to-one correspondence between $\mathbf{C}$ and the plane $\mathbf{R}^2$; we are thereby entitled to call this the Gauss plane or *complex plane*. The field operations in $\mathbf{C}$ carry over to operations on planar vectors. Addition gives nothing new; $z_1 = a_1 + b_1 i$ and $z_2 = a_2 + b_2 i$ add up to

$$z_1 + z_2 = (a_1 + a_2) + (b_1 + b_2)i,$$

and we see that complex addition corresponds to the usual "parallelogram rule" for planar vectors. The product

$$z_1 z_2 = (a_1 a_2 - b_1 b_2) + (a_1 b_2 + a_2 b_1)i$$

is more subtle and needs to be analyzed. To do this, we rewrite z_1 and z_2 in polar coordinates:

$$z_1 = r_1(\cos \theta_1 + i \sin \theta_1),$$
$$z_2 = r_2(\cos \theta_2 + i \sin \theta_2)$$

and compute

$$z_1 z_2 = r_1 r_2(\cos \theta_1 \cos \theta_2 - \sin \theta_1 \sin \theta_2)$$
$$+ r_1 r_2(\cos \theta_1 \sin \theta_2 + \sin \theta_1 \cos \theta_2)i$$
$$= r_1 r_2(\cos(\theta_1 + \theta_2) + i \sin(\theta_1 + \theta_2)),$$

where we used the addition formulae for sine and cosine. This tells us two things: First, introducing the *absolute value*, or modulus, of a complex number $z = a + bi = r(\cos \theta + i \sin \theta)$ as the length of the corresponding vector

$$|z| = \sqrt{a^2 + b^2} = r,$$

we have

$$|z_1 z_2| = |z_1| \cdot |z_2|.$$

Second, introducing the multiple-valued *argument*

$$\arg z = \{\theta + 2k\pi \mid k \in \mathbf{Z}\},$$

we also have

$$\arg(z_1 z_2) = \arg z_1 + \arg z_2$$

(as sets!). The geometry behind this is the following: Consider $\arg : \mathbf{C} \to \mathbf{R}$ as a multiple-valued function on $\mathbf{C} = \mathbf{R}^2$. Its graph is a surface in $\mathbf{R}^3$ that resembles an infinite spiral staircase; multiplying z_1 by z_2 has the effect of walking up and down the staircase (see Color Plate 2b). In particular, consider complex numbers of modulus one:

$$z(\theta) = \cos\theta + i\sin\theta, \quad \theta \in \mathbf{R}.$$

They fill the *unit circle*

$$S^1 = \{z \in \mathbf{C} \mid |z| = 1\}.$$

Since

$$z(\theta_1)z(\theta_2) = z(\theta_1 + \theta_2), \quad \theta_1, \theta_2 \in \mathbf{R},$$

S^1 is a multiplicative subgroup of $\mathbf{C}^\# = \mathbf{C} - \{0\}$. ♡ In fact, as the identity shows, $\theta \mapsto z(\theta)$, $\theta \in \mathbf{R}$, is a homomorphism of the additive group $\mathbf{R}$ onto the multiplicative group S^1 with kernel $2\pi\mathbf{Z}$. Note also that, using the complex exponential function, $z(\theta) = e^{i\theta}$ by the Euler formula (proved in Section 15). Since we are using only the multiplicative property of $z(\theta)$ above, we can postpone the introduction of complex exponentials.

♣ Fix $n \in \mathbf{N}$. The complex numbers

$$z(2k\pi/n), \quad k = 0, 1, \ldots, n-1,$$

are (for $n \geq 3$) the vertices of a *regular n-sided polygon* inscribed in S^1 (with a vertex on the positive first (or real) axis). By the identity above, for fixed n these vertices form a multiplicative subgroup of S^1 (and thereby of $\mathbf{C}^\#$) that is isomorphic with $\mathbf{Z}_n$, the group of congruence classes of integers modulo n. This is a convenient way to view $\mathbf{Z}_n$, since we can use the field operations in $\mathbf{C}$ to discover some of the geometric features of this configuration. Here is one:

Proposition 3.

For $n \geq 2$, we have

$$\sum_{k=0}^{n-1} z\left(\frac{2k\pi}{n}\right) = 0.$$

Remark.

Geometrically, the sum of vectors from the centroid to the vertices of a regular polygon is zero.

Proof.

This is clear for n even, but we give a proof that is independent of the parity of n. First, by complex multiplication,

$$z\left(\frac{2k\pi}{n}\right) = z\left(\frac{2\pi}{n}\right)^k,$$

so that

$$\sum_{k=0}^{n-1} z\left(\frac{2k\pi}{n}\right) = 1 + \omega + \omega^2 + \cdots + \omega^{n-1},$$

where $\omega = z(2\pi/n)$. This looks very familiar! In fact, switching from real to complex, the proof of Proposition 2 in Section 2 gives

$$1 + z + z^2 + \cdots + z^{n-1} = \frac{1 - z^n}{1 - z}, \quad z \neq 1, \quad z \in \mathbf{C}.$$

Substituting $z = \omega$ and noting that $\omega^n = z(2n\pi/n) = z(2\pi) = 1$, the proposition follows. ∎

The first idea used in the proof gives another interpretation of the numbers $z(2k\pi/n)$, $k = 0, \ldots, n - 1$. In fact, raising them to the nth power, we get

$$z\left(\frac{2k\pi}{n}\right)^n = z\left(\frac{2kn\pi}{n}\right) = z(2\pi k) = 1,$$

so that they all satisfy the equation

$$z^n - 1 = 0.$$

By the complex version of the factor theorem, there are at most n roots; so we got them all! We call $z(2k\pi/n)$, $k = 0, \ldots, n - 1$, the *nth roots of unity*. The number $\omega = z(2\pi/n)$ is called a *primitive nth root of unity*. The powers $\omega^k = z(2k\pi/n)$, $k = 0, \ldots, n-1$, give all nth roots of unity. We now begin to believe that complex numbers should suffice, and that after all, the FTA ought to be true!

The nth roots of a complex number z can be written down in a convenient way using the polar form

$$z = |z| \cdot z(\theta), \quad \theta \in \arg z.$$

Indeed, we have

$$\sqrt[n]{z} = \sqrt[n]{|z|}\sqrt[n]{z(\theta + 2k\pi)} = \sqrt[n]{|z|}z(\theta/n)\omega^k, \quad k \in \mathbf{Z}.$$

For $z \neq 0$, these complex numbers are distinct for $k = 0, \ldots, n-1$. Notice that they have the same absolute value and are equally spaced. Geometrically, they are the vertices of a regular n-sided polygon inscribed in a circle with radius $\sqrt[n]{|z|}$ and center at the origin.

Remark.

If $p \geq 3$ is odd, then the solutions of the equation $z^p + 1 = 0$ are the negatives of the pth roots of unity $z = -z(2k\pi/p) = -z(2\pi/p)^k$, $k = 0, \ldots, p-1$. Setting $\omega = z(2\pi/p)$, the factor theorem says that

$$z^p + 1 = (z + 1)(z + \omega) \cdots (z + \omega^{p-1}).$$

Letting $z = a/b$, $a, b \in \mathbf{Z}$, we obtain the identity

$$a^p + b^p = (a + b)(a + \omega b) \cdots (a + \omega^{p-1}b).$$

Going back to Fermat's Last Theorem, we see that if there is an all-positive solution to

$$a^p + b^p = c^p,$$

then

$$c^p = (a + b)(a + \omega b) \cdots (a + \omega^{p-1}b)$$

must hold. $\heartsuit$ The right-hand side is a factorization of c^p with factors not quite in the ring $\mathbf{Z}$ but in the ring $\mathbf{Z}[\omega]$ obtained from $\mathbf{Z}$ by "adjoining" ω. If the factors have no common divisors in this extended ring, then, based on the analogy with the unique factorization in $\mathbf{Z}$, we would think that each factor must be a pth power. Looking back to Pythagorean triples, we see that this would be an essential step toward a resolution of Fermat's Last Theorem. This idea goes back to Kummer.[2] Unfortunately, in $\mathbf{Z}[\omega]$, unique factorization does not hold for all p (in fact, the first prime when it fails is $p = 23$ as was recognized by Cauchy), so he had to adjoin further "ideal numbers" to this ring. In this way, he managed to prove Fermat's Last Theorem for a large class of primes p.

To illustrate how far-reaching the validity of unique factorization in $\mathbf{Z}[\omega]$ is, we now consider the case $n = 4$ ($\omega = i$) and prove that unique factorization in the square lattice $\mathbf{Z}[i]$ of Gaussian integers (cf. Section 2) *implies* that any prime p of the form $4m+1$ is a sum of squares of two integers. This is a result of Fermat already discussed at the end of Section 1. Let $p = 4m + 1$, $m \in \mathbf{N}$, be a prime in $\mathbf{Z}$. By Wilson's criterion for primality (usually proved in an introductory course in number theory), p must divide $(p - 1)! + 1$. We write the latter as

$$(p - 1)! + 1 = (4m)! + 1$$

$$= 1 \cdot 2 \cdots (2m - 1)(2m)(2m + 1)(2m + 2)$$

$$\cdots (4m - 1)(4m) + 1.$$

Modulo p this is equal to

$$1 \cdot 2 \cdots (2m-1)(2m)(-2m)(-2m+1) \cdots (-2)(-1)+1 = (2m)!^2+1.$$

This number lives in $\mathbf{Z}$, but factors in $\mathbf{Z}[i]$ as

$$(2m)!^2 + 1 = ((2m)! + i)((2m)! - i).$$

Since p divides the left-hand side in $\mathbf{Z}$, it also divides the right-hand side in $\mathbf{Z}[i]$. Clearly, p cannot divide the factors. Thus, by a mild generalization of Euclid's characterization of primes (Book IX of the *Elements*), p cannot be a prime in $\mathbf{Z}[i]$! Hence p must have proper divisors. A proper divisor must have the form $a + bi$, where $b \neq 0$, since p is a prime in $\mathbf{Z}$. Since p is real, the conjugate $a - bi$

---

[2]Emil Grosswald's book *Topics from the Theory of Numbers* (Macmillan, New York, 1966) gives an excellent account of Kummer's work.

also divides p. We obtain that $(a + bi)(a - bi) = a^2 + b^2$ divides p. Since $a^2 + b^2$ is real and p is a prime in $\mathbf{Z}$, we arrive at $p = a^2 + b^2$.

Problems

1. Show that the roots r_1, r_2 of a quadratic polynomial $ax^2 + bx + c$ can be obtained in the following geometric way:[3] Consider the points $O = (0, 0)$, $A = (1, 0)$, $B = (1, -b/a)$, and $C = (1 - c/a, -b/a)$. Let the circle with diameter OC intersect the line through A and B at points P and Q. Then r_1 and r_2 are the (signed) lengths of the segments AP and AQ. Is there a generalization of this construction to cubic polynomials?

2. Find all values of $\sqrt[3]{-i}$.

3. Let $z_1, z_2, z_3, z_4 \in \mathbf{C}$ with $z_1 \neq z_2$ and $z_3 \neq z_4$. Show that the line through z_1, z_2 is perpendicular to the line through z_3, z_4 iff $\Re((z_1 - z_2)/(z_3 - z_4)) = 0$.

4. Factor the quartic polynomial $z^4 + 4$ first over $\mathbf{C}$ and then over $\mathbf{R}$.

5. Show that the triangles with vertices z_1, z_2, z_3 and w_1, w_2, w_3 are similar iff the complex determinant

$$\begin{vmatrix} 1 & 1 & 1 \\ z_1 & z_2 & z_3 \\ w_1 & w_2 & w_3 \end{vmatrix}.$$

   is equal to zero.

6. ♡ Show that unique factorization does not hold in $\mathbf{Z}[\sqrt{5}i]$.

7. Use the complex identity $|z_1|^2|z_2|^2 = |z_1 z_2|^2$, $z_1, z_2 \in \mathbf{C}$, to show that if a and b are both sums of squares of integers, then ab is also a sum of squares of integers. ♡ Use the results of Section 1 to prove the following theorem of Fermat: A positive integer a is the sum of two squares of integers iff in the prime factorization of a the primes $4k + 3$ occur with even exponents.

---

[3]Cf. G.H. Hardy, *A Course of Pure Mathematics*, Cambridge University Press, 1960.

6

Quadratic, Cubic, and Quartic Equations

†[1] ♣ Beyond the pure existence of complex roots of polynomials of degree n, guaranteed by the FTA, the next question to ask concerns their constructibility in terms of radicals such as the quadratic formula for $n = 2$. Although the quadratic case was known around 300 B.C. by the Greeks,[2] it was not until François Viète (1540–1603) that the quadratic formula was cast in its present form, due to its creator's insistence on systematic use of letters to represent constants and variables. For later purposes, it is instructive to derive the quadratic formula in a somewhat nontraditional way. In what follows it will be convenient to normalize our polynomials by dividing through by the leading coefficient. We will call a polynomial with leading coefficient 1 *monic*. To emphasize that we are dealing with complex variables, we replace the real variable x by the complex variable z. In particular, we take our quadratic polynomial in

---

[1]This symbol indicates that the material in this section is more technical than the average text.

[2]The fact that extraction of square roots can be used to solve quadratic equations had already been recognized by the Sumerians. Furthermore, an ancient Babylonian tablet (c. 1600 B.C.) states problems that reduce to the solution of quadratic equations.

the form

$$P(z) = z^2 + pz + q = (z - z_1)(z - z_2).$$

Here we have already factored P into root factors. The complex roots z_1, z_2 are to be determined. We will allow the coefficients p and q to be complex numbers. Multiplying out, we obtain

$$z_1 + z_2 = -p \quad \text{and} \quad z_1 z_2 = q.$$

These formulas are symmetric with respect to the interchange $z_1 \leftrightarrow z_2$. Since the *discriminant*

$$\delta = (z_1 - z_2)^2$$

is also symmetric, it is reasonable to expect that we can express δ in terms of the coefficients p and q. This is indeed the case, since

$$\delta = (z_1 + z_2)^2 - 4z_1 z_2 = p^2 - 4q.$$

The two values $\pm\sqrt{\delta} = \pm(z_1 - z_2)$ can be combined with the expression for the sum of the roots above. We obtain

$$2z_1, 2z_2 = -p \pm \sqrt{\delta} = -p \pm \sqrt{p^2 - 4q}.$$

The quadratic formula for P follows. Substituting $p = b/a$ and $q = c/a$, we obtain the quadratic formula for the polynomial $az^2 + bz + c$ discussed in Section 5.

The solution of the cubic equation (with no quadratic term) in terms of radicals was first obtained by del Ferro in 1515. He passed this on to some of his students. In 1535, one of the students, Fiore, challenged Fontana (nicknamed Tartaglia), who had treated some particular cases of cubics, to a public contest of solving cubic equations. Before the contest, Fontana found the solution for general cubics, and inflicted a humiliating defeat on Fiore. Bent on persuasion, Fontana told the trick to Cardano (in a poem), and allegedly swore him to secrecy that he would not reveal it to anyone. Nevertheless, Cardano (with generous references to del Ferro and Fontana) published it in his Ars Magna[3] in 1545, and the formula was subsequently named after him. To derive the Ferro–Fontana–Cardano formula for the roots of a cubic polynomial, we

---

[3] For a brief account on the story, see Oystein Ore's Foreword in the Dover edition of the *Ars Magna*.

will employ a method due to Lagrange. This is somewhat tedious, but less ad hoc than the more traditional approach (cf. Problem 1). The further advantage of presenting the Lagrange method now is that it will reappear in a more subtle setting for the solution of quintics in Section 25.

The first step is the same for all methods. We reduce the general monic cubic equation

$$z^3 + az^2 + bz + c = 0, \qquad a, b, c \in \mathbf{C},$$

to the special cubic equation

$$P(z) = z^3 + pz + q = 0$$

by means of the substitution $z \mapsto z - a/3$. (Once again, a generalization of this seemingly innocent reduction (termed the Tschirnhaus transformation later) will gain primary importance in solving quintic equations. We could also have performed this reduction for quadratic equations, but that would have been the same as completing the square, a standard way to derive the quadratic formula.)

Assume now that z_1, z_2, z_3 are the roots of P:

$$z^3 + pz + q = (z - z_1)(z - z_2)(z - z_3).$$

Multiplying out, we have

$$z_1 + z_2 + z_3 = 0, \qquad z_1z_2 + z_2z_3 + z_3z_1 = p, \qquad z_1z_2z_3 = -q.$$

Remark.

As an application, we show that z_1, z_2, z_3 are the vertices of an equilateral triangle iff

$$z_1^2 + z_2^2 + z_3^2 = z_1z_2 + z_2z_3 + z_3z_1.$$

First, we notice that this equation remains unchanged when z_1, z_2, z_3 are simultaneously subjected to the substitution $z \mapsto z-d$, where d is any complex number. (Geometrically, this corresponds to translation of the triangle by the translation vector $-d$.) Choosing $d = (z_1 + z_2 + z_3)/3$ (the centroid of the triangle), and adjusting the notation, we may thus assume that $z_1 + z_2 + z_3 = 0$ holds. Since

$$(z_1 + z_2 + z_3)^2 = z_1^2 + z_2^2 + z_3^2 + 2(z_1z_2 + z_2z_3 + z_3z_1) = 0,$$

the stated criterion splits into two equations,

$$z_1 + z_2 + z_3 = 0 \quad \text{and} \quad z_1 z_2 + z_2 z_3 + z_3 z_1 = 0.$$

Now consider the monic cubic polynomial with roots z_1, z_2, z_3. Expanding, we have

$$(z - z_1)(z - z_2)(z - z_3) = z^3 + az^2 + bz + c.$$

Our conditions translate into $a = b = 0$. Equivalently, z_1, z_2, z_3 are the solutions of the equation $z^3 + c = 0$. Since these are the three cubic roots of $-c$, they are equally spaced on the circle $|z| = \sqrt[3]{|c|}$. The claim follows. (For another solution to this problem, we could have used the factorization

$$z_1^2 + z_2^2 + z_3^2 - z_1 z_2 - z_2 z_3 - z_3 z_1 = (z_1 + z_2\omega + z_3\omega^2)(z_1 + z_2\omega^2 + z_3\omega),$$

where $\omega = z(2\pi/3)$ is a primitive third root of unity, but we preferred a less ad hoc approach.)

Returning to the main line, once again we consider the discriminant

$$\delta = \prod_{1 \le j < l \le 3} (z_j - z_l)^2 = (z_1 - z_2)^2 (z_2 - z_3)^2 (z_3 - z_1)^2.$$

A mildly unpleasant calculation gives

$$\delta = -4(z_1 z_2 + z_2 z_3 + z_3 z_1)^3 - 27(z_1 z_2 z_3)^2$$

$$= -4p^3 - 27q^2 = -108 \left(\left(\frac{p}{3}\right)^3 + \left(\frac{q}{2}\right)^2 \right).$$

(For the discriminant of the general monic cubic, see Problem 2.) Let $\omega = z(2\pi/3) = (-1 + \sqrt{3})/2$ be a primitive third root of unity, and consider the so-called *Lagrange substitutions*

$$\xi_l = \frac{1}{3} \sum_{j=1}^{3} \omega^{(j-1)l} z_j, \qquad l = 0, 1, 2.$$

Expanding the sums and using $\omega^3 = 1$, we obtain

$$3\xi_0 = z_1 + z_2 + z_3,$$

$$3\xi_1 = z_1 + \omega z_2 + \omega^2 z_3,$$

$$3\xi_2 = z_1 + \omega^2 z_2 + \omega z_3.$$

Since our cubic is reduced, ξ_0 vanishes. Moreover, ξ_1^3 and ξ_2^3 remain unchanged when the roots z_1, z_2, z_3 are subjected to even (in our case cyclic) permutations. Thus, it is reasonable to expect that ξ_1^3 and ξ_2^3 can be expressed as polynomials in p, q, and $\sqrt{\delta}$, since the latter three seem to be more "elementary" and possess the same symmetries. Once again a somewhat tedious calculation gives

$$\xi_1^3 = -\frac{q}{2} + \frac{\sqrt{3}}{18}i\sqrt{\delta} = -\frac{q}{2} + \sqrt{\left(\frac{p}{3}\right)^3 + \left(\frac{q}{2}\right)^2},$$

$$\xi_2^3 = -\frac{q}{2} - \frac{\sqrt{3}}{18}i\sqrt{\delta} = -\frac{q}{2} - \sqrt{\left(\frac{p}{3}\right)^3 + \left(\frac{q}{2}\right)^2}.$$

The second formula follows from the first because

$$9\xi_1\xi_2 = z_1^2 + z_2^2 + z_3^2 + (\omega + \omega^2)(z_1z_2 + z_2z_3 + z_3z_1)$$
$$= (z_1 + z_2 + z_3)^2 - 3(z_1z_2 + z_2z_3 + z_3z_1) = -3p.$$

Here we used Proposition 3 to the effect that $1 + \omega + \omega^2 = 0$.

Summarizing, we obtain

$$\xi_1 = \sqrt[3]{-\frac{q}{2} + \sqrt{\left(\frac{p}{3}\right)^3 + \left(\frac{q}{2}\right)^2}},$$

$$\xi_2 = \sqrt[3]{-\frac{q}{2} - \sqrt{\left(\frac{p}{3}\right)^3 + \left(\frac{q}{2}\right)^2}},$$

where the cubic roots are chosen such that $\xi_1\xi_2 = -p/3$ is satisfied. Finally, the Lagrange substitutions can be solved uniquely for z_1, z_2, z_3 in terms of ξ_1 and ξ_2, since the determinant

$$\begin{vmatrix} 1 & 1 & 1 \\ 1 & \omega & \omega^2 \\ 1 & \omega^2 & \omega \end{vmatrix} = 3(\omega^2 - \omega) = -3\sqrt{3}i$$

is nonzero. The solution is

$$z_j = \sum_{l=1}^{2} \omega^{-(j-1)l}\xi_l, \qquad j = 1, 2, 3.$$

Indeed, substituting we have

$$\sum_{j=1}^{3} \omega^{(j-1)l} z_j = \sum_{j=1}^{3} \left(\sum_{k=1}^{2} \omega^{(j-1)l} \omega^{-(j-1)k} \right) \xi_k$$

$$= \sum_{k=1}^{2} \left(\sum_{j=1}^{3} \omega^{(j-1)(l-k)} \right) \xi_k = 3\xi_l,$$

since the sum in the last set of parentheses is 3 for $k = l$ and zero for $k \neq l$. Putting everything together, we finally obtain the roots of the reduced cubic:

$$z_1 = \xi_1 + \xi_2,$$
$$z_2 = \omega^2 \xi_1 + \omega \xi_2,$$
$$z_3 = \omega \xi_1 + \omega^2 \xi_2,$$

where ξ_1 and ξ_2 are given in terms of p and q above.

Remark.
Recall that in the formulas for ξ_1 and ξ_2 we have to choose the values of the cube roots such that the constraint $\xi_1 \xi_2 = -p/3$ holds. With a fixed choice of cube roots, z_1, z_2, z_3 give the solutions to our cubic equation. Notice, however, that $\xi_l, \omega \xi_l,$ and $\omega^2 \xi_l$ are the 3 cube roots of ξ_l^3, $l = 1, 2$, so that our three solutions can be (and usually are) written more concisely as

$$\sqrt[3]{-\frac{q}{2} + \sqrt{\left(\frac{p}{3}\right)^3 + \left(\frac{q}{2}\right)^2}} + \sqrt[3]{-\frac{q}{2} - \sqrt{\left(\frac{p}{3}\right)^3 + \left(\frac{q}{2}\right)^2}}$$

with the understanding that we choose the cube roots appropriately.

We now turn to quartic equations. Cardano's student Ferrari (1522–1565) settled the problem of solving quartic equations; this is still included in Cardano's *Ars Magna*. As in the cubic case, the solution of the general quartic equation

$$z^4 + az^3 + bz^2 + cz + d = 0, \qquad a, b, c, d \in \mathbf{C},$$

can be reduced to the special case

$$P(z) = z^4 + pz^2 + qz + r = 0,$$

by means of the substitution $z \mapsto z - a/4$. Assuming that z_1, z_2, z_3, z_4 are the roots of this reduced quartic, expanding the root factors, we obtain the following equations:

$$z_1 + z_2 + z_3 + z_4 = 0, \quad z_1z_2 + z_1z_3 + z_1z_4 + z_2z_3 + z_2z_4 + z_3z_4 = p,$$

$$z_1z_2z_3 + z_1z_2z_4 + z_1z_3z_4 + z_2z_3z_4 = -q, \quad z_1z_2z_3z_4 = r.$$

To follow our earlier path, we would need to work out the discriminant

$$\delta = \prod_{1 \le j < k \le 4} (z_j - z_k)^2$$

in terms of p, q, r, but this seems a rather unpleasant task. Instead, based on the analogy with $\sqrt{\delta}$, we look for polynomials in z_1, z_2, z_3, z_4 that possess only partial symmetries when these variables are permuted. We set

$$z_1^* = (z_1 + z_2)(z_3 + z_4),$$

$$z_2^* = (z_1 + z_3)(z_2 + z_4),$$

$$z_3^* = (z_1 + z_4)(z_2 + z_3).$$

Each of these new variables is symmetric only with respect to specific permutations of z_1, z_2, z_3, z_4. The elementary symmetric polynomials $z_1^* + z_2^* + z_3^*$, $z_1^*z_2^* + z_2^*z_3^* + z_3^*z_1^*$ and $z_1^*z_2^*z_3^*$ produced from z_1^*, z_2^*, z_3^*, however, are symmetric with respect to all permutations, and thereby they should be expressible in terms of the coefficients p, q, r! For example, expanding, we obtain

$$z_1^* + z_2^* + z_3^* = 2(z_1z_2 + z_1z_3 + z_1z_4 + z_2z_3 + z_2z_4 + z_3z_4) = 2p.$$

Similarly, we have

$$z_1^*z_2^* + z_2^*z_3^* + z_3^*z_1^* = p^2 - 4r \quad \text{and} \quad z_1^*z_2^*z_3^* = -q^2.$$

We conclude that z_1^*, z_2^*, z_3^* are the three roots of the cubic

$$P^*(z) = z^3 - 2pz^2 + (p^2 - 4r)z + q^2.$$

But to find the roots of a cubic is exactly what we just accomplished! Thus, using our earlier reduction and formulas, z_1^*, z_2^*, z_3^* can be explicitly expressed in terms of p, q, r. Finally, to pass from the roots of P^* to the roots of our original quartic P is now easy. Since P is reduced, we have $z_1 + z_2 + z_3 + z_4 = 0$, and the formulas defining z_1^*, z_2^*, z_3^* can be resolved, yielding

$$z_1 + z_2 = \sqrt{-z_1^*}, \quad z_3 + z_4 = -\sqrt{-z_1^*},$$

$$z_1 + z_3 = \sqrt{-z_2^*}, \quad z_2 + z_4 = -\sqrt{-z_2^*},$$

$$z_1 + z_4 = \sqrt{-z_3^*}, \quad z_2 + z_3 = -\sqrt{-z_3^*}.$$

Once again, we have to fix the ambiguity inherent in the choice of the square roots. As computations confirm, we need

$$\sqrt{-z_1^*}\sqrt{-z_2^*}\sqrt{-z_3^*} = -q$$

to be satisfied. Finally, solving the system above for z_1, z_2, z_3, z_4, we obtain

$$2z_1 = \sqrt{-z_1^*} + \sqrt{-z_2^*} + \sqrt{-z_3^*},$$

$$2z_2 = \sqrt{-z_1^*} - \sqrt{-z_2^*} - \sqrt{-z_3^*},$$

$$2z_3 = -\sqrt{-z_1^*} + \sqrt{-z_2^*} - \sqrt{-z_3^*},$$

$$2z_4 = -\sqrt{-z_1^*} - \sqrt{-z_2^*} + \sqrt{-z_3^*}.$$

One final remark. Since we have

$$z_1^* - z_2^* = -(z_1 - z_4)(z_2 - z_3),$$

$$z_1^* - z_3^* = -(z_1 - z_3)(z_2 - z_4),$$

$$z_2^* - z_3^* = -(z_1 - z_2)(z_3 - z_4),$$

the discriminant of P^* is the same as the discriminant of P! From the explicit formula of the discriminant for cubics (cf. Problem 2), we obtain that the discriminant of P is

$$\delta = 16p^4 r - 4p^3 q^2 - 128p^2 r^2 + 144pq^2 r - 27q^4 + 256r^3.$$

The cubic P^* that helped to solve the quartic equation is called the *resolvent cubic*. For the future discussion of quintics in Section 25 all we need to remember is that a resolvent is a polynomial whose roots are prescribed functions of the roots of the original polynomial.

The solvability of degree-five polynomial equations eluded mathematicians for nearly 300 years, until Abel showed the impossibility of a radical formula in 1824. (Eleven years earlier Ruffini attempted to prove the impossibility but the proof contained several gaps.) The final phase was completed by Galois (and published in 1846 posthumously, 14 years after he was killed in a duel), who gave a group-theoretical criterion as to which equations of a given degree have solutions in terms of radicals.

Problems

1. Find the Ferro–Fontana–Cardano formula for a root of a general cubic polynomial

$$z^3 + az^2 + bz + c$$

following these steps: (a) Make the substitution $z \mapsto z - a/3$ and verify that in terms of z, the polynomial has the form

$$P(z) = z^3 + pz + q.$$

(b) Make the substitution $z = u - v$ and show that the polynomial reduces to

$$P(u - v) = u^3 - v^3 + q - (3uv - p)(u - v).$$

(c) Set $3uv - p = 0$ and $u^3 - v^3 + q = 0$, solve the first equation for v, substitute it into the second equation, and arrive at

$$3^3 u^6 + 3^3 u^3 q - p^3 = 0.$$

(d) Solve this for $w = u^3$ using the quadratic formula and obtain for $z = u - v$,

$$u = \sqrt[3]{-\frac{q}{2} + \sqrt{\left(\frac{q}{2}\right)^2 + \left(\frac{p}{3}\right)^3}}, \qquad v = \sqrt[3]{+\frac{q}{2} + \sqrt{\left(\frac{q}{2}\right)^2 + \left(\frac{p}{3}\right)^3}}.$$

(Note also that the single substitution[4] $z = p/(3v) - v$ reduces the cubic P to a quadratic polynomial in v^3.)

---

[4]This was suggested by Viète and appeared in print posthumously in 1615 in *De aequationum recognitione et emendatione*.

2. Derive the formula

$$\delta = a^2b^2 - 4b^3 - 4a^3c - 27c^2 + 18abc$$

for the discriminant of the general cubic $z^3 + az^2 + bz + c$.

3. Show that the roots of a reduced cubic with real coefficients are all real iff $\delta \geq 0$. Assuming $\delta > 0$, derive a formula for the roots that involves only real numbers.[5] (Hint: Write $\xi_{1,2}^3 = -q/2 \pm i\sqrt{3\delta}/18$ in polar form.)

4. In this (admittedly long) problem we deal with geometric constructions. We start with two points in the plane. These points will be called *constructible*, as will any other points that can be obtained from them by making repeated use of straightedge and compass. The rules for the constructions are the following. (i) The line passing through two constructible points is *constructible*. (ii) A circle with center at a constructible point and passing through another constructible point is *constructible*. (iii) The points of intersection of constructible lines and circles are constructible. (a) Show that the line through a constructible point and perpendicular or parallel to a constructible line is constructible. (b) By marking off the distance between two constructible points on a constructible line starting at a constructible point, prove that we arrive at a constructible point. (c) Set up Cartesian coordinates such that the two initial points correspond to $(0, 0)$ and $(1, 0)$, and the coordinate axes are constructible. Call a number $a \in \mathbf{R}$ *constructible* if $|a|$ is the distance between two constructible points. Show that a point $p = (a, b)$ is constructible iff the components a, b are constructible. (d) Use similar right triangles and the relations $a : 1 = ab : b$ and $1 : a = 1/a : 1$ to prove that the constructible numbers form a subfield of $\mathbf{R}$. As a byproduct, conclude that all rational numbers are constructible. (e) Use Problem 1 of Section 5 to prove that a quadratic polynomial $ax^2 + bx + c$ with constructible coefficients a, b, c has constructible roots r_1 and r_2. (f) ♡ Show that $r \in \mathbf{R}$ is constructible iff r is contained in a subfield K of $\mathbf{R}$, and there is a chain of subfields

$$\mathbf{Q} = K_0 \subset K_1 \subset \cdots \subset K_n = K$$

such that for each $j = 1, \ldots, n$, $K_j = K_{j-1}(\sqrt{r_j})$, where $r_j \in K_{j-1}$ is a positive real number (that is not a square in K_{j-1}). (♠ Conclude that if $r \in \mathbf{R}$ is constructible, then the degree $[\mathbf{Q}(r) : \mathbf{Q}]$ of the field extension $\mathbf{Q}(r)/\mathbf{Q}$ is a power of 2. Warning: The converse is fase! In fact, there are degree-4 field extensions of $\mathbf{Q}$ that contain nonconstructible numbers.) (g) ♡ Show that if a cubic polynomial $x^3 + ax^2 + bx + c$ with rational coefficients has no rational root, then it has no constructible root. (Hint: ¬ Assume that r is a constructible root of the cubic such that the number n in (f) is minimal. Write $r = s + t\sqrt{r_n}$ with $s, t \in K_{n-1}$ and prove that $\tilde{r} = s - t\sqrt{r_n}$ is also a root (cf. Problems 7–8 of Section 4). Verify that the third root of the cubic is in K_{n-1}. ¬) Derive the following results of Wantzel (1837): (i) $\sqrt[3]{2}$ is not

---

[5] The case $\delta > 0$ is historically called "casus irreducibilis," since the Ferro–Fontana–Cardano formula involves complex numbers, while the roots are real.

constructible. (♠ Alternatively, $[\mathbf{Q}(\sqrt[3]{2}) : \mathbf{Q}] = 3$.) ♡ Interpret this result in view of the *Delian problem*:[6] Construct an altar for Apollo twice as large as the existing one without changing its cubic shape. (ii) Use Problem 3 of Section 2 to show that $\cos 20°$ is an irrational root of the polynomial $P(x) = 8x^3 - 6x - 1$, thereby not constructible. Since $\cos 60°$ is constructible, the impossibility of the trisection of a $60°$ angle follows. This problem dates back to the fifth century B.C. in Greek geometry. (h) Show that constructibility of the regular n-sided polygon P_n is equivalent to the constructibility of the primitive nth root of unity $\omega = z(2\pi/n)$. (Note that according to the example in Section 4, ω is a point with algebraic coordinates.)

5. ♠ (a) Construct a regular pentagon in the following algebraic way: The fifth roots of unity, other than 1 itself, satisfy the equation

$$z^4 + z^3 + z^2 + z + 1 = 0.$$

Divide both sides of this equation by z^2, rearrange the terms, and notice that the substitution $w = z + 1/z$ reduces the equation to $w^2 + w - 1 = 0$. Solve this by the quadratic formula and construct the roots (cf. Problem 1 of Section 5). (In Book IV of the *Elements* Euclid constructed a regular pentagon. Simpler constructions were given by Ptolemy, and also by Richmond in 1893.) (b) Apply the argument in (a) to the regular 7-sided polygon. Show that the equation $z^7 = 1$ reduces to the cubic equation

$$w^3 + w^2 - 2w - 1 = 0$$

and $w = 2\cos(360°/7)$ is a solution, thereby irrational. Verify that this equation has no rational roots, so that the regular 7-sided polygon is not constructible.[7]

6. Solve the problem of Zuane de Tonino da Coi (c. 1540): "Divide 10 into 3 parts such that they shall be in continued proportion and that the product of the first two shall be 6." (Hint: $a + b + c = 10$, $a/b = b/c$, and $ab = 6$ give a quartic equation in b.)

---

[6] For a concise treatment of Galois theory including the solutions of some problems of antiquity, see Ian Stewart's *Galois Theory*, Chapman and Hall, 1973.

[7] Gauss proved that a regular n-sided polygon, $n \geq 3$, is constructible iff $n = 2^m p_1 \cdots p_k$, where $p_j = 2^{2^{a_j}} + 1$, $a_j = 0, 1, 2, \ldots, j = 1, \ldots, k$, are distinct Fermat primes. (To be precise, Gauss did not state explicitly that the condition that the Fermat primes are distinct is necessary; this gap was filled by Wantzel in 1837.) For a fairly comprehensive account of geometric constructions, including the construction of Gauss of the regular 17-sided polygon, see Klein, *Famous Problems in Elementary Geometry*, Chelsea, New York, 1955.

7

Stereographic Projection

♣ Consider the complex plane **C** imbedded in **R**^{3} as the plane spanned by the first two axes. Algebraically, we view a complex number $z = a + bi$, $a, b \in \mathbf{R}$, as the 3-vector $(a, b, 0) \in \mathbf{R}^3$. The set of unit vectors in **R**^{3} form the 2-*sphere*:

$$S^2 = \{p \in \mathbf{R}^3 \mid |p| = 1\}.$$

(S^2 intersects **C** in the unit circle S^1.) We now pretend that we are Santa Claus; we sit at the North Pole $N = (0, 0, 1)$ and look down to Earth S^2. We realize that every time we see a location on the transparent Earth, we also see a unique point on **C**. Associating the point on Earth with the point on the complex plane **C** that we see simultaneously gives the *stereographic projection*

$$h_N : S^2 - \{N\} \to \mathbf{C}.$$

Geometrically, given $p \in S^2$, $p \neq N$, $h_N(p)$ is the unique point in **C** such that N, p, and $h_N(p)$ are on the same line (Figure 7.1).

h_N is clearly invertible. Santa Claus aside, the importance of h_N is clear: it allows us to chart maps of various parts of the curved globe on a flat piece of paper.

If we are in Brazil, we may want to use the South Pole $S = (0, 0, -1)$ instead of the North Pole. We arrive at another

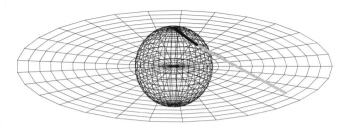

Figure 7.1

stereographic projection

$$h_S : S^2 - \{S\} \to \mathbf{C}.$$

Looking down or up we see that $h_N(S) = h_S(N) = 0$, the origin at $\mathbf{C}$. We can thus form the composition

$$h_N \circ h_S^{-1} : \mathbf{C} - \{0\} \to \mathbf{C} - \{0\}.$$

We claim that

$$(h_N \circ h_S^{-1})(z) = \frac{z}{|z|^2}, \quad 0 \neq z \in \mathbf{C}.$$

Fix a nonzero complex number z. Since all the actions of h_N and h_S are happening in the plane spanned by N, S and z, we arrive at Figure 7.2.

The angles $\angle SNh_S^{-1}(z)$ and $\angle 0zS$ have perpendicular sides, so Euclid's *Elements* tells us that they are equal. Thus, the right triangles

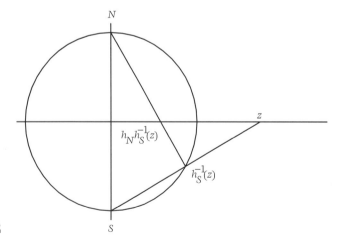

Figure 7.2

$\triangle 0N(h_N \circ h_S^{-1})(z)$ and $\triangle 0zS$ are similar. Again by Euclid, we have

$$|h_N \circ h_S^{-1}(z)| = \frac{1}{|z|},$$

and this gives the required formula.

We need to rewrite $z \to z/|z|^2$ in a more convenient form by introducing *complex conjugation*. Given a complex number $z = a + bi$, its *complex conjugate* $\bar{z}$ is defined as

$$\bar{z} = a - bi.$$

Crossing the Gaussian Bridge we see that conjugation corresponds to reflection in the real axis. As far as our algebra is concerned, we have

$$\overline{z_1 + z_2} = \bar{z}_1 + \bar{z}_2,$$

$$\overline{z_1 \cdot z_2} = \bar{z}_1 \cdot \bar{z}_2,$$

as can be easily verified. The real and imaginary parts of a complex number z can be written in terms of conjugation as

$$\Re(z) = \frac{(z + \bar{z})}{2} \quad \text{and} \quad \Im(z) = \frac{(z - \bar{z})}{2i}.$$

Moreover,

$$z\bar{z} = |z|^2,$$

since $z\bar{z} = (a+bi)(a-bi) = a^2 + b^2$. This Length^2-Identity is important, since it ties complex multiplication to the length of the vector corresponding to z. More generally, the dot product of $z_1, z_2 \in \mathbf{C}$ considered as vectors in $\mathbf{R}^2$, is given by

$$z_1 \cdot z_2 = \frac{(\bar{z}_1 z_2 + z_1 \bar{z}_2)}{2} = \Re(z_1 \bar{z}_2)$$

because $\Re(z_1 \bar{z}_2) = \Re((a_1 + b_1 i)(a_2 - b_2 i)) = a_1 a_2 + b_1 b_2$ (with obvious notations). Cross-dividing, we see that

$$\frac{1}{z} = \frac{\bar{z}}{|z|^2}, \quad z \neq 0.$$

This also provides a convenient formula for complex division. Going back to our stereographic projections, we obtain

$$(h_N \circ h_S^{-1})(z) = 1/\bar{z}, \quad 0 \neq z \in \mathbf{C}.$$

It is rewarding to study the geometry of the stereographic projection more closely. Let $p \in S^2$, $p \neq N$, $q = h_N(p)$, and let l denote the line through these points. Consider the configuration of the three planes V_N, V_p, and $\mathbf{C}$, where V_N is tangent to S^2 at N, and V_p is tangent to S^2 at p. By the geometry of the sphere, the perpendicular bisector of the segment connecting N and p contains the line $V_N \cap V_p$. Since V_N and $\mathbf{C}$ are parallel, the perpendicular bisector of the segment connecting p and $q = h_N(p)$ contains $V_p \cap \mathbf{C}$. Figure 7.3 depicts the situation with an edge-on view of the three planes. A plane W that contains the line l (through N, p, and q) intersects $V_p \cup \mathbf{C}$ in 2 lines meeting at a point $v_0 \in V_p \cap \mathbf{C}$. The configuration consisting of the two half-lines emanating from v_0 and containing p and q and the segment connecting p and q looks like a figure **A** laid on its side (see Figure 7.4). We call this an *A-configuration* and denote it by **A**. The segment connecting p and q is called the *median* of **A**. Since the perpendicular bisector of the median contains v_0, geometrically, an A-configuration is an isosceles triangle whose two sides are extended beyond the vertices of the base. We call these the *extensions* of **A**. The upper extension is tangent to S^2 at its initial point p, and the lower extension is a half-line in $\mathbf{C}$ emanating from q. The inner angles of the two extensions with the median at p and at q are equal. Rotating W around l, we obtain a continuous family of A-configurations whose upper extensions sweep the tangent plane V_p around. Similarly, the lower extensions centered at q sweep $\mathbf{C}$ around.

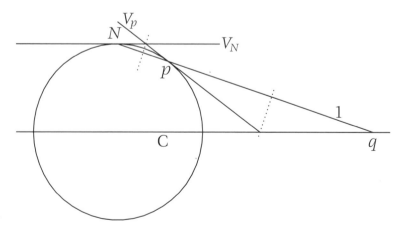

Figure 7.3

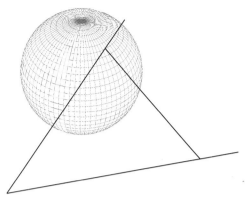

Figure 7.4

Now let $v \in V_p$ be a tangent vector, and let w be the image of v under the (stereographic) projection from N. Then the A-configuration whose upper extension contains v also contains w in its lower extension, and the terminal points of v and w are collinear with N (see Figure 7.5).

Consider now two vectors v_1 and v_2 tangent to S^2 at p, and let w_1 and w_2 be their respective projections from N. The two A-configurations A_1 and A_2 contain v_1, w_1 and v_2, w_2. Moreover, the segment connecting p and q is the common median of A_1 and A_2. By bilateral symmetry with respect to the perpendicular bisector of this common median, the angle between v_1 and v_2 is the same as the angle between w_1 and w_2. We obtain that the sterographic projection is angle-preserving!

We close this section by showing that the stereographic projection h_N is also circle-preserving in the sense that if $C \subset S^2$ is a circle

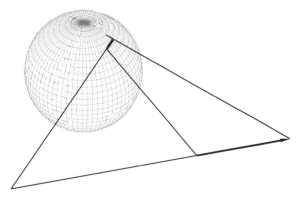

Figure 7.5

that avoids the North Pole N, then $S = h_N(C) \subset \mathbf{C}$ is also a circle. (If C passes through the North Pole, then S is a line. Why?) There are beautiful geometric arguments to prove this.[1] For a change, we give here an algebraic proof. A circle C on S^2 is the intersection of a plane with S^2. Using a, b, c as coordinates, $p = (a, b, c) \in S^2 \subset \mathbf{R}^3$, the equation of the plane is given by

$$\alpha a + \beta b + \gamma c = \delta,$$

where $\alpha, \beta, \gamma, \delta \in \mathbf{R}$ are the coefficients. Since C does not pass through the North Pole, we have $\gamma \neq \delta$. Using the explicit expression of the inverse of h_N (Problem 1(c)), the image $h_N(C)$ is given by

$$\alpha \frac{2\Re(z)}{|z|^2 + 1} + \beta \frac{2\Im(z)}{|z|^2 + 1} + \gamma \frac{|z|^2 - 1}{|z|^2 + 1} = \delta,$$

where we used the complex variable z on $\mathbf{C}$. Letting $z = x + iy$, $x, y \in \mathbf{R}$, we obtain

$$2\alpha x + 2\beta y + \gamma(x^2 + y^2 - 1) = \delta(x^2 + y^2 + 1).$$

Equivalently,

$$(\delta - \gamma)(x^2 + y^2) - 2\alpha x - 2\beta y + \gamma + \delta = 0.$$

This is the equation of a circle, since $\delta - \gamma \neq 0$. The claim follows.

Problems

1. (a) Given $p = (a, b, c) \in S^2 \subset \mathbf{R}^3$, $c \neq 1$, show that

$$h_N(p) = \frac{a + bi}{1 - c} \in \mathbf{C}.$$

(b) Verify the identity

$$h_N(-p) = -1/\overline{h_N(p)}, \quad p \in S^2.$$

(c) Given $z \in \mathbf{C}$, show that

$$h_N^{-1}(z) = \left(\frac{2z}{|z|^2 + 1}, \frac{|z|^2 - 1}{|z|^2 + 1} \right) \in S^2.$$

---

[1] See D. Hilbert and S. Cohn-Vossen, *Geometry and Imagination*, Chelsea, 1952.

(Recall that **C** is imbedded in $\mathbf{R}^3$ as the plane spanned by the first two coordinate axes.)

(d) Let $\omega = z(2\pi/3)$ be a primitive third root of unity. For what value of $r > 0$ are the points 0, r, $r\omega$, and $r\omega^2$ the stereographically projected vertices of a regular tetrahedron inscribed in S^2?

2. Using the formula for h_N in Problem 1 and a similar formula for h_S, work out $h_N \circ h_S^{-1}$ explicitly.

3. Show that $\overline{z_1 z_2} = \bar{z}_1 \bar{z}_2$.

4. Prove that

$$\left| \frac{z - w}{1 - z\bar{w}} \right| < 1$$

if $|z| < 1$ and $|w| < 1$.

Web Site

1. www.geom.umn.edu/~sullivan/java/stereop/

8

Proof of the Fundamental Theorem of Algebra

♣ Consider the polynomial

$$P(z) = c_0 + c_1 z + \ldots + c_n z^n, \quad c_n \neq 0,$$

as a map $P : \mathbf{C} \to \mathbf{C}$. We "pull P up to S^2" by using the stereographic projection $h_N : S^2 - \{N\} \to \mathbf{C}$. Analytically, we define $f : S^2 \to S^2$ by

$$f(p) = \begin{cases} (h_N^{-1} \circ P \circ h_N)(p), & \text{if } p \neq N, \\ N, & \text{if } p = N. \end{cases}$$

f is clearly smooth on $S^2 - \{N\}$ since h_N is smooth. We claim that f is smooth[1] across N. To show this, we "pull f down to S^2" by the stereographic projection $h_S : S^2 - \{S\} \to \mathbf{C}$ and consider

$$Q = h_S \circ f \circ h_S^{-1}.$$

Notice that smoothness of f near N corresponds to smoothness of Q near zero. We now compute

$$Q(z) = (h_S \circ f \circ h_S^{-1})(z)$$

---

[1] See "Smooth Maps" in Appendix D.

$$= [(h_N \circ h_S^{-1})^{-1} \circ P \circ (h_N \circ h_S^{-1})](z)$$

$$= (h_N \circ h_S^{-1})^{-1}(P(1/\bar{z}))$$

$$= \frac{1}{\overline{P(1/\bar{z})}}$$

$$= \frac{z^n}{\bar{c}_0 z^n + \bar{c}_1 z^{n-1} + \cdots + \bar{c}_n}.$$

This is a complex rational function with nonvanishing denominator at $z = 0$. Since any rational function is smooth wherever it is defined, we conclude that Q is smooth near $z = 0$. Thus $f : S^2 \to S^2$ is smooth near N and hence everywhere.

Before going any further, we work out a particular example in order to get a feel for f. Setting $P(z) = z^2$, we first describe the transformation $z \mapsto z^2$ on the complex plane $\mathbf{C}$. Consider first what happens to the unit circle $S^1 \subset \mathbf{C}$: Since S^1 is parametrized by the points

$$z(\theta) = \cos\theta + i\sin\theta, \quad \theta \in \mathbf{R},$$

and $z(\theta)^2 = z(2\theta)$, we see that S^1 is mapped onto itself, and, on S^1, f corresponds to doubling the angle. Topologically, S^1 is stretched to twice its perimeter and then wrapped around itself twice (Figure 8.1).

Going a step further, it is now clear that any circle concentric to S^1 on $\mathbf{C}$ is mapped to another concentric circle (with the radius changing from r to r^2) and, restricted to these, the transformation has the same stretch and wraparound effect. Finally, since under h_N the concentric circles correspond to parallels of latitude on S^2, we see that $f : S^2 \to S^2$ wraps each parallel twice around another parallel and keeps the North and South Poles fixed. What is

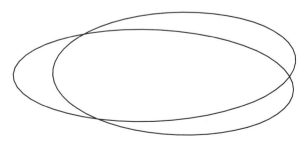

Figure 8.1

important to us is that, apart from the poles (which we will call "singular"), each point $q \in S^2 - \{N, S\}$ has exactly two inverse images; that is, $|f^{-1}(q)| = 2$. (Actually, $f : S^2 \to S^2$ is called a "twofold branched covering with singular points N and S." The wraparound effect or, more precisely, the branching is shown in Color Plate 3a with f being vertical projection to a disk.) Moreover, the singular points are N and S and the latter corresponds (under h_N) to the origin in $\mathbf{C}$ where the derivative of P vanishes.

The description for $P(z) = z^n$, $n \in \mathbf{N}$, is similar with two replaced by n; that is, $|f^{-1}(q)| = n$, for $q \in S^2 - \{N, S\}$.

♠ We now return to the main discussion with a good clue. First we have to settle the problem of singular points. The complex derivative of P is a polynomial of degree $n - 1$:

$$P'(z) = c_1 + 2c_2 z + \cdots + nc_n z^{n-1}.$$

P' vanishes on, at most, $n-1$ points. Let $\sigma \subset S^2$ denote the $(f \circ h_N^{-1})$-image of these points plus N. Then σ consists of at most n points. We now claim that the function $\# : S^2 - \sigma \to \mathbf{N} \cup \{0\}$ defined by

$$\#(q) = |f^{-1}(q)|, \quad q \in S^2 - \sigma,$$

is actually a constant. Since $S^2 - \sigma$ is connected, it is enough to show that $\#$ is locally constant; that is, for every $q \in S^2 - \sigma$, there is an open neighborhood V of q in $S^2 - \sigma$ such that $\#$ is constant on V. Let $q \in S^2 - \sigma$ be fixed and $p \in f^{-1}(q)$. The derivative P' of P at $h_N(p)$ is nonzero (since $p \notin \sigma$), so that P is a local diffeomorphism[2] near $h_N(p)$. Since the stereographic projection h_N is a diffeomorphism, f is a local diffeomorphism at each point of $f^{-1}(q)$. In particular, $f^{-1}(q)$ cannot have any accumulation points, so it must be finite. Let $\#(q) = m$ with $f^{-1}(q) = \{p_1, \ldots, p_m\}$. (Actually, $m \leq n$ because a degree-n polynomial can have at most n roots, but we do not need this additional fact.) For each p_j, $j = 1, \ldots, m$, there exists an open neighborhood V_j of p_j such that $f|V_j : V_j \to f(V_j)$ is a diffeomorphism. By cutting finitely many times, we may assume that the V_j's are mutually disjoint and they all map under f diffeomorphically onto a single V_0, an open neighborhood of q. Finally,

---

[2]We are using here some basic facts in complex calculus. For a quick review, see the beginning of Section 15 and "Smooth Maps" in Appendix D.

we delete from V_0 the *closed* set $f(S^2 - (V_1 \cup \cdots \cup V_m))$ to obtain an open neighborhood V of q. (Notice that $S^2 - (V_1 \cup \cdots \cup V_m)$ is closed in S^2; hence, it is compact. Its f-image is also compact and thus closed in S^2.) Clearly, # is constant ($= m$) on V_0. The claim follows.

The globally constant function # cannot be identically zero, since otherwise f would map into the finite σ and the polynomial P would reduce to a constant. Thus # is at least one on $S^2 - \sigma$. In particular, $S^2 - \sigma$ is contained in the image of f. The latter is closed in S^2 so that the image must be the whole of S^2. We obtain that f is onto. Thus, P must be onto. In particular, $0 \in \mathbf{C}$ is in the image of P. This means that there exists $z \in \mathbf{C}$ with $P(z) = 0$. This is the FTA!

Problems

Fill in the details in the first four problems, which outline additional proofs of the FTA.

1. Let
$$P(z) = c_0 + c_1 z + \cdots + c_n z^n, \quad c_n \neq 0,$$
be a polynomial with complex coefficients.

(a) Show that the nonnegative real function $z \mapsto |P(z)|$ attains its global minimum on $\mathbf{C}$. (Let $R > 0$ and use the triangle inequality to show that, for $|z| > R$, we have
$$|P(z)| > |z|^n \left(|c_n| - \frac{|c_{n-1}|}{R} - \cdots - \frac{|c_0|}{R^n} \right).$$

Choose R large enough so that $|P|$ will be above its greatest lower bound m for $|z| > R$. Refer to closedness and boundedness of the domain and show that $|P|$ attains m on the disk $\{z \mid |z| \leq R\}$.)

(b) Show that if $z_0 \in \mathbf{C}$ is a local minimum of $|P|$, then z_0 is a zero of P. Assume that $P(z_0) \neq 0$ and show that z_0 is not a local minimum of $|P|$ as follows: Consider the polynomial $Q(z) = P(z + z_0)/P(z_0)$ of degree n and with constant term 1 and verify that $|P|$ has no local minimum at z_0 iff $|Q|$ has no local minimum at 0 (iff $|Q|$ takes values < 1 near 0). Let
$$Q(z) = b_0 + b_1 z + \cdots + b_n z^n, \quad b_n \neq 0, \quad b_0 = 1.$$

Let k be the least positive number with $b_k \neq 0$. Let $r > 0$ and use the triangle inequality again to show that for $|z| = r$, we have
$$|Q(z)| \leq |b_n| r^n + \cdots + |b_{k+1}| r^{k+1} + |b_k z^k + 1|.$$

For r small enough, this is dominated by $(1/2)|b_k|r^k + |b_k z^k + 1|$. Choose z on the circle $|z| = r$ such that $b_k z^k + 1 = -|b_k|r^k + 1$. Comparing with the above, for this z, $|Q(z)| \leq 1 - |b_k|r^k/2 < 1$.

2. ♠ ¬ Let P be a nonzero polynomial that has no complex roots. Define f as in the text and observe that the image Y of f does not contain the South Pole S.

   (a) Use compactness of Y to show that S has an open neighborhood disjoint from Y.

   (b) Use the local diffeomorphism property to prove that whenever $P'(z) \neq 0$, the point $f(h_N^{-1}(z))$ is in the interior of Y.

   (c) Exhibit infinitely many boundary points of Y by considering the "southernmost" point in the intersection of Y with any meridian of longitude.

   (d) Conclude that P' vanishes at infinitely many points. ¬

3. ¬ Let $P : \mathbf{C} \to \mathbf{C} - \{0\}$ be as in Problem 2. For each $r > 0$, consider the closed curve $w_r : [0, 2\pi] \to \mathbf{C} - \{0\}$, $w_r(\theta) = P(rz(\theta))$, $0 \leq \theta \leq 2\pi$. Let $M \subset \mathbf{R}^3$ be the graph of the multivalued function arg $: \mathbf{C} - \{0\} \to \mathbf{R}$ with projection $p : M \to \mathbf{C} - \{0\}$.

   (a) Show that w_r can be "lifted" to a curve $\tilde{w}_r : [0, 2\pi] \to M$ satisfying $p \circ \tilde{w}_r = w_r$. (Define $\tilde{w}_r$ locally using a subdivision $0 = \theta_0 < \theta_1 < \ldots < \theta_m = 2\pi$ of $[0, 2\pi]$ into sufficiently small subintervals such that arg is single-valued on each subarc $w_r([\theta_{i-1}, \theta_i])$, $i = 1, \ldots, m$.)

   (b) Verify that the winding number[3] $(1/(2\pi))(\tilde{w}_r(2\pi) - \tilde{w}_r(0))$ of w_r is a nonnegative integer and is independent of the choice of the lift. Note that $\tilde{w}_r$ is unique up to translation with an integer multiple of 2π along the third axis in $M \subset \mathbf{R}^3$.

   (c) Prove that for r large, the winding number is n, the degree of P. (arg w_r is increasing in $\theta \in [0, 2\pi]$ for r large. For this, work out the dot product $(iw_r) \cdot (\partial/\partial\theta)w_r$ and use the first estimate in Problem 1 to conclude that it is positive for r large.)

   (d) Observe that for r small, the winding number is zero.

   (e) Use a continuity argument to show that the winding number is independent of r. ¬

4. Let P be a polynomial of degree n as in Problem 1. Assume first that P has real coefficients. (a) If n is odd, use calculus to show that P has a real root. (In particular, for $n = 3$, this means that $\mathbf{R}^3$ is not an extension field of $\mathbf{R}$, so that, unlike $\mathbf{R}^2$, no multiplication makes $\mathbf{R}^3$ a field.) (b) In general, write $n = 2^m a$, where a is odd and use induction with respect to m to prove that P has at least one complex root. For the general induction step, let $\alpha_1, \ldots, \alpha_n$ denote the roots of P over a splitting field of P. For $k \in \mathbf{Z}$, let Q_k be the polynomial with roots $\alpha_i + \alpha_j + k\alpha_i\alpha_j$, $1 \leq i < j \leq n$, and leading coefficient one. Use the fundamental theorem of symmetric polynomials to show that the coefficients of Q_k are real. Check that the degree of Q_k is $2^{m-1}b$, where b is odd; apply the

---

[3]The winding number can also be defined by the integral $(1/2\pi) \int_{w_r} d\theta$, where θ is the polar angle on $\mathbf{R}^2 - \{0\}$. (Observe that θ is multiple-valued, but $d\theta$ gives a well-defined 1-form on $\mathbf{R}^2 - \{0\}$.)

induction hypothesis, and conclude that $\alpha_i + \alpha_j + k\alpha_i\alpha_j$ is a complex number for some $1 \leq i < j \leq n$. Use the dependence of i and j on k to prove that $\alpha_i\alpha_j \in \mathbf{C}$ and $\alpha_i + \alpha_j \in \mathbf{C}$ for some $1 \leq i < j \leq n$. Apply the quadratic formula to show that $\alpha_i, \alpha_j \in \mathbf{C}$. (c) Extend the results of (a) and (b) to polynomials P with complex coefficients by considering the real polynomial $P\bar{P}$, where $\bar{P}$ is obtained from P by conjugating the coefficients.

5. Show that if a polynomial with real coefficients has a root of the form $a + bi$, then the complex conjugate $a - bi$ is also a root.

Web Site

1. www.cs.amherst.edu/~djv/fta.html

9

Symmetries of Regular Polygons

♣ We now go back again to ancient Greek mathematics, in which regular polygons played a central role. We learned that a regular n-sided polygon, P_n, $n \geq 3$, can be represented by its vertices $z(2k\pi/n)$, $k = 0, 1, \ldots, n-1$, on the complex plane $\mathbf{C}$ (Figure 9.1).

By the multiplicative property of $z(\theta)$, $\theta \in \mathbf{R}$ (see Section 5), we have

$$z(2k\pi/n)z(2l\pi/n) = z(2(k+l)\pi/n), \quad k, l \in \mathbf{Z}.$$

We see that multiplication by $z(2k\pi/n)$ causes the index of the vertex to shift by k. More geometrically, the multiplicative property also tells us that multiplying complex numbers by $z(\theta)$ corresponds to the geometric transformation of counterclockwise rotation R_θ by angle θ around the origin (see Figure 9.2). In particular, the group

Figure 9.1

96

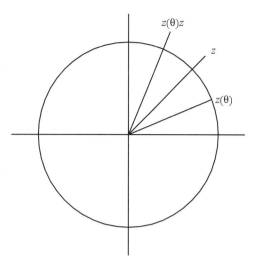

Figure 9.2

of rotations

$$\{R_{2k\pi/n} \mid k = 0, \ldots, n-1\}$$

(isomorphic with $\mathbf{Z}_n$) leaves the regular n-sided polygon invariant. Regularity thus implies that each vertex (and each edge) can be carried into any vertex (and any edge) by a suitable symmetry that leaves the polygon invariant. This seemingly innocent remark gives a profound clue to defining regularity of polyhedra in space.

We now ask the following more general question: What is the largest group of geometric transformations that leaves the regular n-sided polygon invariant? To answer this question, we need to make the term "geometric transformation" precise. Transformation usually refers to a bijection of the ambient space, which, in this case, is the Cartesian plane $\mathbf{R}^2$. (A finer point is whether we should require continuity or differentiability; fortunately, this is not essential here.) Since in Euclidean plane geometry we always work in the concrete model $\mathbf{R}^2$ (and avoid the headache of axiomatic treatment), we have the Euclidean distance function $d : \mathbf{R}^2 \times \mathbf{R}^2 \to \mathbf{R}$,

$$d(p, q) = |p - q|, \quad p, q \in \mathbf{R}^2,$$

and the concepts of angle, area, etc. The term "geometric" stands for transformations that preserve some of these geometric quantities. In our present situation, we require the geometric transformation to preserve d; that is, we consider isometries. Explicitly, a transformation $S : \mathbf{R}^2 \to \mathbf{R}^2$ is an *isometry* if

$$d(S(p), S(q)) = d(p, q), \quad p, q \in \mathbf{R}^2.$$

Euclid's *Elements* tells us that an isometry also preserves angles, areas, etc. The set of all plane isometries form a group denoted by $Iso\,(\mathbf{R}^2)$. (Actually, $\mathbf{R}^2$ equipped with d is usually denoted by E^2. Since we have no fear of confusing $\mathbf{R}^2$ with other (non-Euclidean) models built on $\mathbf{R}^2$, we ignore this finer point.)

We now look at examples of planar isometries.

1. $R_\theta(p)$: rotation with center $p \in \mathbf{R}^2$ and angle $\theta \in \mathbf{R}$;
2. T_v: translation with vector $v \in \mathbf{R}^2$;
3. R_l: reflection in a line $l \subset \mathbf{R}^2$;
4. $G_{l,v}$: glide reflection[1] along a line l with vector v (parallel to l); in fact, $G_{l,v} = T_v \circ R_l = R_l \circ T_v$.

We now claim that every plane isometry is one of these. Although this result is contained in many textbooks, the proof is easy (especially in our model $\mathbf{R}^2$), so we will elaborate on it a little. Before the proof we assemble a few elementary facts.

Given two lines l and l', the composition $R_{l'} \circ R_l$ is a rotation if l and l' intersect and a translation if l and l' are parallel. The rotation angle is twice the signed angle from l to l'; the translation vector is perpendicular to these lines, and its length is twice the signed distance from l to l'. (Thus, rotations and translations are not all that different; consider a rotation and move the center "slowly to infinity"; when the center leaves the plane, the rotation becomes a translation!) Notice that when decomposing a rotation with center at p as a product of reflections R_l and $R_{l'}$, the line l (or l') through p can be chosen arbitrarily. (What is the analogue of this for translations?)

Let $R_{2\alpha}(p)$ and $R_{2\beta}(q)$ be rotations with $p \neq q$ and let l denote the line through p and q. By the above, $R_{2\alpha}(p) = R_{l'} \circ R_l$, where l' is the unique line through p such that the angle from l to l' is α.

[1]You may say, "I do not understand this!" and push this book away closing it. Well, you just performed a glide!

Similarly, $R_{2\beta}(q) = R_l \circ R_{l''}$, where l'' is the line through q such that the angle from l'' to l is β. The composition

$$R_{2\alpha}(p) \circ R_{2\beta}(q) = (R_{l'} \circ R_l) \circ (R_l \circ R_{l''}) = R_{l'} \circ R_{l''}$$

is the product of reflections in the lines l' and l''. If $\alpha + \beta \in \pi\mathbf{Z}$, then l' and l'' are parallel, and the composition is a translation with translation vector twice the vector from q to p. If $\alpha + \beta \notin \pi\mathbf{Z}$, then the lines l' and l'' intersect at a point r, and composition is a rotation with center r. It is convenient to write this rotation as $R_{-2\gamma}(r)$, since then

$$R_{2\alpha}(p) \circ R_{2\beta}(q) \circ R_{2\gamma}(r) = I$$

with $\alpha + \beta + \gamma \in \pi\mathbf{Z}$. This is due to W.F. Donkin (1851). If in the latter argument the roles of $R_{2\alpha}(p)$ and $R_{2\beta}(q)$ are interchanged, then we obtain a rotation $R_{2\gamma}(s) = R_{2\alpha}(p)^{-1} \circ R_{2\beta}(q)^{-1}$ with $s \neq r$. In particular, the commutator

$$R_{2\alpha}(p)^{-1} \circ R_{2\beta}(q)^{-1} \circ R_{2\alpha}(p) \circ R_{2\beta}(q)$$

is a translation. ♡ As a byproduct, we obtain that if a subgroup $G \subset \mathrm{Iso}\,(\mathbf{R}^2)$ contains no translations, then all the rotations in G have the same center.

♣ Finally, note that an isometry that fixes three noncollinear points is the identity, a fact that is easy to show.

We are now ready to prove the claim. Let $S : \mathbf{R}^2 \to \mathbf{R}^2$ be an isometry. Assume first that S is *direct*, that is, *orientation preserving*. We split the argument into two cases according to whether or not S has a fixed point. If S has a fixed point $p \in \mathbf{R}^2$, that is, $S(p) = p$, then choose $q \in \mathbf{R}^2$ different from p. Let θ be the angle $\angle qpS(q)$. The composition $R_\theta(p)^{-1} \circ S$ leaves p and q fixed. It thus fixes every point on the line through p and q. On the other hand, $R_\theta(p)^{-1} \circ S$ is direct (since S is), so that it must be the identity. We obtain $S = R_\theta(p)$. Assume now that S has no fixed points. Let $p \in \mathbf{R}^2$ be arbitrary and consider the vector v emanating from p and terminating in $S(p)$. The composition $(T_v)^{-1} \circ S$ is direct and leaves p fixed. By the first case, it is a rotation $R_\theta(p)$, i.e., $S = T_v \circ R_\theta(p)$. We claim that $R_\theta(p)$ is the identity, i.e., $S = T_v$. ¬ Assume not. Arrange v to be the base of the isosceles triangle with vertex p opposite to v and angle θ at p as Figure 9.3 shows. Clearly, q is a fixed point of $S = T_v \circ R_\theta(p)$. ¬

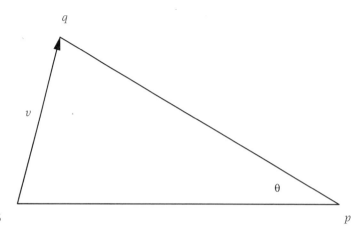

Figure 9.3

Second, assume that S is *opposite*, that is, *orientation reversing*. If S has a fixed point p, then let l be a line through p and consider the composition $R_l \circ S$. This is a direct isometry that fixes p. By the previous case, it must be a rotation $R_\theta(p)$. We obtain $R_l \circ S = R_\theta(p)$, or equivalently, $S = R_l \circ R_\theta(p)$. We now write $R_\theta(p) = R_l \circ R_{l'}$, where l' meets l at p and the angle between l and l' is $\theta/2$. We obtain $S = R_l \circ R_l \circ R_{l'} = R_{l'}$, so that S is a reflection in a line. If S has no fixed point, then let $p \in \mathbf{R}^2$ be arbitrary and denote by $q \in \mathbf{R}^2$ the midpoint of the segment connecting p and $S(p)$. If p, q and $S(q)$ are collinear, it is easy to see that the line l through these points is invariant under S. Thus, the direct isometry $R_l \circ S$ keeps l invariant so that it is a translation T_v with translation vector v parallel to l. Thus $R_l \circ S = T_v$ and so $S = R_l \circ T_v = T_v \circ R_l = G_{l,v}$ is a glide. If p, q and $S(q)$ are not collinear then let r and s denote the orthogonal projections of p and $S(p)$ to the line l through q and $S(q)$ (Figure 9.4).

We claim that $s = S(r)$. The triangle $\triangle qS(p)S(q)$ is isosceles, since $d(p, q) = d(q, S(p)) = d(S(p), S(q))$. Thus, the angles $\angle sqS(p)$ and $\angle sS(q)S(p)$ are equal. We obtain that the triangles $\triangle pqr$ and $\triangle S(p)S(q)s$ are congruent and oppositely oriented. But the same is true for $\triangle pqr$ and $\triangle S(p)S(q)S(r)$, so that $s = S(r)$ follows. Let $G_{l,v}$ be the glide that sends $\triangle pqr$ to $\triangle S(p)S(q)S(r)$, where v has initial point q and terminal point $S(q)$. Then $(G_{l,v})^{-1} \circ S$ fixes p, q, and r, and so it is the identity. $S = G_{l,v}$ follows.

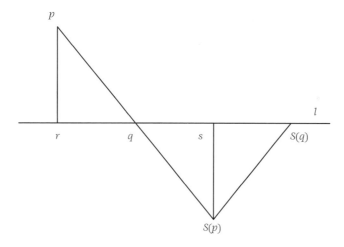

Figure 9.4

Remark.

Since rotations and translations are products of two reflections, as a byproduct of the argument above we obtain that any planar isometry is the product of at most three reflections.

We now go back to our regular polygons. Let X be a set (figure) in the plane and define the symmetry group of X as

$$Symm\,(X) = \{S \in Iso\,(\mathbf{R}^2) \mid S(X) = X\}.$$

Theorem 3.

For $n \geq 3$, Symm (P_n) consists of the rotations

$$R_{2k\pi/n} = R_{2k\pi/n}(0), \quad k = 0, 1, \ldots, n-1,$$

and n reflections $R_{l_1}, \ldots, R_{l_n}$ in the lines $l_1, \ldots, l_n$ joining the origin to the vertices and to the midpoints of the sides.

Proof.

Let $S \in Symm\,(P_n)$. It is clear that S can only be a rotation or a reflection. Indeed, just look at Figure 9.5.

Under S, vertices go to vertices and midpoints of sides go to midpoints of sides; in fact, S is a *permutation* on these two sets. Thus, the origin—the centroid of P_n—is left fixed by S. (For n even, the centroid is the midpoint of a diagonal connecting two vertices. For n odd, the centroid is on a line connecting the midpoint of a

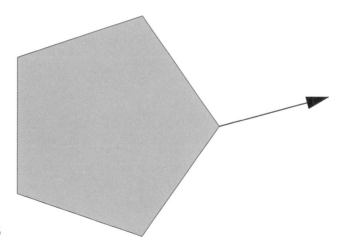

Figure 9.5

side and the opposite vertex, and the center splits this segment in a specified ratio. What is this ratio?) If S is a rotation, then $S = R_\theta = R_\theta(0)$ and $\theta = 2k\pi/n$ clearly follow. If S is reflection in a line l, then l must go through the origin. Again it follows that l is one of the l_i's, $i = 1, \ldots, n$. ∎

Remark.

As seen from the proof, there is a slight distinction between the structure of the lines $l_1, \ldots, l_n$ for n even and n odd. For example, take a look at the triangle P_3 and the square P_4 in Figure 9.6.

We now take an algebraic look at the group $Symm\,(P_n)$. Letting $a = R_{2\pi/n}$, we see that $a^k = R_{2k\pi/n}$, $k = 0, \ldots, n - 1$, so that the rotations generate the cyclic subgroup

$$e = a^0, a, a^2, \ldots, a^{n-1}.$$

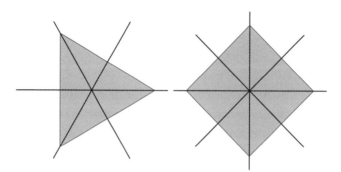

Figure 9.6

We denote this by C_n. Let $b = R_{l_1}$. Then $b^2 = e$ since b is a reflection, and $ba = a^{-1}b$. (This needs verification.) The reflections $b, ab, a^2b, \ldots, a^{n-1}b$ are mutually distinct in $Symm\,(P_n)$, so that they must give $R_{l_1}, \ldots, R_{l_n}$ (in a possibly permuted order). We obtain that $Symm\,(P_n)$ is generated by two elements a and b and relations $a^n = b^2 = e$ and $ba = a^{-1}b$. The elements of $Symm\,(P_n)$ are

$$e, a, a^2, \ldots, a^{n-1}; \quad b, ab, a^2b, \ldots, a^{n-1}b.$$

This is called the *dihedral group* D_n of order $2n$. The name comes from the fact that D_n is the symmetry group of a *dihedron* (as Klein called it), a spherical polyhedron with two hemispheres as faces and n vertices distributed uniformly along the common boundary. (D_n is also the symmetry group of the "reciprocal" spherical polyhedron with two antipodal vertices connected by n semicircles as edges that split the sphere into n congruent spherical wedges as faces.)

Looking back, we see that studying the symmetries of regular polygons leads us to the cyclic group C_n of order n and the dihedral group D_n of order $2n$. It is a remarkable fact that any finite subgroup of $Iso\,(\mathbf{R}^2)$ is isomorphic to one of these.

Theorem 4.

Let $G \subset Iso\,(\mathbf{R}^2)$ be a finite subgroup. Then G fixes a point $p_0 \in \mathbf{R}^2$ and is one of the following:

1. G is a cyclic group of order n, generated by the rotation $R_{2\pi/n}(p_0)$.
2. G is a dihedral group of order $2n$, generated by two elements: $R_{2\pi/n}(p_0)$ and a reflection R_l in a line l through p_0.

Proof.

Let $p \in \mathbf{R}^2$ be any point and consider the *orbit* of G through p:

$$G(p) = \{S(p) \mid S \in G\}.$$

This is a finite subset of $\mathbf{R}^2$ with elements listed as $G(p) = \{p_1, \ldots, p_m\}$. Each element in G is a permutation on this set. Now consider the centroid

$$p_0 = \frac{p_1 + \cdots + p_m}{m}.$$

We claim that p_0 is left fixed by G. Let $S \in G$. We now use the classification of plane isometries above, along with the fact that S permutes $p_1, \ldots, p_m$, to conclude that $S(p_0) = p_0$. To determine the structure of G, we split the proof into two cases.

(1) G contains only direct isometries. Since G fixes p_0, every element in G is a rotation $R_\theta(p_0)$ with center p_0. In what follows, we suppress p_0. Let θ be the smallest positive angle of rotation. We claim that G is generated by R_θ. Since G is finite, it will then follow that G is cyclic. Indeed, let $R_\alpha \in G$ be arbitrary. The division algorithm tells us that

$$\alpha = m\theta + \beta, \quad m \in \mathbf{Z},$$

with remainder $0 \leq \beta < \theta$. Since G is a group, $R_\beta = R_{\alpha - m\theta} = R_\alpha \circ (R_\theta)^{-m} \in G$. Since θ is the smallest positive angle with $R_\theta \in G$ this is possible only if $\beta = 0$. We obtain that $\alpha = m\theta$, and hence

$$R_\alpha = R_{m\theta} = (R_\theta)^m.$$

The rest is clear, since $(R_\theta)^n = I$ for $|G| = n$ so that $\theta = 2\pi/n$.

(2) Assume that G contains an opposite isometry. Any opposite isometry with fixed point p_0 must be a reflection R_l in a line l through p_0. Thus, G contains rotations and reflections, the former being a subgroup of G denoted by G^+. By the previous case, G^+ is generated by a rotation R_θ (with $\theta = 2\pi/n$). Let $R_l \in G$. As in the proof of Theorem 3, we have the following $2n$ elements in G:

$$I, R_\theta, R_\theta^2, \ldots, R_\theta^{n-1},$$

$$R_l, R_\theta \circ R_l, R_\theta^2 \circ R_l, \ldots, R_\theta^{n-1} \circ R_l.$$

These isometries are all distinct, and they form a subgroup G' (of G) isomorphic with D_n. We must show that $G' = G$. It is enough to show that G' contains all reflections in G. Let $R_{l'} \in G$, $l \neq l'$. Since l and l' intersect in p_0, $R_{l'} \circ R_l$ is a rotation in G, hence $R_{l'} \circ R_l = R_\theta^k$ for some $k = 0, \ldots, n-1$. Thus $R_{l'} = R_\theta^k \circ R_l$, and this is listed above. ■

Theorem 4 asserts in particular that the only possible groups of central symmetries in two dimensions are

$$C_1, C_2, C_3, \ldots \quad \text{and} \quad D_1, D_2, D_3, \ldots.$$

Central symmetry frequently occurs in nature. Most of us observed in childhood that snowflakes have sixfold (some threefold) symmetries. Flowers usually have fivefold symmetries, and depending on whether the petals are bilaterally symmetric or not, their symmetry group is D_5 or only C_5. We finish this section with a quotation from Hermann Weyl's *Symmetry* regarding Theorem 4: "Leonardo da Vinci engaged in systematically determining the possible symmetries of a central building and how to attach chapels and niches without destroying the symmetry of the nucleus. In abstract modern terminology, his result is essentially our above table of the possible finite groups of rotations (proper and improper) in two dimensions."

Problems

1. Show that the 3-dimensional cube ($[0, 1]^3 \subset \mathbf{R}^3$) can be sliced by planes to obtain a square, an equilateral triangle, and a regular hexagon.

2. Prove that any two rotations $R_\theta(p)$ and $R_\theta(q)$ with the same angle $\theta \in \mathbf{R}$ are conjugate in *Iso* $(\mathbf{R}^2)$; that is, $R_\theta(q) = T_v \circ R_\theta(p) \circ T_{-v}$, where v is a vector from p to q.

3. (a) Let s_n, $n \geq 3$, denote the side length of P_n, the regular n-sided polygon inscribed in the unit circle. Show that

$$s_{2n} = \sqrt{2 - \sqrt{4 - s_n^2}}.$$

Deduce from this that

$$s_4 = \sqrt{2}, \qquad s_8 = \sqrt{2 - \sqrt{2}}, \qquad s_{16} = \sqrt{2 - \sqrt{2 + \sqrt{2}}}, \ldots.$$

Generalize these to show that

$$s_{2^n} = \sqrt{2 - \sqrt{2 + \sqrt{2 + \cdots + \sqrt{2}}}},$$

with $n - 1$ nested square roots.

(b) Let A_n be the area of P_n. Derive the formula

$$A_{2^{n+1}} = 2^{n-1} s_{2^n} = 2^{n-1} \sqrt{2 - \sqrt{2 + \sqrt{2 + \cdots + \sqrt{2}}}},$$

with $n - 1$ nested square roots. (Hint: Half of the sides of P_n serve as heights of the $2n$ isosceles triangles that make up P_{2n}, so that $A_{2n} = n s_n / 2$.) Conclude that

$$\lim_{n \to \infty} 2^n \sqrt{2 - \sqrt{2 + \sqrt{2 + \cdots + \sqrt{2}}}} = \pi,$$

where in the limit there are n nested square roots. In particular, we have

$$\lim_{n \to \infty} \sqrt{2 + \sqrt{2 + \cdots + \sqrt{2}}} = 2.$$

4. (a) Prove that in the product of three reflections, one can always arrange that one of the reflecting lines is perpendicular to both the others. (b) Derive Theorem 4 without the "orbit argument," using the previously proved fact that all the rotations in G have the same center.

5. The *reciprocal of a point* $p = (a, b) \in \mathbf{R}^2$ to the circle with center at the origin and radius $r > 0$ is the *line* given by the equation $ax + by = r^2$. Show that the reciprocals of the vertices $z(2k\pi/n)$, $k = 0, \ldots, n - 1$, of the regular n-sided polygon P_n to its *inscribed* circle give the sides of another regular n-sided polygon whose vertices are the midpoints of the sides of P_n.

Web Sites

1. www.maa.org

2. aleph0.clarku.edu/~djoyce/java/elements/elements.html

10

Discrete Subgroups of Iso ($\mathbf{R}^2$)

♣ In Section 9, we obtained a classification of all finite subgroups of the group of isometries $Iso\,(\mathbf{R}^2)$ of $\mathbf{R}^2$ by studying symmetry groups of regular polygons. We saw that such subgroups cannot contain translations or glides, a fact that is intimately connected to boundedness of regular polygons. If we want to include translations and glides in our study, we have to start with unbounded plane figures and their symmetry groups. It turns out that classification of these subgroups is difficult unless we assume that the subgroup $G \subset Iso\,(\mathbf{R}^2)$ does not contain rotations of arbitrarily small angle and translations of arbitrarily small vector length. (As we will see later, we do not have to impose any condition on reflections and glides.) Groups $G \subset Iso\,(\mathbf{R}^2)$ satisfying this condition are called *discrete*. In this section we give a complete classification of discrete subgroups of $Iso\,(\mathbf{R}^2)$. Just as cyclic and dihedral groups can be viewed as orientation-preserving and full symmetry groups of regular polygons, we will visualize these groups as symmetries of frieze and wallpaper patterns. Thus, next time you look at a wallpaper pattern, you should be able to write down generators and relations for the corresponding symmetry group!

Let $G \subset Iso\,(\mathbf{R}^2)$ be a discrete group. Assume that G contains a translation $T_v \in G$ that is not the identity ($v \neq 0$). All powers of T_v are then contained in G (by the group property): $T_v^k \in G$, $k \in \mathbf{Z}$. Since $T_v^k = T_{kv}$, we see that all these are mutually distinct. It follows that G must be infinite. (The same conclusion holds for glides, since the square of a glide is a translation.) We see that the presence of translations or glides makes G infinite. The following question arises naturally: If we are able to excise the translations from G, is the remaining "part" of G finite? The significance of an affirmative answer is clear, since we just classified all finite subgroups of Iso ($\mathbf{R}^2$). This gives us a good reason to look at translations first.

Let $\mathcal{T}$ be the group of translations in $\mathbf{R}^2$. It is clearly a subgroup of Iso ($\mathbf{R}^2$). From now on we agree that for a translation $T_v \in \mathcal{T}$, we draw the translation vector v from the origin. Associating to T_v the vector v (just made unique) gives the map

$$\varphi : \mathcal{T} \to \mathbf{R}^2,$$

defined by

$$\varphi(T_v) = v, \quad v \in \mathbf{R}^2.$$

Since

$$T_{v_1} \circ T_{v_2} = T_{v_1+v_2}, \quad v_1, v_2 \in \mathbf{R}^2,$$

and

$$(T_v)^{-1} = T_{-v}, \quad v \in \mathbf{R}^2,$$

we see that φ is an isomorphism. Summarizing, the translations in Iso ($\mathbf{R}^2$) form a subgroup $\mathcal{T}$ that is isomorphic with the additive group $\mathbf{R}^2$.

Let $G \subset Iso$ ($\mathbf{R}^2$) be a discrete group. The translations in G form a subgroup $T = G \cap \mathcal{T}$ of G. Since G is discrete, so is T. The isomorphism $\varphi : \mathcal{T} \to \mathbf{R}^2$ maps T to a subgroup denoted by $L_G \subset \mathbf{R}^2$. This latter group is also discrete in the sense that it does not contain vectors of arbitrarily small length. By definition, L_G is the group of vectors $v \in \mathbf{R}^2$ such that the translation T_v is in G. We now classify the possible choices for L_G.

Theorem 5.

Let L be a discrete subgroup of $\mathbf{R}^2$. Then L is one of the following:

1. $L = \{0\}$;
2. L consists of integer multiples of a nonzero vector $v \in \mathbf{R}^2$:

$$L = \{kv \mid k \in \mathbf{Z}\};$$

3. L consists of integral linear combinations of two linearly independent vectors $v, w \in \mathbf{R}^2$:

$$L = \{kv + lw \mid k, l \in \mathbf{Z}\}.$$

Proof.
We may assume that L contains a nonzero vector $v \in \mathbf{R}^2$. Let $l = \mathbf{R} \cdot v$ be the line through v. Since L is discrete, there is a vector in $l \cap L$ of shortest length. Changing the notation if necessary, we may assume that this vector is v. Let w be any vector in $l \cap L$. We claim that w is an integral multiple of v. Indeed, $w = av$ for some $a \in \mathbf{R}$ since w is in l. Writing

$$a = k + r,$$

where k is an integer and $0 \le r < 1$, we see that $w - kv = (a-k)v = rv$ is in L. On the other hand, if $r \ne 0$, then the length of rv is less than that of v, contradicting the minimality of v. Thus $r = 0$, and $w = kv$, an integer multiple of v. If there are no vectors in L outside of l, then we land in case 2 of the theorem.

Finally, assume that there exists a vector $w \in L$ not in l. The vectors v and w are linearly independent, so that they span a parallelogram P. Since P is bounded, it contains only finitely many elements of L. Among these, there is one whose distance to the line l is positive, but the smallest possible. By changing w (and P), we may assume that this vector is w. We claim now that there are no vectors of L in P except for its vertices. ¬ Assume the contrary and let $z \in L$ be a vector in P. Due to the minimal choice of v and w, this is possible only if z terminates at a point on the opposite side of v or w. In the first case, $z - w \in L$ would be a vector shorter than v; in the second, z would be closer to l than w. ¬ Summarizing, we conclude that there are two linearly independent vectors v and w that span a parallelogram P such that P contains no vectors in L except for its vertices. Clearly, $\{kv + lw \mid k, l \in \mathbf{Z}\}$ is contained in L. To land in case 3 we now claim that every vector z in L is an integral linear combination of v and w. By linear independence, z is certainly a linear combination

$$z = av + bw$$

of v and w with *real* coefficients $a, b \in \mathbf{R}$. We now write

$$a = k + r \quad \text{and} \quad b = l + s,$$

where $k, l \in \mathbf{Z}$ and $0 \leq r, s < 1$. The vector $z - kv - lw = rv + sw$ is in L and is contained in P. The only way this is possible is if $r = s = 0$ holds. Thus $z = kv + lw$, and we are done. ∎

We now return to our discrete group $G \subset Iso\,(\mathbf{R}^2)$ and see that we have three choices for L_G. If $L_G = \{0\}$, then G does not contain any translations (or glides, since the square of a glide is a translation). In this case, G consists of rotations and reflections only. By a result of the previous section, the rotations in G have the same center, say, p_0. Since G is discrete, it follows that G contains only finitely many rotations. If $R_l \in G$ is a reflection, then l must go through p_0, since otherwise, $R_l(p_0)$ would be the center of another rotation in G. Finally, since the composition of two reflections in G is a rotation in G, there may be only at most as many reflections in G as rotations (cf. the proof of Theorem 4). Summarizing, we obtain that if G is a discrete group of isometries with $L_G = \{0\}$ then G must be finite. In the second case T, the group of all translations in G, is generated by T_v, and we begin to suspect that G is the symmetry group of a *frieze pattern*. Finally, in the third case T is generated by T_v and T_w, and T is best viewed by its φ-image $L_G = \{kv + lw \mid k, l \in \mathbf{Z}\}$ in $\mathbf{R}^2$. We say that L_G is a *lattice* in $\mathbf{R}^2$ and G is a (2-dimensional) *crystallographic group*. Since any wallpaper pattern repeats itself in two different directions, we see that their symmetry groups are crystallographic.

We now turn to the process of "excising" the translation part from G. To do this, we need some preparations. Recall that at the discussion of translations we agreed to draw the vectors v from the origin so that the translation T_v by the vector $v \in \mathbf{R}^2$ acts on $p \in \mathbf{R}^2$ by $T_v(p) = p + v$. Now given any *linear* transformation $A : \mathbf{R}^2 \rightarrow \mathbf{R}^2$ (that is, $A(v_1 + v_2) = A(v_1) + A(v_2)$, $v_1, v_2 \in \mathbf{R}^2$, and $A(rv) = rA(v)$, $r \in \mathbf{R}$, $v \in \mathbf{R}^2$), we have the commutation rule

$A \circ T_v = T_{A(v)} \circ A$. Indeed, evaluating the two sides at $p \in \mathbf{R}^2$, we get

$$(A \circ T_v)(p) = A(T_v(p)) = A(p + v) = A(p) + A(v)$$

and

$$(T_{A(v)} \circ A)(p) = T_{A(v)}(A(p)) = A(p) + A(v).$$

Let $O(\mathbf{R}^2)$ denote the group of isometries in $Iso\,(\mathbf{R}^2)$ that leave the origin fixed. $O(\mathbf{R}^2)$ is called the *orthogonal group*. From the classification of the plane isometries, it follows that the elements of $O(\mathbf{R}^2)$ are linear.

Remark.

♠ We saw above that a direct isometry in $O(\mathbf{R}^2)$ is a rotation R_θ. These rotations form the *special orthogonal group* $SO(\mathbf{R}^2)$, a subgroup of $O(\mathbf{R}^2)$. Associating to R_θ the complex number $z(\theta)$ establishes an isomorphism between $SO(\mathbf{R}^2)$ and S^1. Any opposite isometry in $O(\mathbf{R}^2)$ can be written as a rotation followed by conjugation. Thus *topologically* $O(\mathbf{R}^2)$ is the disjoint union of two circles.

♣ Occasionally, it is convenient to introduce superscripts $\pm$ to indicate whether the isometries are direct or opposite. Thus $Iso^+(\mathbf{R}^2)$ denotes the set of direct isometries in $Iso\,(\mathbf{R}^2)$. Note that it is a subgroup, since the composition and inverse of direct isometries are direct. $Iso^-(\mathbf{R}^2)$ is not a subgroup but a topological copy of $Iso^+(\mathbf{R}^2)$.

♡ The elements of $O(\mathbf{R}^2)$ are linear, so that the commutation rule above applies. We now define a homomorphism

$$\psi : Iso\,(\mathbf{R}^2) \to O(\mathbf{R}^2)$$

as follows: Let $S \in Iso\,(\mathbf{R}^2)$ and denote by v the vector that terminates at $S(0)$. The composition $(T_v)^{-1} \circ S$ fixes the origin so that it is an element of $O(\mathbf{R}^2)$. We define $\psi(S) = (T_v)^{-1} \circ S$. To prove that ψ is a homomorphism, we first write $(T_v)^{-1} \circ S = U \in O(\mathbf{R}^2)$, so that $S = T_v \circ U$. This decomposition is unique in the sense that if $S = T_{v'} \circ U'$ with $v' \in \mathbf{R}^2$ and $U' \in O(\mathbf{R}^2)$, then $v = v'$ and $U = U'$. Indeed, $T_v \circ U = T_{v'} \circ U'$ implies that $(T_{v'})^{-1} \circ T_v = U' \circ U^{-1}$. The right-hand side fixes the origin so that the left-hand side, which is a translation, must be the identity. Uniqueness follows.

Using the notation we just introduced, we have $\psi(S) = U$, where $S = T_v \circ U$. Now let $S_1 = T_{v_1} \circ U_1$ and $S_2 = T_{v_2} \circ U_2$, where $v_1, v_2 \in \mathbf{R}^2$ and $U_1, U_2 \in O(\mathbf{R}^2)$. For the homomorphism property, we need to show that $\psi(S_2 \circ S_1) = \psi(S_2) \circ \psi(S_1)$. By definition, $\psi(S_1) = U_1$ and $\psi(S_2) = U_2$, so that the right-hand side is $U_2 \circ U_1$. As for the left-hand side, we first look at the composition

$$S_2 \circ S_1 = T_{v_2} \circ U_2 \circ T_{v_1} \circ U_1.$$

Using the commutation rule for the linear U_2, we have $U_2 \circ T_{v_1} = T_{U_2(v_1)} \circ U_2$. Inserting this, we get

$$S_2 \circ S_1 = T_{v_2} \circ T_{U_2(v_1)} \circ U_2 \circ U_1.$$

Taking ψ of both sides amounts to deleting the translation part:

$$\psi(S_2 \circ S_1) = U_2 \circ U_1.$$

Thus ψ is a homomorphism.

ψ is onto since it is identity on $O(\mathbf{R}^2) \subset Iso\,(\mathbf{R}^2)$. The kernel of ψ consists of translations:

$$\ker \psi = \mathcal{T}.$$

In particular, $\mathcal{T} \subset Iso\,(\mathbf{R}^2)$ is a normal subgroup. Having constructed $\psi : Iso\,(\mathbf{R}^2) \to O(\mathbf{R}^2)$, we return to our discrete group $G \subset Iso\,(\mathbf{R}^2)$. The ψ-image of G is called the *point-group* of G, denoted by $\bar{G} = \psi(G)$. The kernel of $\psi|G$ is all translations in G, that is, T. Thus, we have the following:

$$\psi|G : G \to \bar{G} \subset O(\mathbf{R}^2)$$

and

$$ker(\psi|G) = T.$$

For nontrivial L_G, the point-group $\bar{G}$ interacts with L_G in a beautiful way:

Theorem 6.
   $\bar{G}$ *leaves* L_G *invariant.*

Proof.
Let $U \in \bar{G}$ and $v \in L_G$. We must show that $U(v) \in L_G$. Since $U \in \bar{G}$, there exists $S \in G$, with $S = T_w \circ U$ for some $w \in \mathbf{R}^2$. The assumption $v \in L_G$, means $T_v \in G$, and what we want to conclude, $U(v) \in L_G$, means $T_{U(v)} \in G$. We compute

$$T_{U(v)} = T_{U(v)} \circ T_w \circ (T_w)^{-1}$$
$$= T_w \circ T_{U(v)} \circ (T_w)^{-1}$$
$$= T_w \circ U \circ T_v \circ U^{-1} \circ (T_w)^{-1}$$
$$= S \circ T_v \circ S^{-1} \in G,$$

where the last but one equality is because of the commutation relation

$$U \circ T_v = T_{U(v)} \circ U$$

as established above. The theorem follows.   ∎

$\bar{G}$ is discrete in the sense that is does not contain rotations with arbitrarily small angle. This follows from Theorem 6 if L_G is nontrivial. If L_G is trivial, then by a result of the previous section, G is finite, and so is its (isomorphic) image $\bar{G}$ under ψ. Since $\bar{G}$ is discrete and fixes the origin, it must be finite! Indeed, by now this argument should be standard. Let $R_\theta \in \bar{G}$ with θ being the smallest positive angle. Then any rotation in $\bar{G}$ is a multiple of R_θ. Moreover, using the division algorithm, we have $2\pi = n\theta + r$, $0 \le r < \theta$, $n \in \mathbf{Z}$, and r must reduce to zero because of minimality of θ. Thus $\theta = 2\pi/n$, and the rotations form a cyclic group of order n. Finally, there cannot be infinitely many reflections, since otherwise their axes could get arbitrarily close to each other, and composing any two could give rotations of arbitrarily small angle. We thus accomplished our aim. $\bar{G}$ gives a finite subgroup in $O(\mathbf{R}^2)$ consisting of rotations and reflections only. In particular, if $L_G = \{0\}$—that is, if G contains no nontrivial translations—then the kernel of $\psi|G$ is trivial and so $\psi|G$ maps G isomorphically onto $\bar{G}$. In particular, G is finite. By Theorem 4 of Section 9, G is cyclic or dihedral.

We are now ready to classify the possible frieze patterns, of which there are seven. According to Theorem 6, a frieze group G keeps the line c through L_G invariant, and the group of translations T in G is an infinite cyclic subgroup generated by a shortest translation, say, τ, in the direction of c. The line c is called the "center" of the frieze group. In addition to T, the only nontrivial direct isometries are rotations with angle π, called "half-turns," and their center must be on c. The only possible opposite isometries are reflection to c, reflections to lines perpendicular to c, and glides along c. In the classification below we use the following notations: If G contains a half-turn, we denote its center by $p \in c$. If G does not contain any half-turns, but contains reflections to lines perpendicular to c, the axis of reflection is denoted by l, and p is the intersection point of l and c. Otherwise p is any point on c. Let $p_n = \tau^n(p)$, $n \in \mathbf{Z}$, and $m =$ the midpoint of the segment connecting p_0 and p_1. Finally, let $m_n = \tau^n(m)$, the midpoint of the segment connecting p_n and p_{n+1} (Figure 10.1).

We are now ready to start. First we classify the frieze groups that contain only direct isometries:

1. $G = T = \langle \tau \rangle$, so that G contains[1] no half-turns, reflections or glide reflections.
2. $G = \langle \tau, H_p \rangle$. Aside from translations, G contains the half-turns $\tau^n \circ H_p$. For $n = 2k$ even, $\tau^{2k} \circ H_p$ has center at p_k, and for $n = 2k + 1$ odd, $\tau^{2k+1} \circ H_p$ has center at m_k.

It is not hard to see that these are all the frieze groups that contain only direct isometries. We now allow the presence of opposite isometries.

3. $G = \langle \tau, R_c \rangle$. Since $R_c^2 = I$ and $\tau \circ R_c = R_c \circ \tau$, aside from T, this group consists of glides $\tau^n \circ R_c$ mapping p to p_n.

Figure 10.1

$p \qquad\qquad m \qquad\qquad p_1 \qquad\qquad c$

---

[1] If $\Gamma \subset \textit{Iso}\,(\mathbf{R}^2)$, then $\langle \Gamma \rangle$ denotes the smallest subgroup in $\textit{Iso}\,(\mathbf{R}^2)$ that contains Γ. We say that Γ generates $\langle \Gamma \rangle$ (cf. "Groups" in Appendix B).

4. $G = \langle \tau, R_l \rangle$. Since $R_l^2 = I$ and $R_l \circ \tau = \tau^{-1} \circ R_l$, aside from T, G consists of reflections $\tau^n \circ R_l$. The axes are perpendicular to c, and according to whether $n = 2k$ (even) or $n = 2k + 1$ (odd), the intersections are p_k or m_k.
5. $G = \langle \tau, H_p, R_c \rangle$. We have $H_p \circ R_c = R_l \circ R_c \circ R_c = R_l \in G$. In addition to this and τ, G includes the glides $\tau^n \circ R_c$ (sending p to p_n) and $\tau^n \circ R_l$ discussed above.
6. $G = \langle \tau, H_p, R_{l'} \rangle$. Then l' must intersect c perpendicularly at the midpoint of p and m_k, for some $k \in \mathbf{Z}$.
7. $G = \langle G_{c,v} \rangle$ is generated by the glide $G_{c,v}$ with $G_{c,v}^2 = \tau$.

Figure 10.2 depicts the seven frieze patterns. (The pictures were produced with Kali (see Web Site 2), written by Nina Amenta of the Geometry Center at the University of Minnesota.) Which corresponds to which in the list above?

The fact that the point-group $\bar{G}$ leaves L_G invariant imposes a severe restriction on G if L_G is a lattice, the case we turn to next.

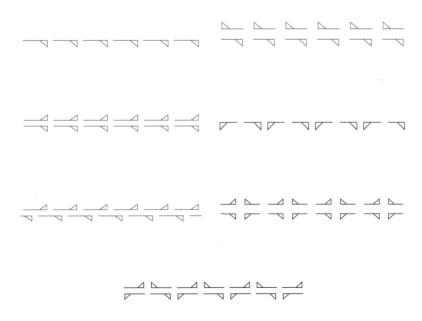

Figure 10.2

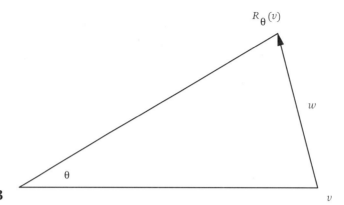

Figure 10.3

Crystallographic Restriction.

Assume that G is crystallographic. Let $\bar{G}$ denote its point-group. Then every rotation in $\bar{G}$ has order 1,2,3,4, or 6, and $\bar{G}$ is C_n or D_n for some $n = 1,2,3,4,$ or 6.

Proof.

As usual, let R_θ be the smallest positive angle rotation in $\bar{G}$, and let v be the smallest length nonzero vector in L_G. Since L_G is $\bar{G}$-invariant, $R_\theta(v) \in L_G$. Consider $w = R_\theta(v) - v \in L_G$ (Figure 10.3).

Since v has minimal length, $|v| \le |w|$. Thus,

$$\theta \ge 2\pi/6,$$

and so R_θ has order ≤ 6. The case $\theta = 2\pi/5$ is ruled out since $R_\theta^2(v) + v$ is shorter than v (Figure 10.4).

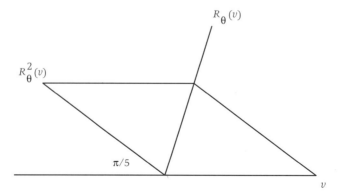

Figure 10.4

The first statement follows. The second follows from the classification of finite subgroups of $Iso\,(\mathbf{R}^2)$ in the previous section.   ∎

Remark.
For an algebraic proof of the crystallographic restriction, consider the trace tr (R_θ) of $R_\theta \in \bar{G}$, $0 < \theta \leq \pi$. With respect to a basis in L_G, the matrix of R_θ has integral entries (Theorem 6). Thus, tr (R_θ) is an integer. On the other hand, with respect to an orthonormal basis, the matrix of R_θ has diagonal entries both equal to $\cos(\theta)$. In particular, tr $(R_\theta) = 2\cos(\theta)$. Thus, $2\cos(\theta)$ is an integer, and this is possible only for $n = 2, 3, 4$, or 6.

EXAMPLE
If $\omega = z(2\pi/n)$ is a primitive nth root of unity, then $\mathbf{Z}[\omega]$ is a lattice iff $n = 3, 4$, or 6. Indeed, the rotation $R_{2\pi/n}$ leaves $\mathbf{Z}[\omega]$ invariant, since it is multiplication by ω. By the crystallographic restriction, $n = 3, 4$, or 6. How do the tesselations look for $n = 3$ and $n = 6$?   □

The absence of order-5 symmetries in a lattice must have puzzled some ancient ornament designers. We quote here from Hermann Weyl's *Symmetry*: "The Arabs fumbled around much with the number 5, but they were of course never able honestly to insert a central symmetry of 5 in their ornamental designs of double infinite rapport. They tried various deceptive compromises, however. One might say that they proved experimentally the impossibility of a pentagon in an ornament."

Armed with the crystallographic restriction, we now have the tedious task of considering all possible scenarios for the point-group $\bar{G}$ and its relation to L_G. This was done in the nineteenth century by Fedorov and rediscovered by Polya and Niggli in 1924. A description of the seventeen crystallographic groups that arise are listed as follows:

Generators for the 17 Crystallographic Groups

1. Two translations.
2. Three half-turns.

3. Two reflections and a translation.
4. Two parallel glides.
5. A reflection and a parallel glide.
6. Reflections to the four sides of the rectangle.
7. A reflection and two half-turns.
8. Two perpendicular glides.
9. Two perpendicular reflections and a half-turn.
10. A half-turn and a quarter-turn.
11. Reflections in the three sides of a $(\pi/4, \pi/4, \pi/2)$ triangle.
12. A reflection and a quarter-turn.
13. Two rotations through $2\pi/3$.
14. A reflection and a rotation through $2\pi/3$.
15. Reflections in the three sides of an equilateral triangle.
16. A half-turn and a rotation through $2\pi/3$.
17. Reflections is in the three sides of a $(\pi/6, \pi/3, \pi/2)$ triangle.

Remark.

The following construction sheds some additional light on the geometry of crystallographic groups. Let G be crystallographic and assume that G contains rotations other than half-turns. Let $R_{2\alpha}(p) \in G$, $0 < \alpha < \pi/2$, be a rotation with integral π/α (cf. the proof of Theorem 4 of Section 9). Let $R_{2\beta}(q) \in G$, $0 < \beta < \pi/2$, be another rotation with integral π/β such that $d(p, q)$ is minimal. ($R_{2\beta}(q)$ exists since G is crystallographic.) Let l denote the line through p and q. Write $R_{2\alpha}(p) = R_{l'} \circ R_l$, where l' meets l at p and the angle from l to l' is α. Similarly, $R_{2\beta}(q) = R_l \circ R_{l''}$, where l'' meets l at q and the angle from l'' to l is β. Since $\alpha + \beta < \pi$, the lines l' and l'' intersect at a point, say, r. In fact, r is the center of the rotation $R_{2\gamma}(r) = (R_{2\alpha}(p) \circ R_{2\beta}(q))^{-1} = R_{l''} \circ R_{l'}$. Since α, β, and γ are the interior angles of the triangle $\triangle pqr$, we have $\alpha + \beta + \gamma = \pi$. On the other hand, since G is discrete, π/γ is rational. It is easy to see that minimality of $d(p, q)$ implies that π/γ is integral. We obtain that

$$\frac{\alpha}{\pi} + \frac{\beta}{\pi} + \frac{\gamma}{\pi} = 1,$$

where the terms on the left-hand side are reciprocals of integers. Since π/α, $\pi/\beta \geq 3$ (and $\pi/\gamma \geq 2$), the only possibilities are $\alpha =$

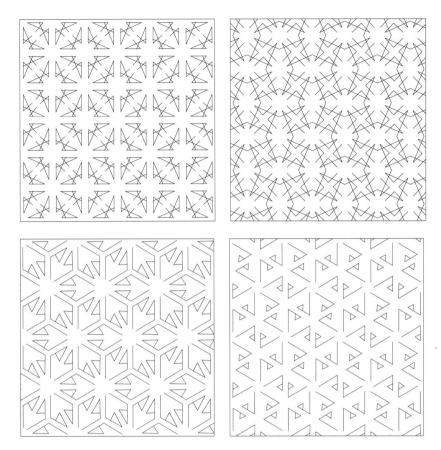

Figure 10.5

$\beta = \gamma = \pi/3$; $\alpha = \beta = \pi/4$, $\gamma = \pi/2$; and $\alpha = \pi/6$, $\beta = \pi/3$, $\gamma = \pi/2$. (Which corresponds to which in the list above?)

As noted above, these groups can be visualized by patterns covering the plane with symmetries prescribed by the acting crystallographic group. Figure 10.5 shows a sample of four patterns (produced with Kali).

Symmetric patterns[2] date back to ancient times. They appear in virtually all cultures; on Greek vases, Roman mosaics, in the thirteenth century Alhambra at Granada, Spain, and on many other Muslim buildings.

---

[2]For a comprehensive introduction see B. Grünbaum and G.C. Shephard, *Tilings and Patterns*, Freeman, 1987.

To get a better view of the repetition patterns, we introduce the concept of fundamental domain. First, given a discrete group $G \subset$ Iso ($\mathbf{R}^2$), a *fundamental set* for G is a subset F of $\mathbf{R}^2$ which contains exactly one point from each orbit

$$G(p) = \{S(p) \mid S \in G\}, \quad p \in \mathbf{R}^2.$$

A *fundamental domain* F_0 for G is a domain (that is, a connected open set) such that there is a fundamental set F between F_0 and its closure[3] $\bar{F}_0$; that is, $F_0 \subset F \subset \bar{F}_0$, and the 2-dimensional area of the boundary $\partial F_0 = \bar{F}_0 - F_0$ is zero.

The simplest example of a fundamental set (domain) is given by the translation group $G = T = \langle T_v, T_w \rangle$. In this case, a fundamental domain F_0 is the open parallelogram spanned by v and w. A fundamental set F is obtained from F_0 by adding the points tv and tw, $0 \leq t < 1$. By the defining property of the fundamental set, the "translates" $S(F)$, $S \in G$, tile[4] or, more sophisticatedly, *tessellate* $\mathbf{R}^2$. (Numerous tessellations appear in Kepler's *Harmonice Mundi*, which appeared in 1619.) If a pattern is inserted in F, translating it with G gives the wallpaper patterns that you see. You are now invited to look for fundamental sets in Figure 10.5!

Problems

1. Prove directly that any plane isometry that fixes the origin is linear.

2. Identify the frieze group that corresponds to the pattern in Figure 10.6.

3. Let $L \subset \mathbf{R}^2$ be a lattice. Show that half-turn around the midpoint of any two points of L is a symmetry of L.

4. Identify the discrete group G generated by the three half-turns around the midpoints of the sides of a triangle.

Figure 10.6

---

[3]See "Topology" in Appendix C.

[4]We assume that the tiles can be turned over; i.e., they are decorated on both sides.

Web Sites

1. www.geom.umn.edu/docs/doyle/mpls/handouts/node30.html

2. www.geom.umn.edu/apps/kali/start.html

3. www.math.toronto.edu/~coxeter/art-math.html

4. www.texas.net/escher/gallery

5. www.suu.edu/WebPages/MuseumGaller/Art101/aj-webpg.htm

6. www.geom.umn.edu/apps/quasitiler/start.html

11

S E C T I O N

Möbius Geometry

♣ Recall from Section 7 that the stereographic projections h_N : $S^2 - \{N\} \to \mathbf{R}^2$ and $h_S : S^2 - \{S\} \to \mathbf{R}^2$ combine to give

$$h_N \circ h_S^{-1} : \mathbf{R}^2 - \{0\} \to \mathbf{R}^2 - \{0\},$$

where

$$(h_N \circ h_S^{-1})(z) = z/|z|^2 = 1/\bar{z}, \quad 0 \neq z \in \mathbf{C} = \mathbf{R}^2.$$

Strictly speaking, this is not a transformation of the plane since it is undefined at $z = 0$. Under h_N, however, it corresponds to the transformation of the unit sphere S^2:

$$h_N^{-1} \circ (h_N \circ h_S^{-1}) \circ h_N = h_S^{-1} \circ h_N : S^2 \to S^2$$

which sends a spherical point $p = (a, b, c)$ to $(a, b, -c)$ so that it is spherical reflection in the equatorial circle S^1 of S^2! That this holds can be seen from Figure 11.1.

Reflections in lines play a central role in Euclidean geometry. In fact, we saw in Section 9 that every isometry of $\mathbf{R}^2$ is the product of at most three reflections. In spherical geometry the ambient space is S^2, lines are great circles of S^2, and reflections are given by spatial reflections in planes in $\mathbf{R}^3$ spanned by great circles. For example, $h_S^{-1} \circ h_N : S^2 \to S^2$ is the restriction to S^2 of the spatial reflection

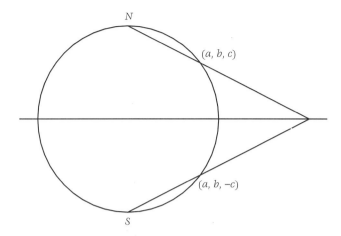

Figure 11.1

$(a, b, c) \mapsto (a, b, -c)$ in the coordinate plane $\mathbf{R}^2 \subset \mathbf{R}^3$ spanned by the first two axes.

We could now go on and study isometries of S^2 and develop spherical geometry in much the same way as we developed plane geometry. Instead, we put a twist on this and insist that we want to view all spherical objects in the plane! "Viewing", of course, means not only the visual perception, but also the description of these objects *in terms of Euclidean plane concepts*. This is possible by the stereographic projection $h_N : S^2 - \{N\} \to \mathbf{R}^2$, which does not quite map the entire sphere to $\mathbf{R}^2$ (this is impossible), but leaves out the North Pole N. Our spherical reflections will thus become "singular" when viewed on $\mathbf{R}^2$. To circumvent this difficulty, we attach an "infinite point" ∞ to $\mathbf{R}^2$ and say that the singular point of the transformation must correspond to ∞. For example, the transformation $z \mapsto 1/\bar{z}$ is singular at the origin but becomes well defined on the extended plane $\hat{\mathbf{R}}^2 = \mathbf{R}^2 \cup \{\infty\}$, where we agree that it should send 0 to ∞. Of course $\hat{\mathbf{R}}^2$ is nothing but S^2, but it is much easier to view the action on $\mathbf{R}^2$ plus one point than on S^2. Our path is now clear; we need to consider all finite compositions of spherical reflections in great circles of S^2 and pull them down to $\mathbf{R}^2$ (via h_N). The transformations of $\mathbf{R}^2$ (or rather $\hat{\mathbf{R}}^2$) obtained this way are named after Möbius. (The concept of reflection in a circle in $\hat{\mathbf{R}}^2$, the case of a single spherical reflection, was invented by Steiner around 1828.)

Although this project is not difficult to carry out, we will pursue a different track and work in $\hat{\mathbf{R}}^2$ from the beginning. Our starting

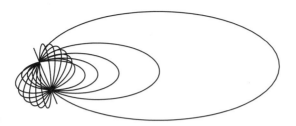

Figure 11.2

point is the following observation: Consider the equatorial circle $S^1 \subset \mathbf{R}^2 \subset \mathbf{R}^3$. Under h_N, it corresponds to itself. Now rotate S^1 around the first axis in $\mathbf{R}^3$ by various angles (Figure 11.2).

Under h_N, the rotated great circles correspond to various circles and, when the great circle passes through N, to a straight line! The conclusion is inevitable. In Möbius geometry we have to treat circles and lines on the same footing. Thus when we talk about circles we really mean circles or lines. For the farsighted this is no problem; a line on the extended plane $\hat{\mathbf{R}}^2$ becomes a circle by closing it up with ∞! Analytically, however, circles and lines have different descriptions:

$$S_r(p_0) = \{p \in \mathbf{R}^2 \mid d(p, p_0) = r\},$$

and

$$l_t(p_0) = \{p \in \mathbf{R}^2 \mid p \cdot p_0 = t\} \cup \{\infty\},$$

where $S_r(p_0) \subset \mathbf{R}^2$ is the usual Euclidean circle with center p_0 and radius r, and $l_t(p_0)$ is the usual Euclidean line (extended with ∞) with normal vector p_0 ($\cdot$ is the dot product). Notice that by fixing p_0 and varying r and t, we obtain concentric circles and parallel lines (Figure 11.3).

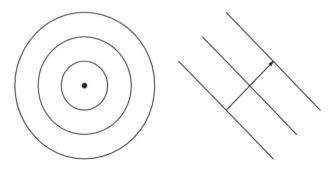

Figure 11.3

To define the reflection $R_S : \hat{\mathbf{R}}^2 \to \hat{\mathbf{R}}^2$ in a circle $S = S_r(p_0)$ in general, we rely on the special case where R_S is obtained from a spherical reflection $R_C : S^2 \to S^2$ in a great circle $C \subset S^2$, i.e., where $R_S = h_N \circ R_C \circ h_N^{-1}$. Since $h_N(C) = S$, this is precisely the case where $S \cap S^1$ contains an antipodal pair of points. At the end of Section 7, we proved that h_N is circle-preserving. Since this is automatically true for isometries (such as R_C), we see that R_S is also circle-preserving in our special case. We now define the reflection R_S in a general circle $S = S_r(p_0)$ by requiring that (i) R_S should be circle-preserving, (ii) it should fix each point of S, and (iii) R_S should interchange p_0 and ∞. That these conditions determine R_S uniquely follows by taking a careful look at the construction in Figure 11.4, in which the point p is chosen outside of S. (The segment connecting p and q is tangent to S at q.) When p is inside S, the points p and $R_S(p)$ should be interchanged. With this in mind, our argument is the following. The line l through p and p_0 should be mapped by R_S to itself, since it contains both p_0 and ∞ (which are interchanged), and it also goes through two diametrically opposite points of S (which stay fixed). Thus if p is on l then so is $R_S(p)$. The circle S' through q, q' and p_0, and the line l' through q, q' (and ∞) are interchanged by R_S since they both contain q, q' (that stay fixed). Thus, if p is outside of S, then $R_S(p)$ must be the common intersection of the lines l and l'. As noted above, the situation is analogous when p is inside S with the roles of p and $R_S(p)$ interchanged.

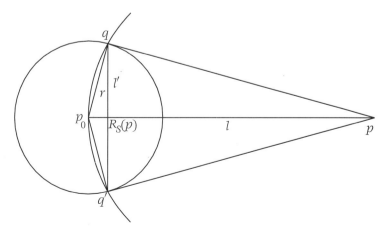

Figure 11.4

The analytical formula for R_S can also be read off from Figure 11.4. We have

$$R_S(p) = p_0 + \left(\frac{r}{d(p, p_0)} \right)^2 (p - p_0), \quad p \in \hat{\mathbf{R}}^2,$$

with p_0 and ∞ corresponding to each other. To see that this is true, we first note that p_0, p, and $R_S(p)$ are collinear, so that $R_S(p) - p_0 = \lambda(p - p_0)$ for some $\lambda \in \mathbf{R}$ to be determined. We have

$$\lambda = \frac{d(R_S(p), p_0)}{d(p, p_0)}.$$

The angles $\angle qpp_0$ and $\angle p_0 q R_S(p)$ are equal (Euclid!), so that the triangles $\triangle qpp_0$ and $\triangle p_0 q R_S(p)$ are similar. Thus,

$$\frac{d(R_S(p), p_0)}{r} = \frac{r}{d(p, p_0)},$$

so that

$$\lambda = \frac{d(R_S(p), p_0)}{d(p, p_0)} = \left(\frac{r}{d(p, p_0)} \right)^2,$$

and the formula for R_S follows when p is outside S. Instead of checking our computations for the case where p is inside S, it suffices to show that R_S^2 is the identity. To work out R_S^2, we first note that

$$R_S(p) - p_0 = \left(\frac{r}{d(p, p_0)} \right)^2 (p - p_0),$$

so that

$$d(R_S(p), p_0) = \left(\frac{r}{d(p, p_0)} \right)^2 d(p, p_0) = \frac{r^2}{d(p, p_0)}.$$

We now compute

$$R_S^2(p) = p_0 + \left(\frac{r}{d(R_S(p), p_0)} \right)^2 (R_S(p) - p_0)$$

$$= p_0 + \left(\frac{r}{r^2/d(p, p_0)} \right)^2 \left(\frac{r}{d(p, p_0)} \right)^2 (p - p_0)$$

$$= p_0 + p - p_0 = p.$$

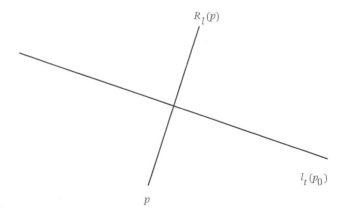

$R_l(p)$

$l_t(p_0)$

p

Figure 11.5

Thus, R_S^2 is the identity, as it should be for a true reflection.

The story is the same for the reflection R_l in the line $l_t(p_0)$ with normal vector p_0. The explicit formula is

$$R_l(p) = p - \frac{2(p \cdot p_0 - t)p_0}{|p_0|^2},$$

and $R_l(\infty) = \infty$ (Figure 11.5).

To show this we note first that $R_l(p) - p$ must be parallel to the normal vector p_0 of $l_t(p_0)$; that is,

$$R_l(p) - p = \lambda p_0.$$

We determine λ using the fact that $l_t(p_0)$ is the perpendicular bisector of the segment connecting p and $R_l(p)$. Analytically, the midpoint

$$\frac{p + R_l(p)}{2}$$

of this segment must be on $l_t(p_0)$:

$$(p + R_l(p)) \cdot p_0 = 2t.$$

On the other hand, $R_l(p) = p + \lambda p_0$, so that

$$\lambda = \frac{2(t - p \cdot p_0)}{|p_0|^2}$$

and the formula for R_l follows.

We now define the Möbius group $M\ddot{o}b\,(\hat{\mathbf{R}}^2)$ as the group of all finite compositions of reflections in circles and lines. Aside from this being a group, one more fact is clear. Our plane isometries are contained in the Möbius group

$$Iso\,(\mathbf{R}^2) \subset M\ddot{o}b\,(\hat{\mathbf{R}}^2).$$

This is because every plane isometry is the composition of (at most three) reflections in lines. (Note also that we automatically extended the plane isometries to $\hat{\mathbf{R}}^2$ by declaring that they should send ∞ to itself.)

Before we go any further, note that any spherical isometry of S^2 is the composition of spherical reflections. (The proof is analogous to the planar case.) Thus, conjugating the group $Iso\,(S^2)$ of all spherical isometries by the stereographic projection h_N, we obtain that $h_N \circ Iso\,(S^2) \circ h_N^{-1}$ is a subgroup of $M\ddot{o}b\,(\hat{\mathbf{C}})$! Finally, notice that $Iso\,(\mathbf{R}^2)$ and $h_N \circ Iso\,(S^2) \circ h_N^{-1}$ intersect in the orthogonal group $O(\mathbf{R}^2)$ (cf. Section 10).

♣ Let $k > 0$. Let R_S° be reflection in $S_1(0)$ and R_S reflection in $S_{\sqrt{k}}(0)$. The composition $R_S \circ R_S^\circ$ works out as follows:

$$(R_S \circ R_S^\circ)(p) = R_S\left(\frac{p}{|p|^2}\right)$$

$$= \left(\frac{\sqrt{k}}{d(0, p/|p|^2)}\right)^2 \frac{p}{|p|^2} = kp.$$

We see that $R_S \circ R_S^\circ$ is nothing but central dilatation with ratio of magnification $k > 0$.

We introduced reflections in circles by the requirement that they should be circle-preseving. This enabled us to derive an explicit formula for R_S as above. It is, however, not clear whether R_S is actually circle-preserving. Our next result states just this.

Theorem 7.

Let S be any Möbius transformation on $\hat{\mathbf{R}}^2$. Then S maps circles to circles.

Proof.

Since isometries and central dilatations send circles to circles, we may assume that S is reflection in a circle $S_r(p_0)$. Since $(T_{p_0})^{-1} \circ S \circ T_{p_0}$ is reflection in the circle $S_r(0)$, we may assume that $p_0 = 0$.

Replacing T_{p_0} in the previous argument with central dilatation with ratio $r > 0$, we may assume that $S = R_S^\circ$, reflection in the unit circle $S_1(0)$. Now let Σ be any circle. Σ can be described by the equation

$$\alpha|p|^2 - 2p \cdot p_0 + \beta = 0, \quad p \in \mathbf{R}^2,$$

where $\alpha, \beta \in \mathbf{R}$ and $p_0 \in \mathbf{R}^2$. The choice $\alpha = 1$ gives the circle with center at p_0, and $\alpha = 0$ defines the line with normal vector p_0. Dividing by $|p|^2$ and rewriting this in terms of $q = R_S^\circ(p) = p/|p|^2$, we obtain

$$\alpha - 2q \cdot p_0 + \beta|q|^2 = 0,$$

and this is the equation of another circle. We are done.[1] ∎

Remark.

♡ Möbius transformations generalize to any dimensions. In fact, the defining formula for the reflection R_S works for $\mathbf{R}^n$ if we think of d as the Euclidean distance function on $\mathbf{R}^n$. Here, $S = S_r(p_0)$ is the $(n-1)$-dimensional sphere with center p_0 and radius r. As an interesting connection between Möbius geometry and the stereographic projection h_N, we note here that, for $n = 3$, h_N is nothing but the restriction of R_S to $S^2 \subset \mathbf{R}^3$, where $S = S_{\sqrt{2}}(N)$ is the sphere with center at the North Pole N and radius $\sqrt{2}$. This can be worked out explicitly by looking at Figure 11.6. (Do you see the Lune of Hippocrates here?) In fact, all we need to show is that $R_S(p)$, $p \in S^2$, is in $\mathbf{R}^2$, or equivalently that $R_S(p)$ and N are orthogonal. Taking dot products and using $|p| = 1$, we compute

$$R_S(p) \cdot N = 1 + \frac{2}{|p - N|^2}(p - N) \cdot N$$

$$= 1 + \frac{2}{2 - 2p \cdot N}(p \cdot N - 1) = 0.$$

---

[1] The first part of the proof was to reduce the case of general Möbius transformations to the single case of reflection in the unit circle. Systematic reduction is a very useful tool in mathematics. Here's a joke: A mathematician and a physicist are asked to solve two problems. The first problem is to uncork a bottle of wine and drink the contents. They solve the problem the same way: They both uncork the bottles and drink the wine. The second problem is the same as the first, but this time the bottles are open. The physicist (without much hesitation) drinks the second bottle of wine. The mathematician puts the cork back in the bottle and says, "Now apply the solution to the first problem!"

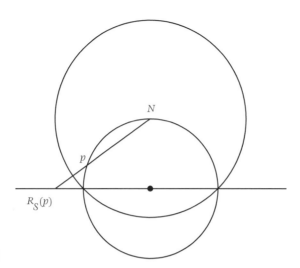

$R_S(p)$

Figure 11.6

We can be even bolder and realize that no computation is needed if we accept the generalization of Theorem 7 to dimension 3, since R_S maps S^2 to a "sphere" that contains $\infty = R_S(N)$, so that (looking at the fixed point set $S^2 \cap S_{\sqrt{2}}(N)$) the image of S^2 under R_S must be $\mathbf{R}^2$.

Problems

1. Show that the Möbius group is generated by isometries, dilatations with center at the origin, and one reflection in a circle.

2. Generalize the observation in the text and show that the composition of reflections in two concentric circles with radii r_1 and r_2 is a central dilatation with ratio of magnification $(r_1/r_2)^2$.

12

SECTION

Complex Linear Fractional Transformations

♣ In Section 11 we created the Möbius group by playing around with stereographic projections, which were a fundamental ingredient in our proof of the FTA. Taking a closer look at the proof, however, we see that we have not attained full understanding of $Möb\,(\mathbf{\hat{R}}^2)$. For example, in the proof of the FTA, it is a crucial fact that reflection to the unit circle can be written not only as $z \mapsto z/|z|^2$, but also as $z \mapsto 1/\bar{z}$ (only the latter gave smoothness of the polynomial P on S^2 across the North Pole). The key to better understanding of the Möbius group is to introduce what is called "complex language in geometry." As usual, we start modestly and express the basic geometric transformations in terms of complex variables:

1. Translation T_v by $v \in \mathbf{C}$ (considered as a vector in $\mathbf{R}^2$) can be defined as $T_v(z) = z + v$.
2. Rotation R_θ with center at the origin and angle $\theta \in \mathbf{R}$ is nothing but multiplication by $z(\theta) = \cos\theta + i\sin\theta$; that is, $R_\theta(z) = z(\theta)z$, $z \in \mathbf{C}$. Now the rotation $R_\theta(p_0)$ with center $p_0 \in \mathbf{C}$ can be written as $R_\theta(p_0) = T_{p_0} \circ R_\theta \circ (T_{p_0})^{-1}$, so that $R_\theta(p_0)(z) = R_\theta(z-p_0)+p_0 = z(\theta)(z - p_0) + p_0$.
3. Reflection R_l in a line can be written in terms of complex variables as follows: Let the line $l = l_t(p_0)$ be given by $\Re(z\bar{p}_0) = t$,

131

$p_0 \neq 0$. (The dot product of two complex numbers z and w considered as plane vectors is given by the real part $\Re(z\bar{w}) = (z\bar{w} + \bar{z}w)/2$. This is what we just used here.) Now, by Section 11, R_l can be written as

$$R_l(z) = z - \frac{2(\Re(z\bar{p}_0) - t)p_0}{|p_0|^2} = z - \frac{z\bar{p}_0 + \bar{z}p_0 - 2t}{\bar{p}_0}$$

$$= -\frac{p_0\bar{z} - 2t}{\bar{p}_0}.$$

4. The complex expression for the glide $G_{l,v} = T_v \circ R_l = R_l \circ T_v$ follows from the expressions of T_v and R_l above.
5. Reflection R_S to the circle $S_r(p_0)$, $r > 0$, is given by

$$R_S(z) = \frac{r^2}{\bar{z} - \bar{p}_0} + p_0 = \frac{p_0\bar{z} + (r^2 - |p_0|^2)}{\bar{z} - \bar{p}_0}.$$

Comparing (3) and (5), we see something in common! They are both of the form

$$z \mapsto \frac{a\bar{z} + b}{c\bar{z} + d}, \quad ad - bc \neq 0.$$

(In the first case the "determinant" $ad - bc$ works out to be $-p_0\bar{p}_0 = -|p_0|^2 \neq 0$, and in the second, $-p_0\bar{p}_0 - (r^2 - |p_0|^2) = -r^2 \neq 0$.)

Theorem 8.

Let $M\ddot{o}b^+(\hat{\mathbf{C}})$ denote the group of direct (orientation-preserving) Möbius transformations of the extended plane $\hat{\mathbf{C}}$. Then $M\ddot{o}b^+(\hat{\mathbf{C}})$ is identical to the group of complex linear fractional transformations

$$z \mapsto \frac{az + b}{cz + d}, \quad ad - bc \neq 0, \quad a, b, c, d, \in \mathbf{C}.$$

Proof.

$M\ddot{o}b^+(\hat{\mathbf{C}})$ is the subgroup of $M\ddot{o}b(\hat{\mathbf{C}})$ consisting of those Möbius transformations that are compositions of an even number of reflections to lines and circles. As noted above, a reflection has the form

$$z \mapsto \frac{a\bar{z} + b}{c\bar{z} + d}, \quad ad - bc \neq 0, \quad a, b, c, d \in \mathbf{C}.$$

The composition of two such reflections (with different a's, b's, c's, and d's) is clearly a linear fractional transformation (cf. also Problem 1), so we have proved that $\text{Möb}^+(\hat{\mathbf{C}})$ is contained in the group of linear fractional transformations.

For the converse, let g be a linear fractional transformation

$$g(z) = \frac{az + b}{cz + d}, \quad ad - bc \neq 0.$$

If $c = 0$, then $g(z) = (a/d)z + (b/d)$, so that if $a = d$, then g is a translation, and if $a \neq d$, then g is the composition of a translation, a rotation, and a central dilatation. This latter statement follows from rewriting g as

$$g(z) = \left(\frac{a}{d}\right)(z - p) + p,$$

where $p = (b/d)/(1 - (a/d))$, and then looking at the complex form of rotations. ($a/d = |a/d| \cdot z(\theta)$, where $\theta = \arg(a/d)$ and $|a/d|$ gives the ratio of magnification.) Thus, we may assume $c \neq 0$. To begin with this case, we first note that $z \mapsto 1/z$ is in $\text{Möb}\,(\hat{\mathbf{C}})$ since it is reflection to the unit circle followed by conjugation (reflection to the real axis). We now rewrite g as

$$g(z) = \frac{bc - ad}{c^2(z + d/c)} + \frac{a}{c}$$

and conclude that g is the composition of a translation, the Möbius transformation $z \mapsto 1/z$, a rotation, a central dilatation, and finally another translation. Thus $g \in \text{Möb}\,(\hat{\mathbf{C}})$, and we are done. $\blacksquare$

Remark.
More geometric insight can be gained by introducing the *isometric circle* S_g of a linear fractional transformation

$$g(z) = \frac{az + b}{cz + d}, \quad ad - bc \neq 0, \quad a, b, c, d \in \mathbf{C},$$

as

$$S_g = \{z \in \mathbf{C} \mid |cz + d| = |ad - bc|^{1/2}\}.$$

(Here we assume that $c \neq 0$.) That S_g is a circle is clear. The name "isometric" comes from the fact that if $z, w \in S_g$, then we have

$$|g(z) - g(w)| = \left| \frac{az + b}{cz + d} - \frac{aw + b}{cw + d} \right|$$

$$= \left| \frac{(ad - bc)(z - w)}{(cz + d)(cw + d)} \right| = |z - w|,$$

so that g behaves on S_g as if it were an isometry! Let R_S denote the reflection in S_g. Since S_g has center $-d/c$ and radius $\sqrt{|\Delta|}/|c|$, where $\Delta = ad - bc$, we have

$$R_S(z) = \frac{|\Delta|/|c|^2}{\bar{z} + \bar{d}/\bar{c}} - \frac{d}{c}$$

$$= \frac{|\Delta|}{c(\overline{cz + d})} - \frac{d}{c}.$$

The composition $g \circ R_S$ is computed as

$$(g \circ R_S)(z) = \frac{aR_S(z) + b}{cR_S(z) + d}$$

$$= \frac{a(cR_S(z) + d) - (ad - bc)}{c(cR_S(z) + d)}$$

$$= \frac{a|\Delta|/(\overline{cz + d}) - \Delta}{c|\Delta|/(\overline{cz + d})}$$

$$= -\frac{\Delta}{c|\Delta|} (\overline{cz + d}) + \frac{a}{c}.$$

This map is of the form $z \mapsto p\bar{z} + q$ with $|p| = 1$. It can be decomposed as

$$z \mapsto \bar{z} \mapsto p\bar{z} \mapsto p\bar{z} + q.$$

The first arrow represents reflection to the first axis; the second, rotation by angle $\arg p$ around the origin; the third, translation by q. The second and third are themselves compositions of two reflections. So we find that $z \mapsto p\bar{z} + q$, $|p| = 1$, is the composition of an odd number of at most five reflections.

Summarizing, we find that $g \circ R_S$ is the composition of an odd number of reflections, so that g is the composition of an even number of reflections.

As a corollary to Theorem 7, we find that linear fractional transformations are circle-preserving. We will return to this later.

Given a direct Möbius transformation represented by the linear fractional transformation

$$z \mapsto \frac{az + b}{cz + d}, \quad ad - bc \neq 0,$$

multiplying a, b, c, d by the *same* complex constant represents the same Möbius transformation. Choosing this complex constant suitably, we can attain

$$ad - bc = 1.$$

We now introduce the *complex special linear group* $SL(2, \mathbf{C})$, consisting of complex 2×2-matrices with determinant 1. $\heartsuit$ By the above, we have

$$M\ddot{o}b^+(\hat{\mathbf{C}}) \cong SL(2, \mathbf{C})/\{\pm I\}.$$

This complicated-looking isomorphism (see Problems 1 and 2) means that each Möbius transformation can be represented by a matrix A in $SL(2, \mathbf{C})$, with A and $-A$ representing the same Möbius transformation. If we write A as

$$A = \begin{bmatrix} a & b \\ c & d \end{bmatrix},$$

then the Möbius transformation is given by the linear fractional transformation

$$z \mapsto \frac{az + b}{cz + d}, \quad z \in \hat{\mathbf{C}}.$$

In what follows, we will not worry about the ambiguity caused by the choice in $\pm A$; our group to study is $SL(2, \mathbf{C})$.

♣ To illustrate how useful it is to represent direct Möbius transformations by linear fractional transformations, we now show that given two triplets of distinct points $z_1, z_2, z_3 \in \hat{\mathbf{C}}$ and $w_1, w_2, w_3 \in \hat{\mathbf{C}}$, there is a unique direct Möbius transformation that carries z_1 to w_1,

z_2 to w_2, and z_3 to w_3. Clearly, it is enough to show this for $w_1 = 1$, $w_2 = 0$, and $w_3 = \infty$. If none of the points z_1, z_2, z_3 is ∞, then the Möbius transformation is given by

$$z \mapsto \frac{z - z_2}{z - z_3} \bigg/ \frac{z_1 - z_2}{z_1 - z_3}$$

If z_1, z_2, or $z_3 = \infty$, then the transformation is given by

$$\frac{z - z_2}{z - z_3}, \quad \frac{z_1 - z_3}{z - z_3}, \quad \frac{z - z_2}{z_1 - z_2}.$$

As for unicity, assume that a direct Möbius transformation

$$z \mapsto \frac{az + b}{cz + d}, \qquad ad - bc = 1,$$

fixes 1, 0, and ∞. Since 0 is fixed, $b = 0$. Since ∞ is fixed, $c = 0$. Now 1 is fixed, so that $1 = (a/d)$ and we end up with the identity.

We finally note that any opposite Möbius transformation can be written as

$$z \mapsto \frac{a\bar{z} + b}{c\bar{z} + d}, \qquad ad - bc = 1, \qquad a, b, c, d \in \mathbf{C}.$$

Indeed, if g is an opposite Möbius transformation, then $z \mapsto g(\bar{z})$ gives an element of $M\ddot{o}b^+(\hat{\mathbf{C}})$, a linear fractional transformation.

One final note: Möbius transformations are not only circle-preserving but angle-preserving as well. What do we mean by this? An angle, after all, consists of two half-lines meeting at a point. An arbitrary transformation maps an angle to two curves that join at the image of the meeting point. The answer, as you know from calculus, is that two curves that meet at a point also have an angle defined by the angle of the corresponding tangent vectors at the meeting point.

We now turn to the proof of preservation of angles (termed *conformality* later) under Möbius transformations. This follows from conformality of the stereographic projection. We give here an independent analytic proof. (For a geometric proof, see Problem 8.) Since isometries preserve angles (congruent triangles in Euclid's *Elements*), we can reduce the case of general Möbius transformations to the case of reflections in circles. Since central dilatations preserve angles (Euclid's *Elements* again on similar triangles), the only case we have to check is reflection to the unit circle $S^1 \subset \mathbf{C}$. This is given by $z \mapsto 1/\bar{z}$, $0 \neq z \in \mathbf{C}$. Even conjugation can be left out, since it is an isometry. Thus, all that is left to check is that $z \mapsto 1/z$ preserves angles. $\diamondsuit$ Let $0 \neq z_0 \in \mathbf{C}$ and $\gamma : (-a, a) \to \mathbf{C}$, $a > 0$, a smooth curve with $\gamma(0) = z_0$. The Möbius transformation $z \mapsto 1/z$ sends the tangent vector $\gamma'(0)$ to the tangent vector $(d/dt)(1/\gamma(t))_{t=0}$. The latter computes as

$$\frac{d}{dt}\left(\frac{1}{\gamma(t)}\right)_{t=0} = -\frac{\gamma'(t)}{\gamma(t)^2}\Big|_{t=0} = -\frac{\gamma'(0)}{z_0^2}$$

since $\gamma(0) = z_0$. Thus $z \mapsto 1/z$ acts on the vector $\gamma'(0)$ by multiplying it with the constant factor $-1/z_0^2$. This is central dilatation by ratio $1/|z_0|^2$ followed by rotation by angle $\arg(-1/z_0^2) = \pi - 2\arg z_0$. Both of these preserve angles, and we are done.

If we accept that Möbius transformations in $\mathbf{R}^3$ are angle-preserving (Problem 8 extended to $\mathbf{R}^3$), then the construction in the remark at the end of Section 11 shows that the stereographic projection is also angle-preserving. This we proved in Section 7 directly.

We are now ready to introduce hyperbolic plane geometry in the next section.

Problems

1. Show that composition of linear fractional transformations corresponds to matrix multiplication in $SL(2, \mathbf{C})$.

2. Check that the inverse of the linear fractional transformation $z \mapsto (az + b)/(cz + d)$, $ad - bc \neq 0$, is $w \mapsto (dw - b)/(-cw + a)$.

3. Find a linear fractional transformation that carries 0 to 1, i to -1, and $-i$ to 0.

4. Let $z_1, z_2, z_3, z_4 \in \hat{\mathbf{C}}$ with z_2, z_3, z_4 distinct.

   (a) Find a linear fractional transformation f that carries z_2 to 1, z_3 to 0, and z_4 to ∞.

   (b) Define the *cross-ratio* (z_1, z_2, z_3, z_4) as $f(z_1) \in \hat{\mathbf{C}}$. Show that the cross-ratio is invariant under any linear fractional transformation; that is, if g is a linear fractional transformation, we have $(g(z_1), g(z_2), g(z_3), g(z_4)) = (z_1, z_2, z_3, z_4)$.

   (c) Prove that $(z_1, z_2, z_3, z_4) \in \mathbf{R}$ iff z_1, z_2, z_3, z_4 lie on a circle or a straight line.

(d) Using (a)–(c), conclude that linear fractional transformations are circle-preserving.

5. Find a linear fractional transformation that maps the upper half-plane $\{z \in \mathbf{C} \mid \Im(z) > 0\}$ to the unit disk $\{z \in \mathbf{C} \mid |z| < 1\}$.

6. Describe the Möbius transformation $z \mapsto 1 + \bar{z}$, $z \in \hat{\mathbf{C}}$, in geometric terms.

7. Show that the non-Möbius transformation $z \mapsto z + 1/z$, $z \in \hat{\mathbf{C}}$, maps circles with center at the origin to ellipses. What is the image of a half-line emanating from the origin?

8. Show that Möbius transformations preserve angles in the following geometric way: (a) Consider a single reflection R_S in a circle $S = S_r(p_0)$. Use the identity

$$d(p, p_0)d(R_S(p), p_0) = r^2$$

(Figure 11.4) to show that R_S maps circles orthogonal to S to themselves.

(b) Given a ray emanating from a point p not in S, construct a circle that is orthogonal to the ray at p, and also orthogonal to S at the intersection points.

(c) Apply (b) to the two sides of an angle, and notice that two intersecting circles meet at the same angle at their two points of intersection.

13

SECTION

"Out of Nothing I Have Created a New Universe" – Bolyai

♣ Plane isometries have one characteristic property that, although obvious, has not been mentioned so far. They map straight lines to straight lines! The more general Möbius transformations do not have this property, but we have seen that they map circles (in the general sense; i.e., circles and lines) to circles. Thus the question naturally arises: Does there exist a "geometry" in which "lines" are circles and "isometries" are Möbius transformations? The answer is not so easy if we think of the strict guidelines, called the Five Postulates, that Euclid gave around 300 B.C. in the *Elements* to obtain a decent geometry. Although we again shy away from the axiomatic treatment, we note here that for any geometry it is essential that through any two distinct points there be a unique line. The famous Fifth Postulate (actually, the equivalent Playfair's axiom),

> Through a given point not on a given line there passes a unique line not intersecting the given line.

is another matter. Even Euclid seemed reluctant to use this in his proofs.

Returning to our circles and Möbius transformations, we now have a general idea how to build a geometry in which "lines" are circles. We immediately encounter difficulty, since looking at the

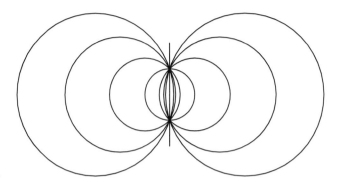

Figure 13.1

pencil of circles in Figure 13.1 we see that the required unicity of the "line" (that is a circle) through two distinct points fails. There are too many 'lines' in our geometry! This is because a circle is determined by three of its distinct points, not two. Thus, to ensure unicity, we have to rule out a lot of circles. This can be done elegantly by imposing a condition on our circles. One simple idea is to fix a point on the plane and require all circles to pass through this point. This point cannot be in the model. Assuming that it is the origin, we perform the transformation $z \mapsto 1/z$, $z \in \hat{\mathbf{C}}$, and realize that our "circles" become straight lines! (This is because this transformation, like any Möbius transformation, is circle-preserving, so that circles through the origin transform into "unbounded circles" that are straight lines.) We thus rediscover Euclidean plane geometry.

A somewhat more sophisticated idea is to fix a line l and consider only those circles that intersect l perpendicularly. The points on this line cannot be in our model. l splits the plane into two half-planes of which we keep only one, denoted by H^2, where the superscript indicates the dimension. That this works can be seen as follows: Take two distinct points p and q in H^2. If the line through p and q is perpendicular to l, then we are done. Otherwise, the perpendicular bisector of p and q intersects l at a point c. Now draw a circle through p and q with center c. Since we keep only H^2 as the point-set of our geometry, we see that the "line" through p and q is, in the first case, the vertical half-line through p and q (ending at a point in l not in H^2) and, in the second case, the semicircle through p and q perpendicular to l at its endpoints. The location of l is irrelevant, so we might just as well take it the real axis. We choose H^2 to be the upper half-plane. We thus arrive at the *half-plane model of hyperbolic geometry*, due to Poincaré. In this model *hyperbolic lines* are Euclidean half-lines or semi-circles perpendicular to the boundary of H^2. The boundary points of H^2 are called *ideal* or (for the obvious reason) points at infinity. It is easy to see that this geometry satisfies the first four postulates of Euclid, along with unicity of the hyperbolic line through two distinct points.

What about the Fifth Postulate? In Figure 13.2 four hyperbolic lines meet at a point, and they are all parallel to the vertical hyperbolic line. The Fifth Postulate thus fails in this model. We have accomplished what mathematicians have been unable to do for almost 2000 years—independence of the Fifth Postulate from the first four! (In fact, we are concentrating here on the explicit model too much. One of the greatest discoveries of Bolyai was the invention of "absolute geometry" in the axiomatic treatment.) This geometry (although not this particular model) was discovered about 1830 by Bolyai (1802–1860) and Lobachevsky (1793–1856), and the story involving Gauss is one of the most controversial pieces in the history of mathematics.[1]

We now return to Möbius transformations. Our path is clear; we need to see which Möbius transformations map H^2 onto itself. Notice that a Möbius transformation of this kind automatically maps the real axis to itself so that, being circle-preserving and conformal (= angle-preserving), it will automatically map our hyperbolic

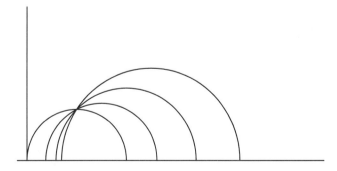

Figure 13.2

---

[1] For a good summary, see M.J. Greenberg, *Euclidean and non-Euclidean Geometries, Development and History*, Freeman, 1993.

lines to hyperbolic lines (since it preserves perpendicularity at the boundary points). We first treat the case of direct Möbius transformations, viewed as elements of $SL(2, \mathbf{C})$.

Theorem 9.

A matrix in $SL(2,\mathbf{C})$ leaves H^2 invariant iff its entries are real; that is, the matrix belongs to $SL(2,\mathbf{R})$.

Proof.

Let g be a linear fractional transformation

$$g(z) = \frac{az + b}{cz + d}, \quad ad - bc = 1, \quad a, b, c, d \in \mathbf{C},$$

and assume that g maps H^2 onto itself. Then g maps the real axis onto itself. In particular, $g^{-1}(0) = r_0$ is real or ∞. We assume that $r_0 \in \mathbf{R}$ since $r_0 = \infty$ can be treated analogously. The composition $g \circ T_{r_0}$ fixes the origin. It has the form

$$(g \circ T_{r_0})(z) = \frac{a(z + r_0) + b}{c(z + r_0) + d} = \frac{az + b + ar_0}{cz + d + cr_0}.$$

The numbers $a, b + ar_0, c, d + cr_0$ are real iff a, b, c, d are. Thus, it is enough to prove the theorem for $g \circ T_{r_0}$. Since this transformation fixes the origin, by changing the notation we can assume that the original linear fractional transformation g fixes the origin. This means that $b = 0$ and

$$g(z) = \frac{az}{cz + d}, \quad ad = 1.$$

First assume that $c = 0$, so that $g(z) = (a/d)z$. Since g maps the real axis onto itself, a/d is real. Since $ad = 1$, it follows that $a/d = a^2$ is real. This means that a is real or purely imaginary. In the first case we are done. In the second, $a = ia_0$ with a_0 real. Thus, $d = 1/a = -i/a_0$ and $g(z) = (a/d)z = -a_0^2 z$. This maps $i \in H^2$ outside of H^2, so that this case is not realized.

Next, we assume that $c \neq 0$. Taking $z = r$ real and working out $ar/(cr + d)$, it follows easily that this latter fraction is real iff a/c and a/d are both real. As before $a/d = a^2$, so that a is real or purely imaginary. If a is real then so are d and c, and we are done. If $a = ia_0$ with a_0 real, then $d = -i/a_0$ and $c = ic_0$ for some c_0 real.

We have

$$g(z) = \frac{ia_0 z}{ic_0 z - i/a_0} = \frac{a_0^2 z}{a_0 c_0 z - 1}.$$

Consider $z = \varepsilon i$, $\varepsilon > 0$, in H^2. We have

$$g(\varepsilon i) = \frac{a_0^2 \varepsilon i}{a_0 c_0 \varepsilon i - 1} = -\frac{a_0^2 \varepsilon i (a_0 c_0 \varepsilon i + 1)}{(a_0 c_0 \varepsilon)^2 + 1}.$$

For ε small enough, this is not in H^2. This case is not realized.

Summarizing, we find that if g maps H^2 onto itself then

$$g(z) = \frac{az + b}{cz + d}, \quad ad - bc = 1,$$

with a, b, c, d real.

Conversely, assume that g is of this form with real coefficients. We claim that $g(H^2) \subset H^2$. (It is clear that g maps the real axis onto itself.) We have

$$\Im(g(z)) = \frac{\Im(z)}{|cz + d|^2}.$$

This is a simple computation. In fact, we have

$$\Im(g(z)) = \Im\left(\frac{az + b}{cz + d}\right) = \Im\left(\frac{(az + b)(c\bar{z} + d)}{|cz + d|^2}\right)$$

$$= \Im\left(\frac{adz + bc\bar{z}}{|cz + d|^2}\right) = \Im\left(\frac{adz - bcz}{|cz + d|^2}\right)$$

$$= \frac{\Im(z)}{|cz + d|^2}.$$

Thus, $\Im(z) > 0$ implies $\Im(g(z)) > 0$; in particular, $g(H^2) \subset H^2$. The theorem follows.   ∎

Corollary.

A general Möbius transformation that maps H^2 onto itself can be represented by

$$z \mapsto \frac{az + b}{cz + d} \quad or \quad z \mapsto \frac{a(-\bar{z}) + b}{c(-\bar{z}) + d}$$

where $a, b, c, d \in \mathbf{R}$.

Proof.

We need to consider only opposite Möbius transformations. If g is opposite, the transformation $z \mapsto g(-\bar{z})$ is direct and maps H^2 onto itself since both g and $z \mapsto -\bar{z}$ do. The corollary follows. (Note that $z \mapsto \bar{z}$ does not work, since it maps H^2 to the lower half-plane.) ∎

Having agreed that the upper half-plane is the model of our new hyperbolic geometry in which the hyperbolic lines are Euclidean half-lines or semi-circles perpendicular to the boundary, we now see that the natural candidate for the group of transformations in this geometry is $SL(2, \mathbf{R})$. The elements of $SL(2, \mathbf{R})$, acting as linear fractional transformations, preserve angles but certainly do not preserve Euclidean distances. The question arises naturally: Does there exist a distance function d_H on H^2 with respect to which the elements of $SL(2, \mathbf{R})$ act as isometries?

We seek a quantity that remains invariant under Möbius transformations. The naive approach is to work out the Euclidean distance $d(g(z), g(w))$ of the image points $g(z)$ and $g(w)$ under

$$g(z) = \frac{az + b}{cz + d}, \quad ad - bc = 1, \quad a, b, c, d, \in \mathbf{R}.$$

Here it is:

$$d(g(z), g(w)) = |g(z) - g(w)| = \left| \frac{az + b}{cz + d} - \frac{aw + b}{cw + d} \right|$$

$$= \left| \frac{(ad - bc)(z - w)}{(cz + d)(cw + d)} \right| = \frac{d(z, w)}{|cz + d||cw + d|}.$$

The Euclidean distance $d(z, w)$ is divided by the "conformality factors" $|cz + d|$ and $|cw + d|$. Notice, however, that these factors also occur in the expressions of $\Im(g(z))$ and $\Im(g(w))$ at the end of the proof of Theorem 9! Dividing, we obtain

$$\frac{d(g(z), g(w))^2}{\Im(g(z))\Im(g(w))} = \frac{d(z, w)^2}{\Im(z)\Im(w)},$$

and we see that this quotient remains invariant! It is now just a matter of scaling to define the *hyperbolic distance function* $d_H : H^2 \times$

$H^2 \to \mathbf{R}$ by

$$\cosh d_H(z, w) = 1 + \frac{d(z, w)^2}{2\Im(z)\Im(w)}.$$

We use here the hyperbolic cosine function: $\cosh t = (e^t + e^{-t})/2$, $t \in \mathbf{R}$. Note that $\cosh : [0, \infty) \to [1, \infty)$ is strictly increasing, so that the formula for d_H makes sense since the right-hand side is ≥ 1. The choice of cosh is not as arbitrary as it seems. As we will show shortly, it is determined by the requirement that measuring hyperbolic distance along a hyperbolic line should be additive in the sense that if z_1, z_2, and z_3 are consecutive points on a hyperbolic line, then $d_H(z_1, z_2) + d_H(z_2, z_3) = d_H(z_1, z_3)$ should hold.

Summarizing, we obtain a model of hyperbolic plane geometry on the upper half-plane H^2 with hyperbolic distance function d_H and group of direct isometries $SL(2, \mathbf{R})/\{\pm I\}$. (Here we inserted $\{\pm I\}$ back for precision.) The hyperbolic lines are Euclidean half-lines and semicircles meeting the boundary of H^2 at right angles. As in Euclidean geometry, the segment on a hyperbolic line between two points is the shortest possible path joining these two points. We emphasize here that "shortest" means with respect to d_H!

By the conformal (angle-preserving) nature of this model, the hyperbolic angles are ordinary Euclidean angles. Let us take a closer look at d_H by traveling along a vertical line toward the boundary of H^2 in Figure 13.3.

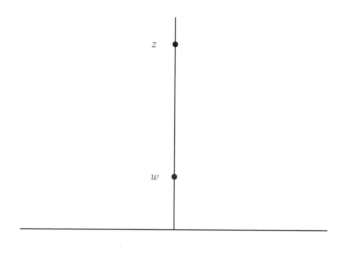

Figure 13.3

Let $z = a + si$ and $w = a + ti$. We have

$$\cosh d_H(z, w) = 1 + \frac{(s - t)^2}{2st} = \frac{s^2 + t^2}{2st} = \frac{1}{2}\left(\frac{s}{t} + \frac{t}{s}\right).$$

Thus,

$$d_H(z, w) = |\log(s/t)|.$$

Note that additivity of d_H along a hyperbolic line is apparent from this formula. If we fix s and let $t \to 0$, we see that $d(z, w)$ tends to ∞ in logarithmic order. More plainly, the distances get more and more distorted as we approach the boundary of H^2.

Another novel feature of hyperbolic geometry is termed as "angle of parallelism." Consider a segment l_0 on a hyperbolic line which, by the abundance of hyperbolic isometries, we may assume to be a vertical line connecting z and w as above. There is a unique hyperbolic line l through w that meets l_0 at a right angle. In our setting, l is nothing but the Euclidean semicircle through w with center at the vertical projection of l_0 to the boundary. Let ω denote the ideal point of l to the right of l_0. Finally, let l' be the unique segment of the hyperbolic line (semicircle) from z to ω. We call l' the *right-sensed parallel* to l (Figure 13.4).

The angle θ at z between l_0 and l' is called the *angle of parallelism*. The length of l_0 (denoted by the same letter) and θ determine each

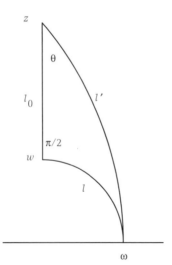

Figure 13.4

other uniquely. Since, unlike Euclidean distances, angles are absolute in both geometries (in the sense that they have absolute unit of measurement), we find that, through the angle of parallelism, hyperbolic distances are also absolute!

Taking a closer inspection of Figure 13.4, we realize that we can actually derive an explicit formula relating θ and l_0. Indeed, inserting some crucial lines, we arrive at the more detailed Figure 13.5.

The Pythagorean Theorem tells us that

$$r^2 = s^2 + (r - t)^2, \quad z = a + si, \quad w = a + ti.$$

Using this, we find

$$\sin \theta = \frac{s}{r} = \frac{2st}{s^2 + t^2}.$$

Comparing this with our earlier computation for d_H, we see that the right-hand side is the reciprocal of cosh of (the hyperbolic length of) l_0! We thus arrive at the *angle of parallelism formula*

$$\sin \theta \cosh l_0 = 1.$$

Encouraged by this, we expect that there must be a formula expressing the area A of a hyperbolic triangle T in terms of its angles α, β, and γ. Here it is:

$$A = \pi - (\alpha + \beta + \gamma).$$

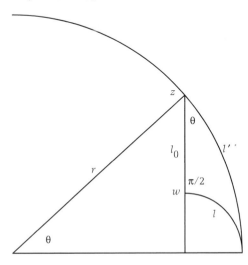

Figure 13.5

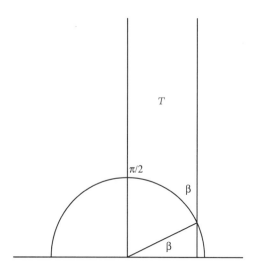

Figure 13.6

The right-hand side is called *angular defect* for obvious reasons. For the proof, we first start with an "asymptotic triangle" with angles $\alpha = \pi/2$ and $\gamma = 0$. As before, we may assume that the sides of the angle γ are vertical, as in Figure 13.6.

We arrange (by an isometry) for the left vertical side to be on the imaginary axis and the finite side to be on the unit circle. Orthogonal angles being equal, a look at Figure 13.6 shows that the right vertical side projects down to the real axis at $\cos \beta$. Using the logarithmic growth formula for hyperbolic distances, we see that the element of arc length in H^2 is $ds = |dz|/\Im(z)$. Hence, the metric is $ds^2 = (dx^2 + dy^2)/y^2$, where $z = x + yi$. By calculus, the area element is thus $dxdy/y^2$. We compute the area A of T as follows:

$$A = \int_0^{\cos \beta} \left(\int_{\sqrt{1-x^2}}^{\infty} \frac{dy}{y^2} \right) dx$$

$$= \int_0^{\cos \beta} \frac{dx}{\sqrt{1 - x^2}}$$

$$= \pi/2 - \beta = \pi - (\pi/2 + \beta).$$

Still in the asymptotic category, it is easy to generalize this to the area formula for a triangle with angles α, β arbitrary and $\gamma = 0$ (we do this by pasting two asymptotic triangles together along a vertical side). Adding areas, we arrive at

$$A = \pi - (\alpha + \beta).$$

Finally, the general formula follows by cutting and pasting; that is, by considering the area of a hyperbolic triangle with nonzero angles as an algebraic sum of the areas of three asymptotic triangles.

As a consequence of the angular defect formula, we find a new non-Euclidean phenomenon: All hyperbolic triangles have area $\leq \pi$.

We derived the angular defect formula as a consequence of the special form of the hyperbolic distance in our model. In an axiomatic setting, the angular defect formula can be *derived* starting from Lobachevsky's angle of parallelism, proving that all trebly asymptotic triangles (triangles with three ideal vertices) are congruent, and finally following[2] the argument of Bolyai (as published in Gauss's collected works).

The angular defect formula is strikingly similar to Albert Girard's spherical excess formula,[3] stating that the area of a spherical triangle in S^2 is

$$\alpha + \beta + \gamma - \pi,$$

where α, β, and γ are the spherical angles of the triangle at the vertices. A proof of this is as follows. The extensions of the sides of the spherical triangle give three great circles that divide the sphere into eight regions, the original triangle whose area we denote by A and its opposite triangle, three spherical triangles with a common side to the original triangle whose areas we denote by X, Y and Z, and their opposites. Let α, β, and γ denote the spherical angles of the original triangle opposite to the sides that are common to the other three triangles with areas X, Y, and Z. Notice that $A + X$ is the area of a spherical wedge with spherical angle α. We thus have $A + X = 2\alpha$. Similarly, $A + Y = 2\beta$ and $A + Z = 2\gamma$. Adding, we obtain $3A + X + Y + Z = 2(\alpha + \beta + \gamma)$. On the other hand,

---

[2]See H.S.M. Coxeter, *Introduction to Geometry*, Wiley, 1969.

[3]This appeared in A. Girard's *Invention nouvelle en algèbre* in 1629.

$2(A + X + Y + Z) = 4\pi$, the area of the entire sphere. Subtracting, the spherical excess formula follows.

As another consequence of the hyperbolic distance formula, we now describe how a *hyperbolic circle* looks. Let $p_0 \in H^2$ and $r > 0$ and work out

$$\{z \in H^2 \mid d_H(z, p_0) = r\}.$$

We can rewrite the defining equation as

$$\cosh d_H(z, p_0) = \cosh r,$$

or equivalently

$$1 + \frac{d(z, p_0)^2}{2\Im(z)\Im(p_0)} = \cosh r.$$

Writing $z = x + yi$ and $p_0 = a + bi$, we have

$$(x - a)^2 + (y - b)^2 = 2(\cosh r - 1)by.$$

This gives

$$(x - a)^2 + (y - b\cosh r)^2 = (b\sinh r)^2,$$

where we used the identity $\cosh^2 r - \sinh^2 r = 1$. But this is the equation of a Euclidean circle! The radius of this circle is $b \sinh r$, and the center has coordinates a and $b \cosh r$. Comparing $(a, b \cosh r)$ with $p_0 = (a, b)$, we see that p_0 is closer to the boundary of H^2. Thus, a hyperbolic circle looks like a Euclidean circle with center moved toward the boundary (Figure 13.7).

We now go back to the roots of this long line of arguments. Remember that the starting point was our study of Möbius transformations—finite compositions of reflections in circles and lines. After much ado, we concluded that analytically they are given by linear fractional transformations (possibly precomposed by conjugation). Restricting to real coefficients, we understood that they act as isometries of the upper half-plane model H^2 of hyperbolic geometry. In the decomposition of a Möbius transformation as a finite composition of reflections, we must take reflections that map H^2 onto itself, and these are exactly the ones with hyperbolic line axes! Thus, for a single reflection, we have the following cases depicted in Figure 13.8.

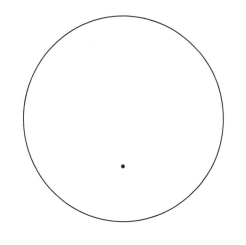

Figure 13.7

Let us explore some of the possible combinations. If we compose two reflections with distinct vertical axes, we get an ordinary Euclidean translation along the real axis such as $z \mapsto z + 1$. The situation is the "same" if we consider reflection to a vertical axis followed by reflection in a circle with one common endpoint (Figure 13.9), or composition of reflections in circles with one common endpoint (Figure 13.10). Notice that parallel vertical lines also have a common endpoint at infinity. Isometries of H^2 of this kind are called *parabolic*.

If the axes intersect in H^2, then the intersection is a fixed point of the composition (Figure 13.11).

These isometries of H^2 are called *elliptic*. Without loss of generality, we may assume that the fixed point is $i \in H^2$. (This is because $SL(2, \mathbf{R})$ carries any point to any point in H^2.) Let g be an elliptic

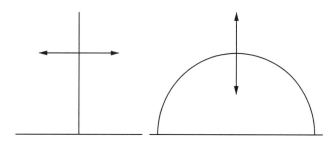

Figure 13.8

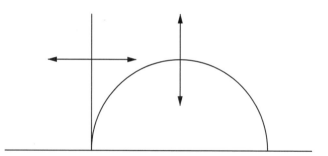

Figure 13.9

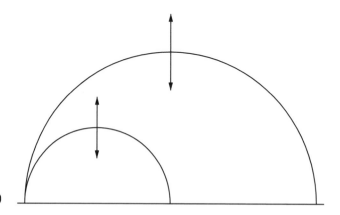

Figure 13.10

isometry with fixed point i. Since g is direct, we can write

$$g(z) = \frac{az + b}{cz + d}, \quad ad - bc = 1, \quad a, b, c, d \in \mathbf{R}.$$

Since i is a fixed point,

$$i = \frac{ai + b}{ci + d}.$$

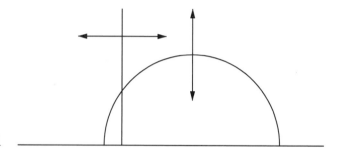

Figure 13.11

Multiplying, we get $a = d$ and $b = -c$. Thus,

$$ad - bc = a^2 + b^2 = 1.$$

This calls for the parameterization

$$a = \cos\theta, \quad b = -\sin\theta, \quad \theta \in \mathbf{R},$$

so that we arrive at

$$g(z) = \frac{\cos\theta \cdot z - \sin\theta}{\sin\theta \cdot z + \cos\theta}.$$

The careless reader may now say, "Well, we have not discovered anything new! After all, the matrix

$$\begin{bmatrix} \cos\theta & -\sin\theta \\ \sin\theta & \cos\theta \end{bmatrix}$$

represents Euclidean rotation on $\mathbf{R}^2$." But this is a grave error, since this matrix is viewed as an element of $SL(2, \mathbf{R})$, and thereby represents a hyperbolic rotation in H^2 given by a linear fractional transformation. If you are still doubtful, substitute $\theta = \pi$ (not 2π!) and verify that g is the identity (and not a half-turn).

Finally, isometries arising from a no-common-endpoint case are called *hyperbolic*. This confusing name has its own advantages. Do not confuse the term "isometry of the hyperbolic plane" with "hyperbolic isometry." A hyperbolic isometry possesses a unique hyperbolic line, called the *translation axis*, perpendicular to both axes (Figure 13.12). Indeed, as the drawing on the left of Figure 13.12 shows, in case one of the axes is vertical with ideal point at $c \in \mathbf{R}$, then the translation axis is the unique Euclidean semicircle with center at c, intersecting the other axis at the point of tangency with the radial line from c. In general (see the drawing at the right of Figure 13.12), one of the axes can be brought to a vertical Euclidean line by a suitable isometry, and the previous construction

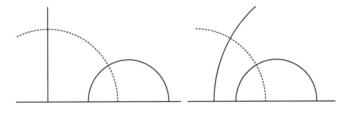

Figure 13.12

applies. Clearly, the translation axis is the unique hyperbolic line that the hyperbolic isometry leaves invariant. A typical example for a hyperbolic isometry is $z \mapsto kz$, $k \neq 1$, with the positive imaginary axis as the translation axis. In fact, *all* hyperbolic isometries are conjugate to $z \mapsto kz$ for some $k \neq 1$, and the conjugation is given by an isometry that carries the translation axis to the positive imaginary axis. Since in this special case the two axes of reflection of the hyperbolic isometry are on concentric Euclidean circles, we have proved the following: Two nonintersecting circles can be brought to a pair of concentric circles by a Möbius transformation. Indeed, if the two circles are disjoint and nonconcentric, then by a suitable Euclidean isometry that carries the line passing through their centers to the real axis, this configuration can be carried into a setting in hyperbolic geometry.

As an application of these constructions, we now realize that we have solved the famous *Apollonius problem*: Find a circle tangent to three given circles. Indeed, first subtract the smallest radius from the three radii to reduce one circle to a point. If the remaining two circles have a common point, then this point can be carried to ∞ by a Möbius transformation, and the Apollonius problem reduces to finding a circle tangent to two lines and passing through a given point. If the reduced circles are disjoint, then they can be brought to two concentric circles, and once again, the Apollonius problem reduces to a trivial one: Find a circle tangent to a pair of concentric circles and passing through a given point. (Notice that the cases where the Apollonius problem has no solution can now be easily listed.) You may feel mildly uncomfortable with the expression "find." To be precise, we really mean here geometric construction as explained in Problem 4 in Section 6. Now realize that all steps that involve Möbius transformations are constructible.

Looking back to the classification of isometries of the hyperbolic plane, we see that parabolic and hyperbolic isometries are of infinite order (that is, all their powers are distinct, or equivalently, their powers generate infinite cyclic groups).

We close this section with a final note: Hyperbolic geometry has several models, and each has its own advantages and disadvantages. One prominent model (also due to Poincaré) is based on the unit disk $D^2 = \{z \in \mathbf{C} \mid |z| < 1\}$. It is easy to derive this from the upper half-plane model H^2. In fact, consider the linear fractional transformation

$$z \to i\frac{i+z}{i-z}.$$

It is easy to see that this transformation sends D^2 onto H^2. (Indeed, this transformation fixes ± 1, sends i to ∞, and sends 0 to i.)

Remark.

♡ Since this formula seems farfetched, we elaborate on it a little. Recall from Section 11 that conjugating spherical isometries of S^2 by the stereographic projection h_N gives Möbius transformations of $\hat{\mathbf{C}}$. We now claim that the linear fractional transformation of D^2 to H^2 above is obtained from the spherical rotation Q around the real axis with angle $\pi/2$. Indeed, h_N^{-1} maps D^2 to the southern hemisphere of S^2, Q rotates this to the "front" hemisphere between the (prime) $0°$ and $180°$ meridians of longitude, and finally, h_N maps this to H^2. To check that $h_N \circ Q \circ h_N^{-1}$ gives the linear fractional transformation above is an easy computation.

♣ Using the linear fractional transformation of D^2 to H^2, we see that whatever we developed in H^2 can now be transported to D^2. We find that hyperbolic lines in D^2 are segments of circles meeting the boundary S^1 of D^2 perpendicularly. In particular, if s is the Euclidean distance beween the origin (the center of D^2) and the "midpoint" of a hyperbolic line, and r is the Euclidean radius of the circular segment representing the hyperbolic line, then perpendicularity at the boundary gives

$$(s+r)^2 = 1 + r^2$$

(Figure 13.13).

Hyperbolic isometries of D^2 are generated by reflections in hyperbolic lines. Analytically, the isometries of D^2 have the form

$$z \to \frac{az + \bar{c}}{cz + \bar{a}} \quad \text{or} \quad z \to \frac{a\bar{z} + \bar{c}}{c\bar{z} + \bar{a}},$$

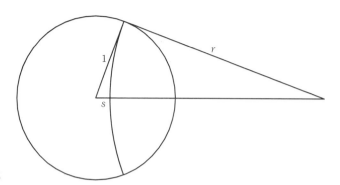

Figure 13.13

where $|a|^2 - |c|^2 = 1$. The hyperbolic metric d_D on D^2 takes the form

$$\tanh\left(\frac{d_D(z, w)}{2}\right) = \frac{d(z, w)}{|1 - z\bar{w}|}, \quad z, w \in D^2,$$

or equivalently,

$$d_D(z, w) = \log \frac{|1 - z\bar{w}| + |z - w|}{|1 - z\bar{w}| - |z - w|}.$$

With these distance formulas, the linear fractional transformation of D^2 to H^2 above becomes an isometry between the two models.

Problems

1. Show that every complex linear fractional transformation of the unit disk D^2 onto itself is of the form $z \mapsto e^{i\theta}(z - w)/(1 - z\bar{w})$, where $w \in D^2$ and $\theta \in \mathbf{R}$ (cf. Problem 4 of Section 7).

2. Solve Monge's problem: Find a circle orthogonal to three given circles.

3. In the upper half-plane H^2, consider two *Euclidean* rays l_1 and l_2 emanating from a boundary point ω.

    (a) Show that all circular segments l_0 with center ω connecting l_1 and l_2 have the same hyperbolic length. (Hint: Use hyperbolic reflections in semi-circles with center at ω.)

    (b) What is the connection between the angle of parallelism for the common hyperbolic length of l_0 in (a) and the angle between l_1 and l_2 at ω?

4. Consider a hyperbolic right triangle with hyperbolic side lengths a, b, and hypotenuse c, and angles α, β, and $\pi/2$.

(a) Prove the Pythagorean Theorem:

$$\cosh a \cosh b = \cosh c.$$

(b) Show that

$$\sinh a \tan \beta = \tanh b.$$

5. Cut a sphere of radius $1/2$ into two hemispheres along the equator and keep the southern hemisphere H. Let H sit on $\mathbf{R}^2$ with the South Pole touching the plane. Let $h : H \to D^2$ be the stereographic projection from the North Pole $N = (0, 0, 1)$, and let $v : H \to D^2_{1/2}$ be the vertical projection to the disk $D^2_{1/2}$ with center at the origin and radius $1/2$. Show that $v \circ h^{-1} : D^2 \to D^2_{1/2}$ maps hyperbolic lines in D^2 to chords in $D^2_{1/2}$. Using this correspondence, develop hyperbolic geometry in $D^2_{1/2}$. (This model is due to F. Klein.)

6. In the upper half-space model

$$H^3 = \{(z, s) \in \mathbf{C} \times \mathbf{R} \mid s > 0\} \subset \mathbf{C} \times \mathbf{R} \cup \{\infty\}$$

of hyperbolic space geometry, hyperbolic lines are Euclidean semicircles or half-lines meeting the ideal boundary

$$\hat{\mathbf{C}} = \{(z, 0) \mid z \in \mathbf{C}\} \cup \{\infty\}$$

at right angles. The hyperbolic distance function $d_H : H^3 \times H^3 \to \mathbf{R}$ is defined by

$$\cosh d_H((z, s), (w, t)) = 1 + \frac{|z - w|^2 + |s - t|^2}{2st}, \qquad (z, s), (w, t) \in H^3.$$

Given a reflection R_S in a circle $S = S_r(p_0)$ on the ideal boundary $\hat{\mathbf{C}}$, define the *Poincaré extension* $\tilde{R}_S$ as reflection to the upper hemisphere (in H^3) with boundary circle S. Given a Möbius transformation g of the ideal boundary $\hat{\mathbf{C}}$, define the *Poincaré extension* $\tilde{g} : H^3 \to H^3$ by decomposing g into a finite composition of reflections and taking the Poincaré extension of each factor.

(a) Show that $\tilde{g}$ is well defined (that is, it does not depend on the particular representation of g as a composition of reflections).

(b) Verify that $\tilde{g}$ is a hyperbolic isometry of H^3. (The Poincaré extension thus defines $M\ddot{o}b^+(\hat{\mathbf{C}})$ as a subgroup of $Iso\,(H^3)$.)

7. Develop hyperbolic space geometry in the 3-dimensional unit ball

$$D^3 = \{p \in \mathbf{R}^3 \mid |p| < 1\}$$

with hyperbolic distance function $d_D : D^3 \times D^3 \to \mathbf{R}$ defined by

$$\cosh(d_D(p, q)) = 1 + \frac{2|p - q|^2}{(1 - |p|^2)(1 - |q|^2)}, \qquad p, q \in D^3.$$

14

S E C T I O N

Fuchsian Groups

♡ We are now ready to discuss the final task in hyperbolic geometry: classification of discrete subgroups of the group of direct hyperbolic isometries in $SL(2, \mathbf{R})$. Unlike the case of crystallographic groups, we do not consider opposite Möbius transformations here. However, this does not prevent us from constructing these discrete subgroups by means of opposite isometries (such as groups generated by reflections in the sides of a hyperbolic triangle; see the discussion of triangle groups at the end of this section). Here we are entering an area in which full understanding has not been obtained, although a vast number of results are known. To be modest, we will only give a few illustrations that reveal some subtleties of the subject.

First some definitions. Given a linear fractional transformation

$$g(z) = \frac{az + b}{cz + d}, \quad ad - bc = 1, \quad a, b, c, d \in \mathbf{C},$$

we introduce its *norm*:

$$|g| = \sqrt{|a|^2 + |b|^2 + |c|^2 + |d|^2}.$$

This is well defined since the right-hand side is unchanged if we replace a, b, c, d by their negatives. (Remember the ambiguity

$\{\pm I\}$.) A subgroup $G \subset M\ddot{o}b^+(\hat{\mathbf{C}})$ is *discrete* if for any $n > 0$, the set

$$\{g \in G \mid |g| < n\}$$

is finite. A discrete group G, considered as a subgroup in $SL(2, \mathbf{C})$ is said to be *Kleinian*.

The most typical example of a Kleinian group is the *modular group* $SL(2, \mathbf{Z})$ consisting of linear fractional transformations

$$z \to \frac{az + b}{cz + d}, \quad ad - bc = 1,$$

with integral coefficients $a, b, c, d \in \mathbf{Z}$.

A Kleinian group that leaves a half-plane or a disk invariant is called *Fuchsian*.[1] In the case of a Fuchsian group we may assume that the invariant half-plane is H^2 or the invariant disk is D^2. In the first case (and this is where we give most of the examples), we thus have a discrete subgroup of $SL(2, \mathbf{R})$. As expected, the theory of Fuchsian groups is very well developed; much less is known about Kleinian groups.

We wish to visualize Fuchsian groups as hyperbolic tilings, or *tesssellations*, of H^2. We thus have to introduce the notion of fundamental set (domain) in much the same way as we did for crystallographic groups on $\mathbf{R}^2$. Given a Fuchsian group $G \subset SL(2, \mathbf{R})$, we say that $F \subset H^2$ is a *fundamental set* if F meets each orbit

$$G(z) = \{g(z) \mid g \in G\}, \quad z \in H^2,$$

exactly once. A *fundamental domain* for G is a domain F_0 in H^2 such that there is a fundamental set F for G between F_0 and $\bar{F}_0$ and $\partial F_0 = \bar{F}_0 - F_0$ has zero dimensional area.[2]

The simplest Fuchsian groups, G, are cyclic; that is, they are generated by a single linear fractional transformation $g \in SL(2, \mathbf{R})$. We write this as $G = \langle g \rangle$. Depending on whether g is parabolic, hyperbolic, or elliptic, we arrive at the following examples:

---

[1] The terms 'Kleinian' and 'Fuchsian' are due to Poincaré.

[2] A word of caution. The closure here is taken in H^2, and by "area" we mean hyperbolic area.

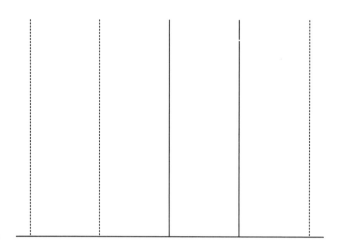

Figure 14.1

EXAMPLE 1

Let g be the parabolic isometry

$$g(z) = z + 1, \quad z \in H^2.$$

A fundamental domain F_0 for $G = \langle g \rangle$ is given by $0 < \Re(z) < 1$, shown in Figure 14.1.

Figure 14.2 shows the transformed fundamental domain and tiling on D^2.   □

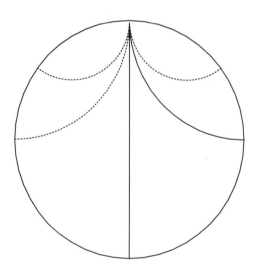

Figure 14.2

Remark.
It is instructive to look at the transformation g from D^2 to H^2 via $h_N \circ Q \circ h_N^{-1}$, where Q is a quarter-turn around the real axis (cf. Section 13).

EXAMPLE 2
Let $k > 1$ and consider the hyperbolic isometry

$$g(z) = kz, \quad z \in H^2.$$

A fundamental domain for $G = \langle g \rangle$ is given by

$$F_0 = \{z \in H^2 \mid 1 < |z| < k\}$$

(Figure 14.3).

The corresponding fundamental domain and tiling on D^2 is shown in Figure 14.4.   □

EXAMPLE 3
Let g be elliptic with fixed point at $i \in H^2$. As noted above, g can be written as

$$g(z) = \frac{\cos\theta \cdot z - \sin\theta}{\sin\theta \cdot z + \cos\theta}, \quad z \in H^2,$$

and discreteness implies that $\theta = \pi/n$ for some $n \geq 2$. The tiling for $G = \langle g \rangle$ is shown in Figure 14.5 for H^2 and in Figure 14.6 for D^2 (noting that i corresponds to the origin).   □

Before we leave the cyclic Fuchsian groups, here is one more simple example, in which G is generated by an elliptic and a hyperbolic isometry.

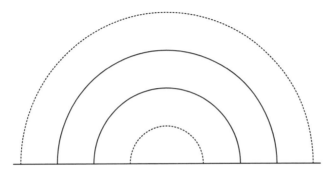

Figure 14.3

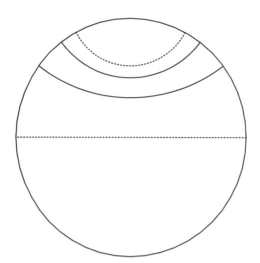

Figure 14.4

Figure 14.5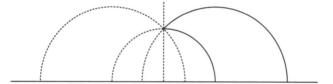

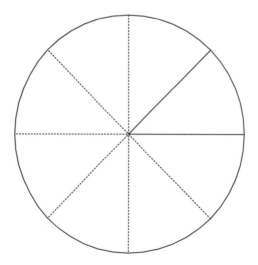

Figure 14.6

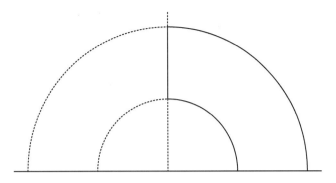

Figure 14.7

EXAMPLE 4

Let $k > 1$ and $G = \langle g_1, g_2 \rangle$, where

$$g_1(z) = -1/z \quad \text{and} \quad g_2(z) = kz, \quad z \in H^2.$$

A fundamental domain and the tiling is shown in Figure 14.7.   □

Examples 1 to 4 (and their conjugates in $SL(2, \mathbf{R})$) make up what we call *elementary Fuchsian groups*. From now on we concentrate on nonelementary Fuchsian groups.

EXAMPLE 5

Let $G = SL(2, \mathbf{Z})$ be the modular group. We claim that F_0 defined by

$$|z| > 1 \quad \text{and} \quad |\Re(z)| < 1/2$$

is a fundamental domain (Figure 14.8).

Let g be an arbitrary element in G and write

$$g(z) = \frac{az + b}{cz + d}, \quad ad - bc = 1, \quad a, b, c, d \in \mathbf{Z}.$$

Assume that $z \in F_0$, i.e., the defining inequalities for F_0 hold. We compute

$$|cz + d|^2 = c^2|z|^2 + 2\Re(z)cd + d^2 > c^2 + d^2 - |cd|$$

$$= (|c| - |d|)^2 + |cd|.$$

The lower bound here is a nonnegative integer and is zero iff $c = d = 0$. This cannot happen since $ad - bc = 1$. Thus,

$$|cz + d|^2 > 1, \quad z \in F_0.$$

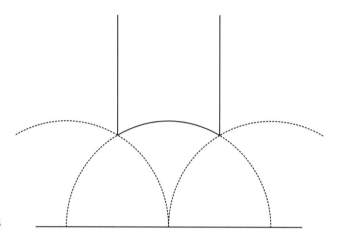

Figure 14.8

Using this, we have

$$\Im(g(z)) = \frac{\Im(z)}{|cz + d|^2} < \Im(z), \quad z \in F_0,$$

where we used the formula for $\Im(g(z))$ in the proof of Theorem 9. We now show that F_0 contains at most one point from each orbit of G. ¬ Assuming the contrary, we can find $z \in F_0$ and $g \in G$ such that $g(z) \in F_0$. By the computation above,

$$\Im(g(z)) < \Im(z).$$

Replacing z by $g(z)$ and g by g^{-1} in this argument, we obtain

$$\Im(g^{-1}(g(z))) = \Im(z) < \Im(g(z)),$$

and this contradicts the inequality we just obtained! ¬

The parabolic isometry $z \mapsto z + 1$ and the elliptic "half-turn" $z \mapsto -1/z$ are in G, and by applying them, we can easily move any point in H^2 to $\bar{F}_0$. Thus, F_0 is a fundamental domain for G. (As a by-product, we obtain that $SL(2, \mathbf{Z})$ is generated by $z \mapsto z+1$ and $z \mapsto -1/z, z \in H^2$.) Notice that F_0 is a hyperbolic triangle (in the asymptotic sense with one vertex at infinity). A fundamental set F with $F_0 \subset F \subset \bar{F}$ is given by the inequalities $-1/2 < \Re(z) \le 1/2$, $|z| \ge 1$, and $\Re(z) \ge 0$ if $|z| = 1$. Geometrically, we add to F_0 the right half of the boundary circle and the right vertical side of F_0. Thus, the modular group tiles H^2 with triangles (Figure 14.9). □

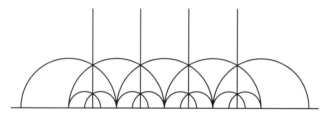

Figure 14.9

Remark.

At the end of Section 2, we introduced the modular group $SL(2, \mathbf{Z})$ by studying various bases in a lattice in $\mathbf{R}^2 = \mathbf{C}^2$ (cf. Problem 15). Since we constructed a fundamental set F for $SL(2, \mathbf{Z})$ above, it follows that for all the bases $\{v, w\}$ of a given lattice $L \subset \mathbf{C}$ there is a unique ratio $\tau = w/v$ in F. A basis $\{v, w\}$ with this property is called *canonical*. Given a ratio $\tau \in F$ corresponding to a canonical basis, there is a choice of two, four, or six canonical bases with this ratio. Indeed, $\{-v, -w\}$ is always canonical, and more canonical bases occur if τ is the fixed point of an element in $SL(2, \mathbf{Z})$. This happens only if $\tau = i$ ($z \mapsto -1/z$) and $\tau = \omega = e^{2\pi i/3}$ ($z \mapsto -(z+1)/z, -1/(z+1)$).

The complexity of Fuchsian groups increases as we look at more and more "random" examples such as the following:

Example 6
Let $G = \langle g_1, g_2 \rangle$, where

$$g_1(z) = \frac{3z + 4}{2z + 3} \quad \text{and} \quad g_2(z) = 2z, \quad z \in H^2.$$

A fundamental domain for G is shown in Figure 14.10. $\qquad \square$

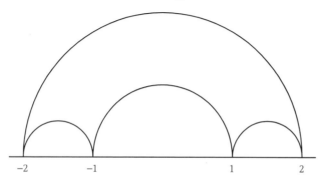

Figure 14.10

A group G of isometries of H^2 is said to be a *triangle group of type* (α, β, γ) if G is generated by reflections in the sides of a hyperbolic triangle with angles α, β, and γ. Here $\alpha, \beta, \gamma \geq 0$, and we have

$$\alpha + \beta + \gamma < \pi.$$

Notice that G cannot be Fuchsian, since it contains opposite isometries. We remedy this by considering the subgroup G^+ of direct isometries in G. The elements of G^+ are compositions of elements in G with an even number of factors. We call G^+ a *conformal group of type* (α, β, γ). For example, if l, l', and l'' denote the sides of a hyperbolic triangle, then $R_{l'} \circ R_l$, $R_l \circ R_{l''}$ (and $R_{l''} \circ R_{l'}$) all belong to (in fact, generate!) G^+. Let the triangle be as shown in Figure 14.11. $R_{l'} \circ R_l$ fixes p so that it is either elliptic (with rotation angle 2α, $\alpha > 0$) or parabolic ($\alpha = 0$). Similarly, $R_l \circ R_{l''}$ fixes q and is elliptic or parabolic depending on whether $\beta > 0$ or $\beta = 0$. Thus G^+ is generated by two isometries g_1, g_2, each being parabolic or elliptic. If G^+ is discrete, then every elliptic element must have finite order in G^+. It follows that if α, β, and γ are positive, then $\pi/\alpha, \pi/\beta$, and π/γ are rational.

EXAMPLE 7

Let T be the asymptotic triangle in H^2 with vertices $z(\pi/3)$, ∞, and i as shown in Figure 14.12. The triangle group G corresponding to T

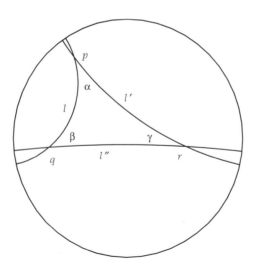

Figure 14.11

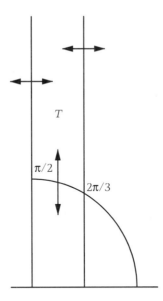

Figure 14.12

is of type $(\pi/3, 0, \pi/2)$. Hyperbolic reflections in the sides of T give $z \mapsto -\bar{z}, z \mapsto 1/\bar{z}$, and $z \mapsto -\bar{z}+1, z \in H^2$. Taking compositions of these, we see that the corresponding conformal group G^+ of type $(\pi/3, 0, \pi/2)$ is generated by the translation $z \mapsto z + 1$ and the half-turn $z \mapsto -1/z, z \in H^2$. In perfect analogy with Example 5, we obtain that G^+ is the modular group $SL(2, \mathbf{Z})$. □

Poincaré proved that a triangle group of type (α, β, γ) is discrete if $\pi/\alpha, \pi/\beta$, and π/γ are integers ≥ 3 (possibly ∞) with

$$\frac{\alpha}{\pi} + \frac{\beta}{\pi} + \frac{\gamma}{\pi} < 1$$

(cf. the remark at the end of Section 10). It can also be proved that the corresponding conformal group of type (α, β, γ) tiles the hyperbolic plane H^2. Thus, the situation is radically different from the Euclidean case of crystallographic groups, where there are only finitely many (in fact, 17) different tilings. You are now invited

Figure 14.13

to look at Figure 14.13, depicting a hyperbolic tiling of a triangle group[3] of type $(\pi/6, \pi/4, \pi/2)$, and discover various elliptic isometries of the corresponding conformal group.

Our last two examples involve regular n-sided hyperbolic polygons P in D^2 with centroid at the origin. Let c be the center of a circle that contains a side of P, m the midpoint of the side, and v a vertex of P adjacent to m. Let r be the Euclidean radius of the circle and s the Euclidean distance between the origin 0 and m. Figure 14.14 depicts the case $n = 3$. Finally, let $\alpha = \angle 0vm$. Since $\angle v0m = \pi/n$, the (Euclidean) law of sines gives

$$\frac{s+r}{\cos \alpha} = \frac{r}{\sin(\pi/n)},$$

where we used the identity $\sin(\alpha + \pi/2) = \cos \alpha$. As noted at the end of Section 13, we have

$$(s+r)^2 = 1 + r^2.$$

---

[3]See H.S.M. Coxeter, *Introduction to Geometry*, Copyright 1969 by John Wiley & Sons, Inc. Reprinted by permission of John Wiley & Sons, Inc.

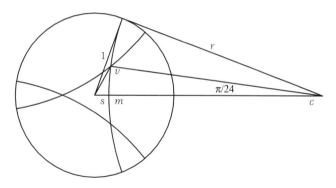

Figure 14.14

Combining these, we obtain

$$r = \frac{\sin(\pi/n)}{\sqrt{\cos^2 \alpha - \sin^2(\pi/n)}}$$

and

$$s = \frac{\cos \alpha - \sin(\pi/n)}{\sqrt{\cos^2 \alpha - \sin^2(\pi/n)}}.$$

EXAMPLE 8
A triangle group G of type $(\pi/4, \pi/4, \pi/4)$ is generated by reflections in the sides of a hyperbolic (equilateral) triangle with $\alpha = \pi/8$ as shown in Figure 14.14. The formulas above reduce to

$$r = \frac{\sqrt{3}}{\sqrt{\sqrt{2} - 1}} \quad \text{and} \quad s = \frac{\sqrt{2 + \sqrt{2}} - \sqrt{3}}{\sqrt{\sqrt{2} - 1}}.$$

With these, we have

$$v = s + r(1 + z(23\pi/24)).$$

The corresponding conformal group G^+ of type $(\pi/4, \pi/4, \pi/4)$ is generated by elliptic quarter-turns around the vertices v and $z(2\pi/3)v$. What is a suitable fundamental domain for G^+?   □

EXAMPLE 9
Finally, we discuss tilings of H^2 with fundamental polygons of $4p$ sides, $p \geq 2$, and Fuchsian groups generated by $2p$ hyperbolic elements. We consider in full detail the case $p = 2$ of a Fuchsian group

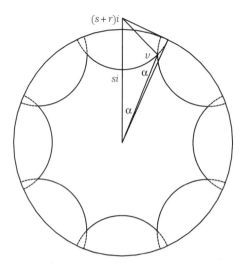

Figure 14.15

generated by four hyperbolic elements. A fundamental polygon is the hyperbolic octagon depicted in Figure 14.15. Each hyperbolic isometry will map a side of the octagon to the opposite side. We first work out the hyperbolic isometry g that maps the bottom side of the octagon to the top side. Let $si \in D^2$ be the midpoint of the top side, and v an adjacent vertex of the octagon. The hyperbolic triangle with vertices 0, si and v has angle $\alpha = \pi/8$ at 0 and v. This is because D^2 is tesselated by images of the octagon under the Fuchsian group. Extending a side to a circle of radius r, the formulas above reduce to

$$r = \frac{\sin \alpha}{\sqrt{\cos(2\alpha)}} \quad \text{and} \quad s = \frac{\cos \alpha - \sin \alpha}{\sqrt{\cos(2\alpha)}}.$$

The hyperbolic distance of si from the origin computes as

$$d_D(si, 0) = \log \frac{1 + s}{1 - s} = \log \frac{\cos \alpha + \sqrt{\cos(2\alpha)}}{\sin \alpha}$$

(see Section 13).

As in Example 2, we represent g on H^2 by $z \mapsto kz$, $z \in H^2$, and the normalization in $SL(2, \mathbf{R})$ implies that the matrix associated to g is

$$\begin{bmatrix} \sqrt{k} & 0 \\ 0 & 1/\sqrt{k} \end{bmatrix}.$$

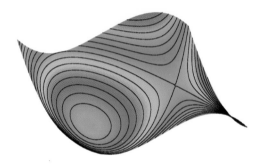

COLOR PLATE 1A

COLOR PLATE 1B

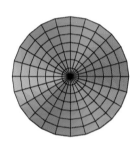

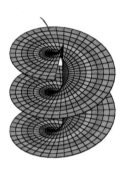

COLOR PLATE 2A

COLOR PLATE 2B

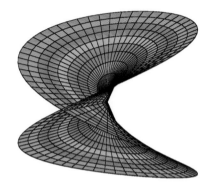

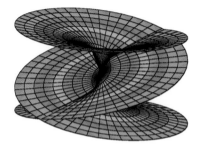

COLOR PLATE 3A

COLOR PLATE 3B

COLOR PLATE 4A

COLOR PLATE 4B

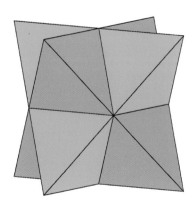

COLOR PLATE 5A

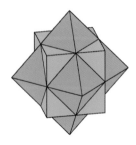

COLOR PLATE 5B

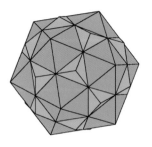

COLOR PLATE 5C

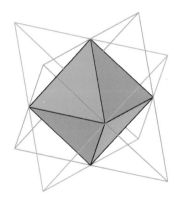

COLOR PLATE 6

COLOR PLATE 7

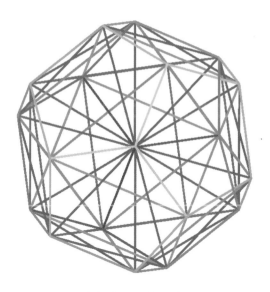

COLOR PLATE 8

COLOR PLATE 9

COLOR PLATE 10

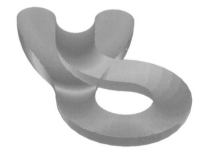

COLOR PLATE 11

COLOR PLATE 12

On D^2, g maps $-si$, the midpoint of the bottom side, to si, the midpoint of the top side. By symmetry, the hyperbolic distance between these points is $2d_D(si, 0)$. On H^2, the hyperbolic distance between i and ki is $\log k$ (see Section 13). It follows that

$$\log k = 2d_D(si, 0),$$

so that

$$\sqrt{k} = \frac{\cos \alpha + \sqrt{\cos(2\alpha)}}{\sin \alpha}.$$

The formula representing g as a linear fractional transformation follows. By using symmetries of the octagon, we see that g, conjugated by rotations around the origin with angles $\pi/4$, $\pi/2$, and $3\pi/4$, defines the three other hyperbolic isometries.

Replacing $\alpha = \pi/8$ by $\alpha = \pi/4p$, $p \geq 2$, we obtain a Fuchsian group generated by $2p$ hyperbolic elements with fundamental domain a regular $4p$-sided hyperbolic polygon. $\qquad\square$

Problems

1. Show that an isometry of the hyperbolic plane H^2 is conjugate in $SL(2, \mathbf{Z})$ to
   (a) $z \mapsto z + 1$ iff it is parabolic;
   (b) to $z \mapsto (\cos \theta \cdot z - \sin \theta)/(\sin \theta \cdot z + \cos \theta)$, $\theta \in \mathbf{R}$, iff it is elliptic;
   (c) $z \mapsto k \cdot z$, $k \neq 1$, iff it is hyperbolic.

2. Characterize the parabolic, elliptic, and hyperbolic isometries of the hyperbolic plane in terms of their fixed points on $H^2 \cup \mathbf{R} \cup \infty$.

3. Let $g(z) = (az + b)/(cz + d)$, $ad - bc = 1$, $a, b, c, d \in \mathbf{R}$, be an isometry of H^2. Show that $\text{trace}^2(g) = (a + d)^2$ is well defined (that is, it depends only on g and not on a, b, c, d). Prove that g is parabolic iff $\text{trace}^2(g) = 4$, elliptic iff $0 \leq \text{trace}^2(g) < 4$ and hyperbolic iff $\text{trace}^2(g) > 4$. (The latter two relations imply that $\text{trace}^2(g)$ is real.)

4. Use Problems 2–3 to conclude that an isometry g of H^2 has no fixed point in H^2 iff $\text{trace}^2(g) \geq 4$.

5. Let g be a linear fractional transformation that satisfies $g^n(z) = z$ for some $n \geq 2$. Show that g is elliptic.

6. Let g_1 and g_2 be isometries of H^2 such that $g_1 \circ g_2 = g_2 \circ g_1$. Show that g_1 parabolic implies that g_2 is also parabolic.

7. Let T_1 and T_2 be adjacent asymptotic hyperbolic triangles in H^2 with vertices $z(\pi/3), \infty, i$ and $z(\pi/3), 1, \infty$ (see Example 7). Show that the conformal groups corresponding to T_1 and T_2 are both equal to $SL(2, \mathbf{Z})$. (Thus, incongruent triangles can define the same conformal group.)

8. Work out the vertices of a regular hyperbolic pentagon in D^2 (with centroid at the origin) whose consecutive sides are perpendicular. Generalize this to regular n-sided polygons in D^2 for $n \geq 5$.

15

SECTION

Riemann Surfaces

◇ Having completed our long journey from Euclidean plane geometry through crystallographic groups, Möbius transformations, and hyperbolic plane geometry, we are now ready to gain a geometric insight into the structure of Fuchsian groups on H^2. Before this, however, we need a bit of complex calculus. Just like linear fractional transformations, complex functions are usually defined on open sets U of the complex plane $\mathbf{C}$ and have their values in $\mathbf{C}$. Given a complex valued function $f : U \to \mathbf{C}$, we say that f is *differentiable (in the complex sense)* at $z_0 \in U$ if the limit

$$\lim_{z \to z_0} \frac{f(z) - f(z_0)}{z - z_0}$$

exists. In this case, we call the limit the *derivative* of f at z_0 and denote it by $f'(z_0)$. We say that f is *differentiable (in the complex sense)* on U if $f'(z_0)$ exists for all $z_0 \in U$. One of the most stunning novelties of complex calculus is that the existence of the first derivative of f on U implies the existence of all higher derivatives $f^{(n)}$, $n \in \mathbf{N}$, on U! (This is clearly nonsense in real calculus; for example, $f : \mathbf{R} \to \mathbf{R}$ defined by

$$f(x) = \begin{cases} x^n, & \text{if } x \geq 0 \\ -x^n, & \text{if } x < 0 \end{cases}$$

has derivatives up to order $n - 1$ at $x_0 = 0$, but

$$f^{(n)}(x) = \begin{cases} n!, & \text{if } x \geq 0 \\ -n!, & \text{if } x < 0 \end{cases}$$

is discontinuous at $x_0 = 0$.) As a matter of fact, even more is true! If f' exists on U, then f is *analytic* on U; that is, given $z_0 \in U$, f can be expanded into a power series

$$f(z) = \sum_{n=0}^{\infty} c_n (z - z_0)^n$$

that is absolutely convergent in a neighborhood of z_0 (contained in U). (♠ Like almost everything in complex calculus, this is a consequence of the Cauchy formula.) ◇ The coefficients of the expansion have no other choice but to be equal to

$$c_n = \frac{f^{(n)}(z_0)}{n!},$$

so that the power series is actually Taylor. (Once analyticity is accepted, this follows since an absolutely convergent series can be differentiated term by term.) From now on we use the term *analytic* for a complex function $f : U \to \mathbf{C}$ whose derivative exists on U. The usual rules of differentiation are valid, and the proofs are the same as in the real case. In particular, every rational function (such as a linear fractional transformation)

$$\frac{a_0 + a_1 z + \cdots + a_n z^n}{b_0 + b_1 z + \cdots + b_m z^m},$$

$$a_n \neq 0 \neq b_m, \quad a_0, \ldots, a_n; b_0, \ldots, b_m \in \mathbf{C},$$

is analytic everywhere except at finitely many ($\leq m$) points where the denominator vanishes. Moreover, and this is very important for Riemann surfaces, composition of analytic functions is analytic.

    Complex functions $f : U \to \mathbf{C}$ are often thought of as being locally defined transformations of the complex plane. As such, the existence of the derivative of f must carry a geometric meaning. Let us eleborate on this a little. Let f be analytic on U and assume that $f'(z_0) \neq 0$ at $z_0 \in U$. Let $\gamma : (-a, a) \to \mathbf{C}$, $a > 0$, be a smooth curve through z_0 with $\gamma(0) = z_0$. Consider the image $f \circ \gamma$. Differentiating, we get

$$(f \circ \gamma)'(0) = f'(\gamma(0)) \cdot \gamma'(0) = f'(z_0)\gamma'(0).$$

This means that

$$|(f \circ \gamma)'(0)| = |f'(z_0)| \cdot |\gamma'(0)|$$

and

$$\arg(f \circ \gamma)'(0) = \arg f'(z_0) + \arg \gamma'(0).$$

Thus, f acts on tangent vectors at z_0 by multiplying them by the constant $|f'(z_0)|$ and rotating them by $\arg f'(z_0)$. In particular, f preserves (signed) angles, a property that we note by saying that f is *conformal*. Summarizing, we obtain that an analytic function is conformal[1] where its derivative does not vanish! Notice that this argument applies to linear fractional transformations and was given earlier for Möbius transformations.

We begin to suspect that the big leap from first-order differentiability to analyticity will rule out a lot of (otherwise nice) functions. The simplest example is *conjugation* $z \mapsto \bar{z}$, $z \in \mathbf{C}$. We claim that conjugation is *nowhere* differentiable. In fact, setting $z - z_0 = r(\cos\theta + i\sin\theta)$ and keeping θ fixed, we have

$$\lim_{z \to z_0} \frac{\bar{z} - \bar{z}_0}{z - z_0} = \lim_{r \to 0} \frac{r(\cos\theta - i\sin\theta)}{r(\cos\theta + i\sin\theta)}$$

$$= \frac{\cos\theta - i\sin\theta}{\cos\theta + i\sin\theta} = \cos(2\theta) - i\sin(2\theta),$$

and for different arguments θ, this takes different values! Thus, the limit and hence the derivative do not exist.

Using this, a number of nondifferentiable complex functions can be manufactured; $f(z) = 1/\bar{z}$ will appear shortly. (♠ You know, of course, the underlying theme that differentiability in complex sense is smoothness *plus the Cauchy-Riemann equations*. These we avoided as part of our desperate effort to keep the length and level of the exposition to a minimum.)

---

[1]The connection between complex differentiability and conformality was first recognized by Gauss in 1825.

◇ Let $f : U \to \mathbf{C}$ be analytic on U and $f'(z_0) = 0$ at $z_0 \in U$. The Taylor expansion of f at z_0 implies that

$$f(z) - f(z_0) = (z - z_0)^m g(z)$$

for some $m \geq 2$, where g is analytic and nonzero on an open neighborhood $U_0 \subset U$ of z_0. Choosing a suitable branch of the mth root of g, we have

$$f(z) - f(z_0) = \left((z - z_0) \sqrt[m]{g(z)} \right)^m .$$

In other words, f is the mth power of an analytic function h defined on U_0 by

$$h(z) = (z - z_0) \sqrt[m]{g(z)}, \quad z \in U_0.$$

Notice that $h(z_0) = 0$ and $h'(z_0) \neq 0$, so that h establishes a conformal equivalence between a neighborhood of z_0 (in U_0) and a neighborhood of the origin.

It is now time to introduce the most basic elementary function in complex calculus: the exponential function. We define it by the power series

$$e^z = \sum_{n=0}^{\infty} \frac{z^n}{n!},$$

which is clearly convergent on the entire complex plane. In particular, $(e^z)' = e^z$. When $z = r$ is real, this reduces to the ordinary exponential function. To see what happens in the imaginary direction, we take $z = i\theta$, $\theta \in \mathbf{R}$. We compute

$$e^{i\theta} = \sum_{n=0}^{\infty} \frac{i^n \theta^n}{n!}$$

$$= \sum_{k=0}^{\infty} \frac{i^{2k} \theta^{2k}}{(2k)!} + \sum_{k=0}^{\infty} \frac{i^{2k+1} \theta^{2k+1}}{(2k+1)!}$$

$$= \sum_{k=0}^{\infty} \frac{(-1)^k \theta^{2k}}{(2k)!} + i \sum_{k=0}^{\infty} \frac{(-1)^k \theta^{2k+1}}{(2k+1)!}$$

$$= \cos \theta + i \sin \theta.$$

Notice that in the second equality we split the infinite sum into two sums; one running on even indices ($n = 2k$), the other on odd indices ($n = 2k+1$). This is legitimate, since the series is absolutely convergent. Then we used the fact that $i^{2k} = (i^2)^k = (-1)^k$, and finally we remembered the Taylor expansion of sine and cosine. We arrive at the famous Euler formula

$$e^{i\theta} = \cos\theta + i\sin\theta, \quad \theta \in \mathbf{R}.$$

This is our old friend $z(\theta)$, whom we met a long time ago when discussing complex arithmetic! Multiplying through by r, we obtain the exponential form of a complex number:

$$z = re^{i\theta}, \quad |z| = r, \quad \arg z = \theta + 2k\pi, \quad k \in \mathbf{Z}.$$

$\theta = \pi$ in the Euler formula gives the equation

$$e^{i\pi} = -1,$$

connecting the three most prominent numbers π, e, and i of mathematics. This appears in Euler's *Introductio*, published in Lausanne in 1748. Aside from its commercial value shown on the T-shirts of mathematics students, "We are all number $-e^{i\pi}$!", its significance can hardly be underrated. Without much explanation, we humbly recite the words of Benjamin Pierce (1809–1880) to his students:

> Gentlemen, that is surely true, it is absolutely paradoxical; we cannot understand it, and we don't know what it means, but we have proved it, and therefore we know it must be the truth.

The natural extension of the relation

$$e^x = \lim_{n\to\infty} \left(1 + \frac{x}{n}\right)^n$$

to complex exponents can easily be understood. For purely imaginary exponents Euler's formula implies

$$e^{i\theta} = \lim_{n\to\infty} \left(1 + i\frac{\theta}{n}\right)^n.$$

Indeed, the argument of the complex number $1+i\theta/n$ is $\tan^{-1}(\theta/n)$, so that the argument of the right-hand side is $\lim_{n\to\infty} n\tan^{-1}(\theta/n) = \theta$. The absolute value of the right-hand side is unity, since

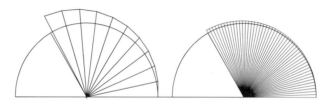

Figure 15.1

$\lim_{n\to\infty}(1 + \theta^2/n^2)^{n/2} = 1$ (a simple consequence of the binomial formula). There is a beautiful geometric interpretation of this. For fixed n, the points $(1 + i\theta/n)^k$, $k = 0, \ldots, n$, are vertices of n right triangles arranged in a fanlike pattern as shown in Figure 15.1 for $n = 10$ and $n = 50$. Complex multiplication by $(1 + i\theta/n)$ amounts to multiplication by $\sqrt{1 + \theta^2/n^2}$ and rotation by the angle $\tan^{-1}(\theta/n)$. Thus, the hypotenuse of each triangle in the fan is the base of the next.

Remark.

The equation $1 + e^{i\pi} = 0$ was used by Lindemann to prove transcendentality of π (cf. Section 4). In fact, he proved[2] that in an equation of the form

$$c_0 + c_1 e^{a_1} + \cdots + c_n e^{a_n} = 0,$$

the coefficients $c_0, \ldots c_n$ and the exponents $a_1, \ldots, a_n$ cannot all be complex algebraic numbers.

The exponential function satisfies the usual exponential identity

$$e^{z_1 + z_2} = e^{z_1} \cdot e^{z_2}, \quad z_1, z_2 \in \mathbf{C},$$

which follows easily from trigonometric identities. (If you try to prove this from the Taylor series definition, you have to work a little harder and use the binomial formula.) In particular, e^z is periodic with period $2\pi i$:

$$e^{z + 2k\pi i} = e^z, \quad k \in \mathbf{Z}.$$

The exponential map $\exp : \mathbf{C} \to \mathbf{C}$ has image $\mathbf{C} - \{0\}$, as can easily be seen by writing an image point in exponential form. In fact,

---

[2]For a lively account, see F. Klein. *Famous Problems of Elementary Geometry*, Chelsea, New York, 1955.

given $0 \neq w \in \mathbf{C}$, the complex numbers

$$\log |w| + i \arg w$$

($\arg w$ has infinitely many values!) all map under exp to w. This is thus the inverse of exp, legitimately called the complex logarithm $\log w$ of w. It is now the multiple valuedness of log that gives rise to Riemann surfaces. In fact, attempting to make log single valued is the same as trying to make the exponential map one-to-one. This is what we will explain next.

Consider the complex plane $\mathbf{C}$. The periodicity formula for exp tells us that z and $z + 2k\pi i$, $k \in \mathbf{Z}$, are mapped to the same point by exp. Thus, we should not consider these two points different in $\mathbf{C}$! Identifying them means rolling $\mathbf{C}$ into a cylinder in the imaginary direction. (This can be demonstrated easily by pouring paint on the Chinese rug at home and rolling it out.) We obtain that exp actually maps the cylinder to the punctured plane (see Color Plate 2a).

The rulings of the cylinder are mapped to rays emanating from the origin. (This is because the rulings can be parametrized by $t \mapsto t + \theta i$ with $\theta \in \mathbf{R}$ fixed, and the images are parametrized by $t \mapsto e^t \cdot e^{i\theta}$.) Similarly, circles on the cylinder map to concentric circles around the origin. We now see that the exponential map establishes a one-to-one correspondence between the cylinder and the punctured plane. We will say later that the cylinder and the punctured complex plane are *conformally equivalent* Riemann surfaces. We now notice that the set of points

$$z + 2k\pi i \in \mathbf{C}, \quad k \in \mathbf{Z},$$

that are to be identified form the orbit[3] of the first frieze group generated by the translation $T_{2\pi i}$!

It is clear how to generalize this to obtain more subtle Riemann surfaces. We consider discrete groups on $\mathbf{C} = \mathbf{R}^2$ or Fuchsian groups on H^2, and in each case, we identify points that are on the same orbit of the acting group. The concept of fundamental set (domain), which we used to visualize wallpaper patterns, now gains primary importance! By its very definition, it contains exactly

---

[3]For group actions, refer to the end of "Groups" in Appendix B.

one point from each orbit, so that as a point-set, it is in one-to-one correspondence with the Riemann surface. In our cases, the fundamental domain is a Euclidean or hyperbolic polygon, so that when taking the closure of our fundamental set, the points that are on the same orbit (and thereby are to be identified) appear on the sides of this polygon. Thus, to obtain our Riemann surface topologically, we need to paste the polygon's sides together in a certain manner as prescribed by the acting group. This prescription is called *side-pairing transformation*.

Enough of these generalities. Let us now consider crystallographic groups on $\mathbf{C} = \mathbf{R}^2$. We first consider translation groups. Any such crystallographic group G is generated by two translations T_v and T_w with v and w linearly independent. A fundamental domain F_0 for G is a parallelogram with vertices $0, v, w, v+w$ (Figure 15.2).

The boundary ∂F_0 consists of the four sides of the parallelogram. The restriction of the action of the group G to these sides gives the side-pairing transformations. It is clear that under T_w, tv, $0 \leq t \leq 1$, gets identified with $T_w(tv) = tv + w$ (Figure 15.3).

Similarly, under T_v, tw, $0 \leq t \leq 1$, gets identified with $T_v(tw) = tw + v$ (Figure 15.4).

Thus, the Riemann surface is obtained by pasting together the base and top sides and the left and right sides. We obtain what is called a *complex torus* (Figure 15.5).

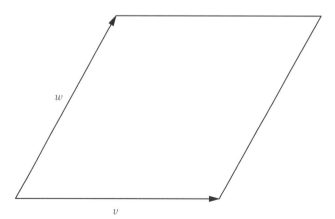

Figure 15.2

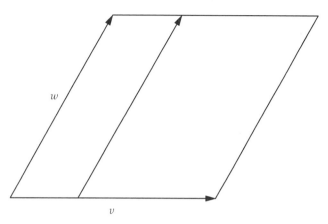

Figure 15.3

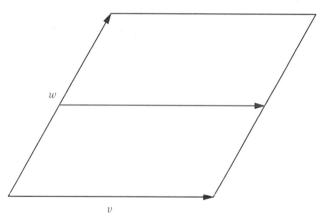

Figure 15.4

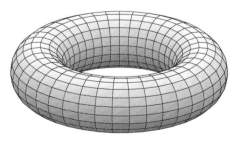

Figure 15.5

You might say that all these tori (the plural of torus) look alike, so why did not we just take v and w to be orthogonal unit vectors spanning the integer lattice $\mathbf{Z}^2$ in $\mathbf{R}^2$? The answer depends on what we mean by "alike". A topologist would certainly consider all of them the same, since they are actually homeomorphic.[4] But from the analyst's point of view, they may be different, since the conformal structure (the way we determine angles) should depend on the vectors v and w. This we will make more precise shortly.

Looking back, we see that we used the term *Riemann surface* a number of times without actually defining what it is. We now have enough intuition to fill the gap properly. To tell you the truth, in the proof of the FTA we came very close to this concept! Recall the stereographic projections $h_N : S^2 - \{N\} \to \mathbf{C}$ and $h_S : S^2 - \{S\} \to \mathbf{C}$ and their connecting relation

$$(h_N \circ h_S^{-1})(z) = 1/\bar{z}, \quad 0 \neq z \in \mathbf{C}.$$

Putting h_N and h_S on symmetric footing, in the proof of the FTA they were used to "view" the map $f : S^2 \to S^2$ in two different ways. One was the original polynomial

$$P = h_N \circ f \circ h_N^{-1} : \mathbf{C} \to \mathbf{C},$$

and the other,

$$Q = h_S \circ f \circ h_S^{-1} : \mathbf{C} \to \mathbf{C},$$

was a rational function. h_N and h_S both have the "defect" that their domains do not cover the entire sphere, but the union of these domains is S^2, and so P and Q describe f completely. We will call h_N and h_S *coordinate charts* for S^2.

There is, however, one minor technical difficulty. To define the notion of analyticity of functions defined on open sets of S^2, we would use h_N and h_S to pull the functions down to open sets of $\mathbf{C}$ and verify analyticity there. Take, for example, the function h_N. To pull this down to $\mathbf{C}$ we use h_N itself and obtain $h_N \circ h_N^{-1} : \mathbf{C} \to \mathbf{C}$, the identity. If we use h_S, however, we obtain $h_N \circ h_S^{-1} : \mathbf{C} \to \mathbf{C}$, and this is not analytic, since $z \to \bar{z}$ is nowhere differentiable! The

---

[4]See "Topology" in Appendix C.

problem, of course, is the presence of conjugation, and it is easily remedied by taking $\bar{h}_S$ instead of h_S. This gives

$$(h_N \circ \bar{h}_S^{-1})(z) = \frac{1}{z}, \quad 0 \neq z \in \mathbf{C},$$

and now analyticity of a function on S^2 no longer depends on whether it is viewed by h_N or $\bar{h}_S$! (Replacing h_S with $\bar{h}_S$ is natural, since $z \mapsto 1/z$ is the simplest direct Möbius transformation that is not an isometry. Notice also the role of this transformation in the proof of Theorem 8 of Section 12.) Now the general definition:

♠ A connected (Hausdorff) topological space[5] M is a Riemann surface if M is equipped with a family

$$\{\varphi_j : U_j \to \mathbf{C} \mid j \in \mathbf{N}\}$$

called the *atlas* (each $\varphi_j : U_j \to \mathbf{C}$ is called a *coordinate chart* of M) such that

1. $\cup_{j \in \mathbf{N}} U_j = M$;
2. Each φ_j is a homeomorphism of U_j to an open set of the complex plane $\mathbf{C}$;
3. If $U = U_k \cap U_j$ is nonempty, then

$$\varphi_k \circ \varphi_j^{-1} : \varphi_j(U) \to \varphi_k(U)$$

is an analytic map between open sets of the complex plane $\mathbf{C}$ (Figure 15.6).

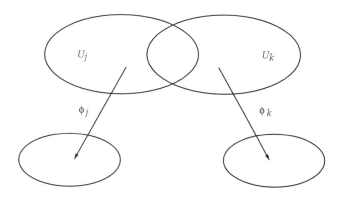

Figure 15.6

---

[5]See "Topology" in Appendix C.

Remark.

We can use a subset of the set of positive integers as the index set for the atlas, which is not required in the usual definition of a Riemann surface. It is, however, a deeper result that every Riemann surface carries a *countable* atlas, so that this choice can always be made. Looking back, we see that

$$\{h_N : S^2 - \{N\} \to \mathbf{C}, \quad \bar{h}_S : S^2 - \{S\} \to \mathbf{C}\}$$

is an atlas for S^2, so that S^2 is a Riemann surface.

Analyticity is a local property, so that any open subset of a Riemann surface is also a Riemann surface. Thus, the complex plane $\mathbf{C}$, the punctured complex plane $\mathbf{C} - \{0\}$, etc. are Riemann surfaces.

The Riemann surface structure on the cylinder and tori are derived from the Riemann surface structure of the complex plane $\mathbf{C}$ in the following way: Recall that they are defined by identifying the points that are on the same orbit of the acting (translation) group G. The identification space, that is, the space of orbits $\mathbf{C}/G$, is a topological space under the quotient topology. Actually, the topology on $\mathbf{C}/G$ is defined to make continuous the natural projection $\pi : \mathbf{C} \to \mathbf{C}/G$ associating to $z \in \mathbf{C}$ its orbit $G(z) = \{g(z) \,|\, g \in G\}$. Now let $z_0 \in \mathbf{C}$ and consider the open disk $D_{r_0}(z_0) = \{z \in \mathbf{C} \,|\, |z - z_0| < r_0\}$ of radius r_0 and center z_0. The restriction $\pi | D_{r_0}(z_0)$ is one-to-one onto an open subset U_0 of $\mathbf{C}/G$, provided that r_0 is small. In our explicit cases, this happens if r_0 is less than half of the minimum translation length in G. Now define $\varphi_0 = (\pi | D_{r_0}(z_0))^{-1} : U_0 \to D_{r_0}(z_0)$. The definition of quotient space topology translates into φ_0 being a homeomorphism. It is clear that we can choose small disks $D_{r_j}(z_j)$, $j \in \mathbf{N}$, as above, such that they all cover $\mathbf{C}$. Finally,

$$\varphi_k \circ \varphi_j^{-1} = (\pi | D_{r_k}(z_k))^{-1} \circ (\pi | D_{r_j}(z_j))$$

is the restriction of an element in G (a translation) and thereby analytic. The cylinder and all complex tori thus become Riemann surfaces.

We may still feel uneasy about the initial choice of the group G acting on $\mathbf{C}$, since it consists of translations only. As a matter of fact, we may think that less trivial choices of G may lead to more subtle

Riemann surfaces $\mathbf{C}/G$. That this is not the case for the complex plane $\mathbf{C}$ is one result in the theory of Riemann surfaces.

In seeking new domains, we may also consider the extended complex plane $\hat{\mathbf{C}}$ or, what is the same, the sphere S^2. The question is the same: "Does there exist a discrete group G acting on S^2 that gives a Riemann surface S^2/G?" As far as the group G is concerned, the answer is certainly yes; we just have to remember the proof of the FTA, where the map $z \to z^n$ induced an n-fold wrap of S^2 to itself, leaving the North and South Poles fixed. It is not hard to see that this map can be thought of as the projection $\pi : S^2 \to S^2/G$, where G is generated by the rotation $R_{2\pi/n}$ around the origin in $\mathbf{C} \subset \hat{\mathbf{C}}$. As far as the Riemann surface S^2/G is concerned, it is again a result in the theory of Riemann surfaces that we do not get anything other than S^2. Summarizing, the only Riemann surfaces that are quotients of S^2 or $\mathbf{C}$ are the sphere, the complex plane, the cylinder, and the complex tori.

We now take the general approach a little further and define analyticity of a map $f : M \to N$ between Riemann surfaces M and N. Given $p_0 \in M$, we say that f is (complex) differentiable at p_0 if the composition

$$\psi_k \circ f \circ \varphi_j^{-1}$$

shown in Figure 15.7 is (complex) differentiable at $\varphi_j(p_0)$. Here $\varphi_j : U_j \to \mathbf{C}$ is a chart covering p_0; that is, $p_0 \in U_j$, and $\psi_k : V_k \to \mathbf{C}$ is a chart covering $f(p_0)$, so that the composition is defined near $\varphi_j(p_0)$. (To be perfectly precise, the composition is defined on $\varphi_j(U_j \cap f^{-1}(V_k))$, but this is really too much distraction.) We also see that we defined the concept of Riemann surfaces in just such a

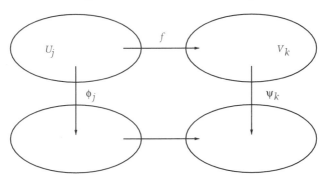

Figure 15.7

way as to make this definition independent of the choice of charts! (You are invited to choose other charts, work out the compositions, and verify independence. Advice: Draw a detailed picture rather than engage in gory computational details.)

If $f : M \to N$ is differentiable at each point of M, we say that f is *analytic*. As an example, it is clear from the way we defined $\mathbf{C}/G$ for G a discrete translation group on $\mathbf{C}$, that the projection $\pi : \mathbf{C} \to \mathbf{C}/G$ is analytic. At the other extreme, an invertible analytic map (whose inverse is also analytic) is called a *conformal equivalence*. (The name clearly comes from the fact that if f is invertible, then its derivative is everywhere nonzero, so that f viewed as a transformation is conformal.) Finally, two Riemann surfaces are called *conformally equivalent* if there is a conformal equivalence between them. Our prominent example: The cylinder and the punctured complex plane are conformally equivalent Riemann surfaces.

If $f : M \to N$ is an analytic map between Riemann surfaces and $p_0 \in M$, then the nonvanishing of the derivative of the local representation $\psi_k \circ f \circ \phi_j^{-1}$ at p_0 is independent of the choice of the charts ϕ_j and ψ_k. In this case, f is a local conformal equivalence between some open neighborhoods U_0 of p_0 and V_0 of $f(p_0)$. If the derivative of $\psi_k \circ f \circ \phi_j^{-1}$ vanishes at p_0, then as the Taylor expansion shows (cf. the argument above for a complex function f), there are local charts ϕ_j and ψ_k such that $\psi_k \circ f \circ \phi_j^{-1}$ is the mth power function for some m. This is exactly the case when f has a branch point at p_0 with branch number $m - 1$. (In fact, this can be taken as the definition of the branch point.) We now see that we used this (somewhat) intuitively in the proof of the FTA in Section 8!

Liouville's theorem in complex calculus implies that the entire complex plane $\mathbf{C}$ is not conformally equivalent to D^2. (Another proof is based on Schwarz's lemma, which asserts that the conformal self-maps of D^2 are linear fractional transformations, and thereby they have the form given in Problem 1 of Section 13. They can be parametrized by three real parameters: $\Re(w)$, $\Im(w)$, and θ. In contrast, the linear transformations $z \mapsto az + b$, $a, b \in \mathbf{C}$, form a 4-parameter family of conformal self-maps of the complex plane.) On the other hand, the linear fractional transformation $z \mapsto i(i + z)/(i - z)$ restricted to the unit disk D^2 establishes a conformal equivalence between D^2 and the upper half-plane H^2. More generally, the Riemann mapping theorem states that any simply connected[6] domain in $\mathbf{C}$ that is not the whole plane is conformally equivalent to D^2.

The question about the tori being "alike" can now be reformulated rigorously: "Which tori are conformally equivalent?" This is a question of Riemann moduli, an advanced topic.[7]

A beautiful result of complex function theory asserts that two complex tori $T_1 = \mathbf{C}/G_1$ and $T_2 = \mathbf{C}/G_2$ with $G_1 = \langle T_{v_1}, T_{w_1} \rangle$ and $G_2 = \langle T_{v_2}, T_{w_2} \rangle$ are conformally equivalent iff w_1/v_1 and w_2/v_2 are on the same orbit under the modular group $SL(2, \mathbf{Z})$. To give a sketch proof, we notice first that up to conformal equivalence, a torus can be realized as $\mathbf{C}/\langle T_1, T_\tau \rangle$, where $\tau \in H^2$. Assume now that we have a conformal equivalence $f : \mathbf{C}/\langle T_1, T_{\tau_1} \rangle \to \mathbf{C}/\langle T_1, T_{\tau_2} \rangle$ between two tori given by τ_1 and τ_2 in H^2. Since both tori are obtained from fundamental parallelograms in $\mathbf{C}$ by side-pairing transformations, it is clear that f can be "lifted up" to a conformal equivalence $\tilde{f} : \mathbf{C} \to \mathbf{C}$ satisfying $\tilde{f}(0) = 0$ and the relation $\pi_2 \circ \tilde{f} = f \circ \pi_1$, where $\pi_1 : \mathbf{C} \to \mathbf{C}/\langle T_1, T_{\tau_1} \rangle$ and $\pi_2 : \mathbf{C} \to \mathbf{C}/\langle T_1, T_{\tau_2} \rangle$ are natural projections. Complex calculus tells us that a conformal equivalence of $\mathbf{C}$ is linear, so that we have

$$\tilde{f}(z) = \alpha z, \quad z \in \mathbf{C},$$

for some $\alpha \in \mathbf{C}$. (This follows since a conformal equivalence of $\mathbf{C}$ has a removable singularity at infinity, so that it extends to a conformal equivalence of the Riemann sphere.) The commutation relation for f and $\tilde{f}$ above implies that $\tilde{f}$ maps any linear combination of 1 and τ_1 with integer coefficients to a linear combination of 1 and τ_2 with integer coefficients. We thus have

$$\tilde{f}(1) = \alpha = a + b\tau_2,$$

$$\tilde{f}(\tau_1) = \alpha\tau_1 = c + d\tau_2,$$

---

[6]In topology, the definition of simply connectedness requires a short detour into homotopy theory. Fortunately, simply connectedness of a domain in the complex plane $\mathbf{C}$ is equivalent to connectedness of its complement in the extended plane $\hat{\mathbf{C}}$.

[7]See H. Farkas and I. Kra, *Riemann Surfaces*, Springer, 1980.

for some $a, b, c, d \in \mathbf{Z}$. Since $\tilde{f}$ is invertible, we have $ad - bc = \pm 1$. Solving for τ_1, we obtain

$$\tau_1 = \frac{c + d\tau_2}{a + b\tau_2},$$

so that τ_1 is in the $SL(2, \mathbf{Z})$-orbit of τ_2. (Note that $ad - bc = 1$, since both τ_1 and τ_2 are in H^2.)

♡ We turn now to the most important case of Fuchsian groups acting on the hyperbolic plane H^2. Discreteness of the Fuchsian group G on H^2 implies that H^2/G is a Riemann surface with analytic projection $\pi : H^2 \to H^2/G$. The proof of this is a souped up version of the one we just did for discrete translation groups for $\mathbf{C}$. We omit the somewhat technical details. It is more important for us that a rich source of Riemann surfaces can be obtained this way. (In fact, all Riemann surfaces arise as quotients; this is "uniformization," a more advanced topic.) Instead, we look at the examples of Fuchsian groups obtained in the previous section and find the corresponding Riemann surface by looking at the side-pairing transformations on the fundamental hyperbolic polygon.

EXAMPLE 1

$G = \langle g \rangle$ with $g(z) = z + 1$, $z \in H^2$. The modified exponential map

$$z \to \exp(2\pi i z), \quad z \in H^2,$$

is invariant under G, so that it projects down to H^2/G and gives a conformal equivalence between H^2/G and the punctured disk $D^2 - \{0\}$ (Figure 15.8). This example gives us the clue that parabolic elements in G are responsible for punctures in H^2/G. (In general,

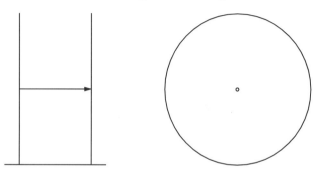

Figure 15.8

there is a one-to-one correspondence between the punctures of H^2/G and the conjugacy classes of parabolic elements in G.)   □

EXAMPLE 2

Let $k > 1$ and $G = \langle g \rangle$ with $g(z) = kz$, $z \in H^2$. The Riemann surface H^2/G is conformally equivalent to the annulus $A_r = \{z \in \mathbf{C} \mid 1 < |z| < r\}$, where $r = e^{2\pi^2/\log k}$. In fact, the map $z \mapsto \exp(-2\pi i \log z/\log k)$ is analytic on H^2, and, being invariant under G, it projects down to H^2/G and gives the conformal equivalence of H^2/G and A_r. This representation of the annulus can be used to show that two annuli, A_{r_1} and A_{r_2}, are conformally equivalent iff $r_1 = r_2$. Indeed, a conformal equivalence $f : A_{r_1} \to A_{r_2}$ can be lifted up to a conformal equivalence $\tilde{f} : H^2 \to H^2$ that satisfies the commutation relation

$$\tilde{f}(k_1 z) = k_2 \tilde{f}(z), \quad z \in H^2,$$

where $k_1 = e^{2\pi^2/\log r_1}$ and $k_2 = e^{2\pi^2/\log r_2}$. Complex calculus (the Schwarz lemma) tells us that a conformal equivalence of H^2 is necessarily Möbius, so that

$$\tilde{f}(z) = \frac{az + b}{cz + d}, \quad ad - bc = 1, \quad a, b, c, d \in \mathbf{R}.$$

Combining this with the commutation relation above, we see that $b = 0$, $k_1 = k_2$, and hence $r_1 = r_2$.   □

EXAMPLE 3

Let $n \in \mathbf{N}$ and $G = \langle g \rangle$ with

$$g(z) = \frac{\cos(\pi/n) \cdot z - \sin(\pi/n)}{\sin(\pi/n) \cdot z + \cos(\pi/n)}, \quad z \in H^2.$$

This is best viewed on D^2 where G is generated by the Euclidean rotation $R_{2\pi/n}$ around the origin (Figure 15.9).

The map $z \mapsto z^n$ is invariant under $\langle R_{2\pi/n} \rangle$, projects down to the quotient, and defines a conformal equivalence between H^2/G and the unit disk D^2.   □

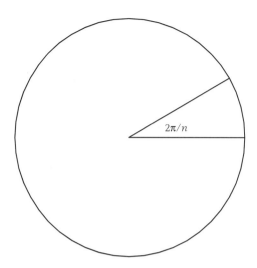

$2\pi/n$

Figure 15.9

EXAMPLE 4

Let $k > 1$ and $G = \langle g_1, g_2 \rangle$ with $g_1(z) = -1/z$ and $g_2(z) = kz$, $z \in H^2$. A somewhat involved argument shows that H^2/G is conformally equivalent to the unit disk. □

EXAMPLE 5

Let $G = SL(2, \mathbf{Z})$ be the modular group. The Riemann surface H^2/G is conformally equivalent to $\mathbf{C}$. This example is important for the Riemann moduli of tori. In fact, we now see that $H^2/SL(2, \mathbf{Z})$ and thereby $\mathbf{C}$ "parametrizes" the set of conformally inequivalent tori. □

EXAMPLE 6

Let $G = \langle g_1, g_2 \rangle$, where

$$g_1(z) = \frac{3z + 4}{2z + 3} \quad \text{and} \quad g_2(z) = 2z, \quad z \in H^2.$$

Looking at the fundamental set, we see that H^2/G is conformally equivalent to the punctured sphere; that is, to $\mathbf{C}$. □

Finally, if a Fuchsian group G has a fundamental polygon with $4p$ sides (and G identifies the opposite sides), then H^2/G is conformally equivalent to a compact genus p Riemann surface, or more plainly, a "torus with p holes." As an example, use the four

Figure 15.10

hyperbolic side-pairing transformations for the hyperbolic octagon at the end of Section 14 as a fundamental domain of a Fuchsian group and realize that H^2/G is a genus 2 Riemann surface (Figure 15.10).

To close this section, we make a note on the origins of the theory of Riemann surfaces. We saw that the exponential map $\exp : \mathbf{C} \to \mathbf{C}$ gives rise to a conformal equivalence between the cylinder and the punctured plane, with inverse being the complex logarithm

$$\log w = \log |w| + i \arg w, \quad w \neq 0.$$

If we reject the cylinder as range and keep log to be defined on the punctured plane $\mathbf{C} - \{0\}$, then it is inevitably multiple valued. How can we get around this difficulty? Well, since log is multiple valued, the domain $\mathbf{C} - \{0\}$ has to be replaced by a Riemann surface on which it becomes single valued. The new domain has to have infinitely many layers of $\mathbf{C} - \{0\}$, so that it must look like an infinite staircase denoted by St_∞ (see Color Plate 2b). With this, we see that $\exp : \mathbf{C} \to St_\infty$ is a conformal equivalence and that the Riemann surfaces are simply connected (no holes!).

The situation is similar for the power function $z \to z^n$. In this case, the inverse $w \to \sqrt[n]{w}$ is n-valued (FTA!), so that the finite staircase St_n will do. Color Plates 3a–b depict the cases $n = 2$ and $n = 3$.

(Notice that St_2 and St_3 seem to have self-intersections. But remember, this is due to our limited 3-dimensional vision; as a matter of fact, St_n, the graph of $z \to z^n$, lives in $\mathbf{C} \times \mathbf{C} = \mathbf{R}^4$, and for a surface in 4 dimensions there is plenty of room to avoid self-intersections!)

♠ Given a monic complex polynomial P of degree n, by cutting and pasting, a compact Riemann surface M_P can be constructed[8] on which $\sqrt{P}$ is *single-valued* and analytic. P can be assumed to have distinct roots, since the square root of a double root factor is single-valued. M_P comes equipped with an analytic projection $\pi : M_P \rightarrow \hat{\mathbf{C}}$, and π is a twofold branched covering with branch points above the roots of P (and ∞ for n odd). If $z \in \mathbf{C}$ is away from the roots of P, then the two points $\pi^{-1}(z)$ correspond to $\pm\sqrt{P(z)}$. We will construct M_P explicitly for $n \leq 4$. From this the general case will follow easily. If P is monic and linear with root $a \in \mathbf{C}$, then the Riemann surface M_P of $P(z) = \sqrt{z - a}$ is essentially St_2 (See Color Plate 3a) centered at a. M_P is obtained by stacking up two copies of $\hat{\mathbf{C}}$, making in each copy a (say) radial cut from a to ∞, and finally, pasting[9] the four edges crosswise. The map $\pi : M_P \rightarrow \hat{\mathbf{C}}$ corresponds to vertical projection, and this is a twofold branched covering with branch points above a and ∞. The function $1/\sqrt{P}$ has a simple pole above a and a simple zero above ∞. Notice that M_P is conformally equivalent to $\hat{\mathbf{C}}$, and with the identification $M_P = \hat{\mathbf{C}}$, π becomes the map $z \mapsto (z - a)^2$, $z \in \hat{\mathbf{C}}$. We now realize that we met this a long time ago in the proof of the FTA!

The situation for quadratic P with distinct roots $a, b \in \mathbf{C}$ is similar. The Riemann surface M_P for $P(z) = \sqrt{(z - a)(z - b)}$ is obtained by cutting the two copies of $\hat{\mathbf{C}}$ by the line segment (actually, any smooth curve) connecting a and b, and pasting the four edges crosswise. The map π has branch points above a and b. The double-valued $\sqrt{P}$ lifted up along $\pi : M_P \rightarrow \hat{\mathbf{C}}$ becomes single-valued on M_P, since the winding number (cf. Problem 3 of Section 8) of a closed curve that avoids the cuts is the same with respect to the points above a and b. The function $1/\sqrt{P}$ has simple poles at the points above a and b and simple zeros at the *two* points above ∞. Once again, M_P is conformally equivalent to $\hat{\mathbf{C}}$. We also see that the quadratic case can be reduced to the linear case by sending b to ∞ by a linear fractional transformation. Analyti-

---

[8]In the rest of this section, we describe Siegel's approach to the Weierstrass $\wp$-function. For details, see C.L. Siegel, *Topics in Complex Function Theory*, Vol. I, Wiley-Interscience, New York, 1969.

[9]For pasting topological spaces in general, see "Topology" in Appendix C.

cally, the substitution $z \mapsto 1/z + b$ transforms $\sqrt{(z-a)(z-b)}$ to $(1/z)\sqrt{b-a}\sqrt{z + 1/(b-a)}$.

It is rewarding to take a closer look at the particular case $P(z) = z^2 - 1$. Let the cuts be the line segments that connect ± 1 in the two copies of $\hat{\mathbf{C}}$. Since $\sqrt{P}$ becomes single-valued on M_P, we can consider the line integral

$$I(C) = \int_C \frac{dz}{\sqrt{P(z)}},$$

where the curve C emanates from a fixed point $p_0 \in M_P$ and terminates at a variable point $p \in M_P$. Although M_P is simply connected, the integral I depends on C due to the simple poles of $1/\sqrt{P}$ above a and b. (To be precise, by the monodromy theorem, the line integral $I(C)$ with respect to a curve C that avoids a, b, and ∞ depends only on the homotopy class of C in $M_P - \{a, b, \infty\}$.) To make the integral I path-independent, we have to construct a Riemann surface M above M_P by taking infinitely many copies of M_P and then cutting and pasting, following the recipe that the integration prescribes (counting the residues à la Cauchy). The prescription in question becomes more transparent when we realize that the inverse of the complex sine function is the antiderivative of $1/\sqrt{1-z^2} = i/\sqrt{z^2-1}$, so that M should be the Riemann surface of $\sin^{-1}$! The sine function itself is given by the Euler formula for complex exponents:

$$\sin z = \frac{e^{iz} - e^{-iz}}{2i}.$$

This is one-to-one on any vertical strip $(k - 1/2)\pi < \Re(z) < (k + 1/2)\pi$, $k \in \mathbf{Z}$, and each strip is mapped onto the whole complex plane with cuts $(-\infty, -1)$ and $(1, \infty)$ along the real axis (cf. Problem 2). The line $\Re(z) = (k + 1/2)\pi$ corresponds to the two edges of the positive cut if k is even, and to the negative cut if k is odd. Thus, the Riemann surface on which the inverse of sine is single-valued is obtained from infinitely many copies of the complex plane with the cuts $(-\infty, -1)$ and $(1, \infty)$ as above, and the even and odd layers connect each other alternately.

If P is cubic with three distinct roots $a, b, c \in \mathbf{C}$, then we group a, b, c, and ∞ into two pairs, connect them by disjoint smooth curves, cut the two copies of $\hat{\mathbf{C}}$ along these curves, and join the edges crosswise. This time M_P becomes conformally equivalent to a complex torus. (This can be seen most easily by arranging the cuts in different hemispheres of $S^2 = \hat{\mathbf{C}}$, cutting one of the copies of S^2 along the equator, pasting first along the branch cuts crosswise, and finally pasting the cut equators back together.) As before, $\pi : M_P \to \hat{\mathbf{C}}$ is an analytic twofold branched covering with branch points above a, b, c, and ∞, and $\sqrt{P}$ is single-valued on M_P. The function $1/\sqrt{P}$ has simple poles at a, b, c and a triple zero at ∞. By a linear change of the variables, we can put P in the *classical Weierstrass form*

$$P(z) = 4z^3 - g_2 z - g_3.$$

Since P has distinct roots, the discriminant δ of P (cf. Section 6) is nonzero:

$$\delta = \frac{1}{16}(g_2^3 - 27g_3^2) \neq 0.$$

The line integral

$$I(C) = \int_C \frac{dz}{\sqrt{4z^3 - g_2 z - g_3}}$$

is called an *elliptic integral of the first kind*. As before, the monodromy theorem says that for a curve C that avoids a, b, c, the line integral $I(C)$ depends only on the homotopy class of C in $M_P - \{a, b, c\}$. In addition, being a complex torus, M_P itself has nontrivial topology. Instead of cutting and pasting we unify these path-dependencies into a single concept. We collect the values of the line integral $I(C)$ for all *closed* curves C (based at a fixed point p_0 away from a, b, c) and call them *periods*. By additivity of the integral, the set L_P of all periods forms an additive subgroup of $\mathbf{C}$. In fact, L_P is a 2-dimensional lattice in $\mathbf{C}$. (This is because the fundamental group $\pi_1(M_P, p_0)$ is $\mathbf{Z}^2$.) We call L_P the *period lattice*. We see that in order to make the line integral $I(C)$ dependent on the terminal point of C only, we need to consider the value of $I(C)$ *modulo the period lattice* L_P. In other words, I effects a conformal equivalence from the Riemann surface M_P to the complex torus $\mathbf{C}/L_P$. The inverse of this map is the *Weierstrass $\wp$-function*, which is best viewed as an analytic map lifted from $\mathbf{C}/L_P$ to $\mathbf{C}$. By definition, $\wp$ satisfies the differential equation

$$(\wp')^2 = 4\wp^3 - g_2\wp - g_3, \quad w \in \mathbf{C}.$$

(What is the analogue for the sine function?) Moreover, $\wp$ is doubly periodic with periods in L_P:

$$\wp(w + \omega) = \wp(w), \quad w \in \mathbf{C}, \quad \omega \in L_P.$$

This is all very elegant but does not give $\wp$ in an explicit form. To get to this, we first notice that $\wp$ must have poles (Liouville's theorem). In fact, $\wp$ is the simplest doubly periodic function. After a detailed analysis we arrive at the partial fractions expansion

$$\wp(w) = \frac{1}{w^2} + \sum_{\omega \in L_P, \omega \neq 0} \left(\frac{1}{(w - \omega)^2} - \frac{1}{\omega^2} \right).$$

Notice that $\wp$ has double poles at the lattice points in L_P, and the difference is needed to ensure convergence. (The series converges absolutely and uniformly on any compact subset in $\mathbf{C} - L_P$.) Finally, we note that the lattice L_P itself determines the coefficients g_2 and g_3 in the classical Weierstrass form above. In fact, we have

$$g_2 = 60 \sum_{\omega \in L_P, \omega \neq 0} \frac{1}{\omega^4} \quad \text{and} \quad g_3 = 140 \sum_{\omega \in L_P, \omega \neq 0} \frac{1}{\omega^6}$$

(cf. Problem 3).

We now take a closer look at the differential equation that $\wp$ satisfies. This looks very familiar! In fact, we immediately notice that, for any $w \in \mathbf{C}$, the pair $(\wp(w), \wp'(w))$ satisfies the equation

$$y^2 = P(x) = 4x^2 - g_2 x - g_3.$$

This is our old friend the elliptic curve C_f, $f(x, y) = y^2 - P(x)$, which we studied in Section 3! The only difference is that here x and y are complex variables, so that our elliptic curve sits in $\mathbf{C}^2$ (rather than in $\mathbf{R}^2$). In fact, we also need to recall the points at infinity that needed to be attached to $\mathbf{R}^2$. In our case, an ideal point is a pencil of complex lines in $\mathbf{C}^2$, and, in analogy with the real case, $\mathbf{C}^2$ with the ideal points becomes the *complex projective plane* $\mathbf{C}P^2$. Thus the equation above defines a *complex* elliptic curve $C_f(\mathbf{C})$ in $\mathbf{C}P^2$, and we have an analytic map

$$(\wp, \wp') : \mathbf{C}/L_P \to C_f(\mathbf{C}) \subset \mathbf{C}P^2.$$

It is not hard to show that this map is a conformal equivalence. With this, we obtain that $C_f(\mathbf{C})$ is a complex torus. We now realize that Figures 3.10 to 3.17 depict various slices of this torus; the only visual problem is the absence of the vertical infinity!

We want to assert that our conformal equivalence is actually an algebraic isomorphism. We have a little technical trouble here. It is clear how to add points in $\mathbf{C}/L_P$, but we defined addition on elliptic curves only for real coordinates and not, in general, for points on $C_f(\mathbf{C})$. This problem can be easily resolved. All we need to do is to work out the algebraic formulas for the geometric rule for the addition (the chord method), and the formulas will automatically extend to the complex case (cf. Problem 17 of Section 3). Notice also that the algebraic isomorphism between $\mathbf{C}/L_P$ and $C_f(\mathbf{C})$ is nothing but the classical addition formula for the $\wp$-function:

$$\wp(w_1 + w_2) = -\wp(w_1) - \wp(w_2) + \frac{1}{4}\left(\frac{\wp'(w_1) - \wp'(w_2)}{\wp(w_1) - \wp(w_2)}\right)^2.$$

Remark.

In Section 3, we defined addition of points on an elliptic curve using geometry (chord method). In an analytical approach[10] we could first define the Weierstrass $\wp$-function, then use $\wp$ to establish the conformal equivalence of an elliptic curve (over $\mathbf{C}$) with a complex torus (by choosing the lattice suitably), and finally define addition on the elliptic curve by carrying over the obvious addition on the torus to the curve, or equivalently, declaring the conformal equivalence to be an algebraic isomorphism.

As expected, the case of a quartic polynomial P with distinct roots $a, b, c, d \in \mathbf{C}$ can be reduced to the cubic case. The complex algebraic curve defined by $y^2 = P(x)$ is birationally equivalent to an elliptic curve.

Without going into details, we mention yet another connection. The elliptic integral for a quartic P can be put into the *Legendre*

---

[10]This is followed in Koblitz, *Introduction to Elliptic Curves and Modular Forms*, Springer, 1993.

form

$$I(C) = \int_C \frac{dz}{\sqrt{(1 - z^2)(1 - kz^2)}},$$

and this gives the Schwarz–Christoffel formula for a conformal map of the rectangle with vertices $(\pm 1, \pm k)$ onto the upper half-plane H^2!

The construction of the Riemann surface M_P for P a monic polynomial of any degree n (with distinct roots) can be easily generalized from the particular cases above. As before, we group the roots in pairs with ∞ added if n is odd, connect the pairs of points with disjoint smooth curves, make the $[(n + 1)/2]$ cuts on each copy of $\hat{\mathbf{C}}$ along the curves, and, finally, join the corresponding cuts crosswise. The Riemann surface M_P is conformally equivalent to a torus with $[(n - 1)/2]$ holes.

Problems

1. Derive Euler's formula for complex exponents using $e^z = \lim_{n \to \infty}(1 + z/n)^n$, where z is a complex number.

2. Prove the basic identities for the complex sine function and derive its mapping properties stated in the text.

3. Use (the derivative of) the geometric series formula for $1/(w - \omega)^2 - 1/\omega^2$ to derive the expansion

$$\wp(w) = \frac{1}{w^2} + 3G_4 w^2 + 5G_6 w^4 + 7G_8 w^6 + \cdots,$$

where

$$G_k = \sum_{\omega \in L_P, \omega \neq 0} \frac{1}{\omega^k}.$$

Substitute this into the differential equation of $\wp$ to obtain the formulas for g_2 and g_3 stated in the text.

4. Work out the side pairing transformations for the hyperbolic octagon in the last example of Section 14 and verify that pasting[11] gives the two-holed torus.

---

[11] For a different cutting-and-pasting construction of hyperbolic octagons and dodecagons, cf. D. Hilbert and S. Cohn-Vossen, *Geometry and Imagination*, Chelsea, New York, 1952.

Web Site

1. www.geom.umn.edu/~banchoff/script/CFGPow.html

16 | General Surfaces

♡ All Riemann surfaces can be listed as either S^2, $\mathbf{C}/G$ with G a translation group acting on $\mathbf{C}$ or H^2/G with G a Fuchsian group acting on H^2. The meager possibilities for S^2 and $\mathbf{C}$ tempt us to think that we may obtain surfaces more general than Riemann surfaces by relaxing some of the conditions on the acting discrete group G. What should we expect to give up to arrive at these more general surfaces?

To answer this question, we need to reconsider the definition of a Riemann surface. The main restriction there came from requiring the chart-changing transformation $\varphi_k \circ \varphi_j^{-1}$, $j, k \in \mathbf{N}$, to be differentiable *in the complex sense* since, as we have seen, there are many transformations that are nice (in the real sense) but fail to be analytic. For example, any Riemann surface carries an orientation given a Riemann surface, we know how to rotate positively around a point. This is because the local orientations given by the charts φ_j patch up to a global orientation, since changing charts amounts to performing $\varphi_k \circ \varphi_j^{-1}$, and this, being conformal, preserves the local orientations.

♠ It is now clear how to define the concept of a general surface. Just repeat the definition of a Riemann surface (replacing $\mathbf{C}$ by $\mathbf{R}^2$), and instead of saying that each $\varphi_k \circ \varphi_j^{-1}$, $j, k \in \mathbf{N}$, is analytic, we just require this to be differentiable[1] *in the real sense*! We now expect to obtain interesting new surfaces of the form $\mathbf{R}^2/G$, where G is a discrete group of diffeomorphisms acting on $\mathbf{R}^2$. The classification of these quotients is possible but still a formidable task.

Since we wish to stay in geometry, we require the elements of G to be isometries. In addition, we will assume that each element of G (that is not the identity) acts on $\mathbf{R}^2$ without fixed points. If you worry about this condition being too restrictive, recall $z \rightarrow z^n$, $z \in \mathbf{C}$, from the proof of the FTA, where the corresponding quotient did not give anything new but S^2! Your worry, however, is not unfounded. Excluding transformations with fixed points makes the projection map $\pi : \mathbf{R}^2 \rightarrow \mathbf{R}^2/G$ a simple covering, while at the fixed points of the elements in G, π would "branch over"—a much more interesting phenomenon. Our reason for leaving out the branched coverings is mostly practical; otherwise we would never get out of two dimensions and on to later parts of the Glimpses!

We now discuss examples. First assume that G is one of the seven frieze groups (see Section 10). Since G can only contain translations and glides, this leaves us only the first and seventh types. The first frieze group is generated by a translation, and the resulting surface $\mathbf{R}^2/G$ is a cylinder, a Riemann surface discussed in Section 15. Assume now that G is of the seventh type; that is, it is generated by a single glide $G_{l,v}$. We may assume that l is the first axis and v is 2π times the first unit vector: $v = (2\pi, 0) \in \mathbf{R}^2$. A fundamental domain F_0 for G is the vertical strip $(0, 2\pi) \times \mathbf{R}$, and the side-pairing transformation

$$G_{l,v}|\{0\} \times \mathbf{R} : \{0\} \times \mathbf{R} \rightarrow \{2\pi\} \times \mathbf{R}$$

is given by

$$G_{l,v}(0, y) = (2\pi, -y), \quad y \in \mathbf{R}.$$

The quotient $\mathbf{R}^2/G$ is called the *infinite Möbius band*. It can be visualized in the following way: Consider the unit circle $S^1 \subset \mathbf{R}^2 \subset \mathbf{R}^3$ (in the plane spanned by the first two axes) and the line parallel to the third axis in $\mathbf{R}^3$ through $(1, 0) \in S^1$. If you slide the line along S^1 by keeping it perpendicular to $\mathbf{R}^2$ all the time, it sweeps

---

[1] See "Smooth Maps" in Appendix D.

an ordinary cylinder. Now slide the line along S^1 with unit speed
and, while sliding, rotate it by half of that speed. (Do not worry
about self-intersections at this point; in fact, if you rotate the line in
an extra 2-dimensional plane perpendicular to $S^1 \subset \mathbf{R}^2$, then there
will be no self-intersections, and the Möbius band will be imbedded
in $\mathbf{R}^2 \times \mathbf{R}^2 = \mathbf{R}^4$!) Upon going around S^1 once, we complete a half-
turn of the line. Now, what the line sweeps is the Möbius band. It
is much easier to visualize this when we consider the action of the
glide only on a finite strip $\mathbf{R} \times (-h/2, h/2)$ of height $h > 0$. We
obtain the *finite Möbius band*.

A note about orientability: We observed above that for a Riemann
surface, changing the charts from $\varphi_j : U_j \to \mathbf{C}$ to $\varphi_k : U_k \to \mathbf{C}$
amounts to performing $\varphi_k \circ \varphi_j^{-1}$, and this, being complex differentiable, is always orientation preserving. In the case of general
surfaces, we see that the surface is *orientable* if there exists an atlas
in which every chart-changing transformation $\varphi_k \circ \varphi_j^{-1}$ is orientation preserving[2]. It is easy to see that such an atlas cannot exist on
a Möbius band, and therefore it is not orientable. The same applies
to all surfaces that contain the Möbius band, and this observation
is sufficient for all the examples that follow.

A note before we go any further: The square of a glide is a translation, and the quotient of $\mathbf{R}^2$ by a group generated by a single
translation is the cylinder. Thus, the map

$$\mathbf{R}^2/\langle G_{l,v}^2 \rangle \to \mathbf{R}^2/\langle G_{l,v} \rangle$$

that associates to the orbit $\langle G_{l,v}^2 \rangle(p)$ the orbit $\langle G_{l,s} \rangle(p)$ is two-to-one
(that is, every point on the range has exactly two inverse images).
We obtain that the cylinder is a twofold cover of the Möbius band!

♡ We now turn to crystallographic groups acting on $\mathbf{R}^2$. To obtain
a surface that has not been listed so far, we assume that G contains a
glide. A quick look at the seventeen crystallographic groups shows
that we are left with only the case when G is generated by two
parallel glide reflections. (Note that the case of two perpendicular
glides cannot occur. In fact, if $G = \langle G_{l_1,v_1}, G_{l_2,v_2} \rangle$ with $v_1 \cdot v_2 = 0$, then
$(v_2 - v_1)/2$ is a fixed point of the composition $G_{l_2,v_2} \circ G_{l_1,v_1} = T_{v_2} \circ R_{l_2} \circ$
$R_{l_1} \circ T_{v_1}$, since $R_{l_2} \circ R_{l_1} = H$ is a half-turn.) We set $G = \langle G_{l_1,v_1}, G_{l_2,v_2} \rangle$,

---

[2]See "Smooth Maps" in Appendix D.

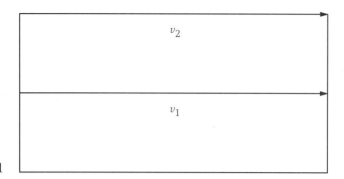

Figure 16.1

where $l_1 = \mathbf{R} \times \{0\}$, $l_2 = \mathbf{R} \times \{h\}$, $h > 0$, and $v_1 = v_2 = (2\pi, 0)$. A fundamental domain is given by $F_0 = (0, 2\pi) \times (-h, h)$ (Figure 16.1). The side-pairing transformations are illustrated in Figure 16.2.

The first side-pairing transformation is the glide G_{l_1,v_1}, and the second is the translation $G_{l_2,v_2}^{-1} \circ G_{l_1,v_1}$. The resulting quotient $\mathbf{C}/G$ is called the *Klein bottle*, denoted by K^2. We claim that K^2 is obtained by pasting two copies of the Möbius band together along their boundary circle (of perimeter 4π). Indeed, cut the fundamental domain F horizontally along the lines $\mathbf{R} \times \{h/2\}$ and $\mathbf{R} \times \{-h/2\}$ (Figure 16.3).

This cut is a topological circle on K^2 since $(0, h/2)$ is identified with $(2\pi, -h/2)$ and $(0, -h/2)$ is identified with $(2\pi, h/2)$. The middle portion gives a Möbius band. The upper and lower portions (identified by $G_{l_2,v_2}^{-1} \circ G_{l_1,v_1}$) first give a rectangle (Figure 16.4), the two vertical sides are identified again by a glide (from appropriate restrictions of G_{l_1,v_1}), and this gives another Möbius band! A somewhat more visual picture of K^2 is shown in Color Plate 4a.

Here, instead of rotating a straight segment, we rotated two halves of the lemniscate (Figure 16.5) to obtain two topological copies of the Möbius band. Then pasting is no problem! Notice that since K^2 contains a Möbius band (actually, it contains

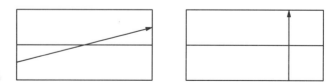

Figure 16.2

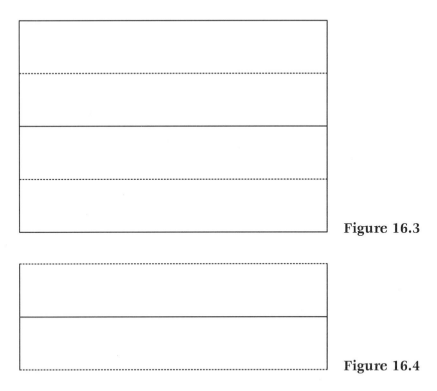

Figure 16.3

Figure 16.4

two), it is nonorientable. Notice also that K^2 can be covered by a torus with a twofold covering. In fact, the torus in question is $\mathbf{R}^2/\langle G^{-1}_{l_2,v_2} \circ G_{l_1,v_1}, G^2_{l_1,v_1} \rangle$, and the fundamental domain of the torus cover is obtained by "doubling" F in the horizontal direction.

$\diamond$ Finally, we consider discrete groups acting on the sphere S^2. Our condition that the isometries act on S^2 without fixed points imposes a severe restriction. The following theorem is essentially due to Euler:

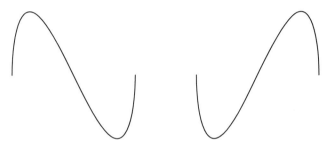

Figure 16.5

Theorem 10.

Let $S:S^2 \to S^2$ be a nontrivial isometry. If S has a fixed point on S^2, then S is the restriction of a spatial reflection or a spatial rotation. If S has no fixed point on S^2, then there exists $p_0 \in S^2$ such that $S(p_0) = -p_0$, S leaves the great circle C orthogonal to p_0 invariant, and S acts on C as a rotation.

Proof.

First note that S is the restriction of an orthogonal transformation $U : \mathbf{R}^3 \to \mathbf{R}^3$. This follows from the following argument: Let $p_1, p_2, p_3 \in S^2$ be points not on the same great circle. Then there exists an orthogonal transformation $U : \mathbf{R}^3 \to \mathbf{R}^3$ such that $U(p_l) = S(p_l)$, $l = 1, 2, 3$. (This is because the spherical triangles $\Delta p_1 p_2 p_3$ and $\Delta S(p_1)S(p_2)S(p_3)$ are congruent.) The composition $U^{-1} \circ S$ is an isometry on S^2 and fixes p_l, $l = 1, 2, 3$. As in the Euclidean case, it follows that $U^{-1} \circ S$ is the identity, so that $S = U$ on S^2.

U is represented by a 3×3 orthogonal matrix. To look for eigenvectors p and eigenvalues λ for U, we solve the equation $U(p) = \lambda \cdot p$. We know from linear algebra that λ satisfies the characteristic equation $\det(U - \lambda I) = 0$. Since U is a 3×3 matrix, this is a cubic polynomial in λ. Every cubic (in fact, odd degree) polynomial $P(\lambda)$ has a real root λ_0 (see Problem 4 of Section 8). Let $p_0 \in \mathbf{R}^3$ be an eigenvector of U corresponding to the eigenvalue λ_0. Since U preserves lengths, we have $\lambda_0 = \pm 1$, so that $U(p_0) = \pm p_0$. Let $C \subset S^2$ be the great circle perpendicular to p_0. C is the intersection of the plane $p_0^\perp$ perpendicular to p_0 and S^2. We now claim that U leaves $p_0^\perp$ invariant. This follows from orthogonality. In fact, if w is perpendicular to p_0, then $U(w)$ is perpendicular to $U(p_0) = \pm p_0$, and the claim follows. U restricted to $p_0^\perp$ is a linear plane isometry, so it must be a rotation or a reflection. By looking at the possible combinations, we see that the theorem follows. ∎

Let G be a discrete group of isometries of S^2 and assume that each nonidentity element of G has no fixed point. We consider only the simplest case in which G is cyclic and generated by a single element $g \in G$. Discreteness, along with Theorem 10, implies that g is a spatial rotation with angle $2\pi/n$, $n \neq 2$, followed by spatial reflection in the plane of the rotation (perpendicular to the axis of the rotation). It is now a simple fact that S^2/G is topologically the same for all n. (Look at a fundamental domain bounded by two meridians of longitude!) We set $n = 2$. Then g becomes the *antipodal map* $-I : S^2 \to S^2$, $-I(p) = -p$, and $G = \{\pm I\}$. The quotient $S^2/\{\pm I\}$ is called the *real projective plane* denoted by $\mathbf{R}P^2$.

In the standard model for the real projective plane, *projective points* are interpreted as lines through the origin in $\mathbf{R}^3$, and *projective lines* as planes containing the origin of $\mathbf{R}^3$. Since the origin is a multiple intersection point, it is deleted from the model. Since every two projective lines intersect, we obtain a model for *elliptic geometry*. Algebraically, we let $\sim$ be the equivalence relation on $\mathbf{R}^3 - \{0\}$ defined by $p_1 \sim p_2$ iff $p_2 = tp_1$ for some nonzero real t. The equivalence class containing $p \in \mathbf{R}^3 - \{0\}$ is the projective point that corresponds to the line $\mathbf{R} \cdot p$ passing through p and with the origin deleted. If $p = (a, b, c)$, then this projective point is classically denoted by $[a : b : c]$. We also say that a, b, c are the *homogeneous coordinates* of the projective point with the understanding that for t nonzero, ta, tb, tc are also projective coordinates of the same projective point. By definition, the set $\mathbf{R}^3 - \{0\}/ \sim$ of equivalence classes is the real projective plane $\mathbf{R}P^2$. Associating to a nonzero point the equivalence class it is contained in gives the natural projection $\mathbf{R}^3 - \{0\} \to \mathbf{R}P^2$.

Since we have no space-time here to explore the sublime beauty of projective geometry, we will understand $\mathbf{R}P^2$ in topological terms only. The idea is to replace $\mathbf{R}^3 - \{0\}$ by the unit sphere S^2 and to consider the intersections of projective points and lines with S^2. Each projective point intersects S^2 at an antipodal pair of points. Moreover, knowing this pair, one can reconstruct the projective point by considering the Euclidean line through them. The intersection of a projective line with S^2 is a great circle (which we see as a better representative of a line than a plane anyway). Every projective line is thus a topological circle.[3]

---

[3]I heard the following story from a reliable source: A desperate student asked a professor what he could do for a passing grade in geometry. "Draw a projective line," was the answer. The student took the chalk and started drawing a horizontal line. "Go on," said the professor when he got to the end of the chalkboard. So

Since every projective point corresponds to a pair of antipodal points of S^2, a topological model of the projective plane is obtained by identifying the antipodal points with each other: $\mathbf{R}P^2 = S^2/\{\pm I\}$. We also see that the identification projection $\pi : S^2 \to \mathbf{R}P^2$ is a twofold cover.

How can we visualize this? Think of S^2 as being the Earth and divide it into three parts with the Arctic and Antarctic Circles. Between these parallels of latitude lies a spherical belt that we further divide by the 0° and 180° meridians of longitude. Since the two spherical caps and the two halves of the belt are identified under $-I$, we keep only one of each (see Color Plate 4b). The longitudinal sides of the half-belt are identified under $-I$, and we obtain a Möbius band. The cap is attached to this. Thus, $\mathbf{R}P^2$ is a Möbius band and a disk pasted together along their boundaries! In particular, $\mathbf{R}P^2$ is nonorientable.

The real projective plane has another classical model, based on an extension of the Euclidean plane by adding a set of so-called *ideal points*. To describe these, consider the equivalence relation of parallelism on the set of all straight lines in $\mathbf{R}^2$. We call an equivalence class (that is, a pencil of parallel lines) an ideal point. We define $\mathbf{R}P^2$ as the union of $\mathbf{R}^2$ and the set of all ideal points. Thus, a projective point is either an ordinary point in $\mathbf{R}^2$ or an ideal point given by a pencil of parallel lines. A projective line is the union of points on a line l plus the ideal point given by the pencil of lines parallel to l. There is a single ideal line, filled by all ideal points. Incidence is defined by (set-theoretical) inclusion. How does the algebraic description fit in with the geometric description of $\mathbf{R}P^2$ as the extension of $\mathbf{R}^2$ with ideal points corresponding to pencils of parallel lines? If $c \neq 0$, then $[a : b : c]$ and $[a/c : b/c : 1]$ denote the same projective point. Hence, adjusting the notation (or setting $c = 1$), the Euclidean point $(a, b) \in \mathbf{R}^2$ can be made to correspond to the projective point $[a : b : 1] \in \mathbf{R}P^2$, and this correspondence is one-to-one with the plane of ordinary points in $\mathbf{R}P^2$. Thus, the ordinary points in $\mathbf{R}P^2$ are exactly those that have nonzero third homogeneous coordinates. For an ideal point $[a : b : 0] \in \mathbf{R}P^2$, we

---

he continued drawing on the wall, went out of the classroom, down the hallway and out to the street. By the time he got back from his roundabout tour, the professor had already marked in the passing grade.

either have $a \neq 0$ and $[a : b : 0] = [1 : m : 0]$, $m = b/a$, so that this may be thought to represent the pencil of parallel lines with common slope m, or $a = 0$ and $[0 : b : 0] = [0 : 1 : 0]$, which represents the pencil of vertical lines.

It is clear that two distinct projective lines always intersect. (In axiomatic treatment, this is called "the elliptic axiom.") Indeed, they either meet at an ordinary point or (their Euclidean restictions) are parallel, in which case they meet at the common ideal point given by these lines.

This model is the same as the topological model given by $\mathbf{R}P^2 = S^2/\{\pm I\}$. Indeed, cut a unit sphere into two hemispheres along the equator and keep only the southern hemisphere H. Let H sit on $\mathbf{R}^2$ with the south pole S touching the plane (see Figure 16.6). Apply stereographic projection from the center O of H. The points in $\mathbf{R}^2$ correspond to points of the interior of H (i.e., H without the boundary equatorial circle). Each line in $\mathbf{R}^2$ corresponds to a great semicircle on H ending at two antipodal points of the equatorial circle. These endpoints, identified by the antipodal map, give the single ideal point of the projective extension of the line. Thus, the ideal points correspond to our horizontal view, and the boundary equatorial circle (on H modulo the antipodal map) gives the ideal projective line. You are now invited to check the basic properties of this model.

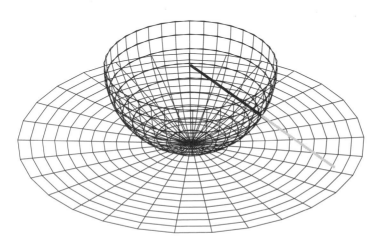

Figure 16.6

Problems

1. Fill in the details in the following argument, which gives another proof of Theorem 10.

   (a) Show that if S leaves a great circle C on S^2 invariant, then each of the two $p_0 \in S^2$ perpendicular to C satisfies $S(p_0) = \pm p_0$.

   (b) To construct C, consider the function $f : S^2 \to \mathbf{R}$ defined by $f(p) =$ the spherical distance between p and $S(p)$, $p \in S^2$. Assume that $0 < f < \pi$, since otherwise Theorem 10 follows. Let $q_0 \in S^2$ be a point where f attains its (positive) minimum. Use the spherical triangle inequality to show that the great circle C through q_0 and $S(q_0)$ is invariant under S.

2. Generalize the previous problem to show that any orthogonal transformation $U : \mathbf{R}^n \to \mathbf{R}^n$ can be diagonalized with diagonal 2×2 blocks

$$\begin{bmatrix} \cos(\theta) & -\sin(\theta) \\ \sin(\theta) & \cos(\theta) \end{bmatrix}$$

   corresponding to planar rotation with angle θ (and, for n odd, by a single 1×1 block with unit entry).

3. Define and study $Iso\,(\mathbf{R}P^2)$.

4. Make a topological model of the Klein bottle K^2 from a finite cylinder bent into a half-torus by pasting the two boundary circles together appropriately.

5. Define the *real projective n-space* $\mathbf{R}P^n$ using homogeneous coordinates. (a) Show that $\mathbf{R}P^1$ is homeomorphic to S^1. (b) Imbed $\mathbf{R}P^{n-1}$ into $\mathbf{R}P^n$ using the inclusion $\mathbf{R}^n \subset \mathbf{R}^{n+1}$ given by augmenting n-vectors with an extra zero coordinate. What is the difference $\mathbf{R}P^n - \mathbf{R}P^{n-1}$?

Web Site

1. vision.stanford.edu/~birch/projective/node3.html

17

The Five Platonic Solids

SECTION

♣ A triangle in the plane can be thought of as the intersection of three half-planes whose boundary lines are extensions of the sides of the triangle. More generally, a *convex polygon* is defined as a bounded region in the plane that is the intersection of finitely many half-planes. A convex[1] polygon has the property that for each pair p_1, p_2 of points of the polygon, the segment connecting p_1 and p_2 is entirely contained in the polygon (Figure 17.1).

The regular *n*-sided polygon is a primary example of a convex polygon. It is distinguished among all convex polygons by the fact that all its sides and angles are congruent.

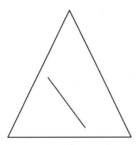 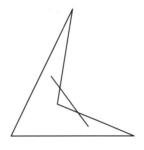

Figure 17.1

---

[1]Nonconvex objects surround us. Next time you eat breakfast, take a closer look at your croissant or bagel.

In a similar vein, in space, a *convex polyhedron K* is defined as a bounded region in $\mathbf{R}^3$ that is the intersection of finitely many half-spaces. The notion of convexity carries over to three (and in fact any) dimensions. The part of the boundary plane of the half-space participating in the intersection that is common with the polyhedron is called a *face* of K. Any common side of two faces is an *edge*. The endpoints of the edges are the *vertices* of K. Regularity of convex polyhedra, although simple in appearance, is not so easy to define. It is clear that the faces of a regular polyhedron must be made up of regular polygons all congruent to each other. That this is not enough for regularity is clear when one considers a double pentagonal pyramid called *Bimbo's lozenge* (Figure 17.2). At two vertices five equilateral triangles meet, while only four meet at the remaining five vertices.

To define regularity, we turn to group theory. First we define *Iso* ($\mathbf{R}^3$), the group of isometries of $\mathbf{R}^3$ (in much the same way as we did for $\mathbf{R}^2$); a transformation $S : \mathbf{R}^3 \to \mathbf{R}^3$ belongs to *Iso* ($\mathbf{R}^3$) if S preserves spatial distances:

$$d(S(p), S(q)) = d(p, q), \quad p, q \in \mathbf{R}^3,$$

where $d : \mathbf{R}^3 \times \mathbf{R}^3 \to \mathbf{R}$ is the Euclidean distance function:

$$d(p, q) = |p - q|, \quad p, q \in \mathbf{R}^3.$$

As in two dimensions, we can easily classify all spatial isometries. Those that have a fixed point have been described in Theorem 10 (see Section 16). In fact, this result can be rephrased by saying that any element of *Iso* ($\mathbf{R}^3$) that leaves a point fixed is either a spatial reflection or a *rotatory reflection*, a rotation followed by re-

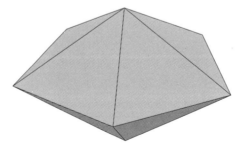

Figure 17.2

flection in the plane perpendicular to the rotation axis. (What is a rotatory half-turn?) Notice that Theorem 10 is actually stated in a bit stronger setting; for the conclusions all we need is the restriction of the spatial isometry S to the unit sphere $S^2(p_0)$ with center p_0, a fixed point of S. As in the plane, a spatial isometry S with no fixed points is the composition of a spatial translation T_v and another spatial isometry U that leaves the origin fixed; $S = T_v \circ U$, where $v = S(0)$. (U is linear, but we do not need this additional fact here.) Since we are in space, the possible outcomes of the composition of T_v and U depend on how the translation vector v relates to the reflection plane or the rotation axis for U; think of the motion of the frisbee or uncorking a bottle! If U is a spatial reflection, then S is a glide reflection, a reflection in a plane followed by a translation with vector parallel to the reflection plane. (Indeed, this follows by writing $v = v_1 + v_2$ and $T_v = T_{v_1} \circ T_{v_2}$, where v_1 is parallel and v_2 perpendicular to the reflection plane. Notice the presence of the two-dimensional statement that any plane reflection in a line followed by a translation with translation vector perpendicular to the reflection axis is another reflection; see Section 9.) If U is a rotation, then $S = T_v \circ U$ is a *screw displacement*, a rotation followed by a translation along the rotation axis. (Once again this follows by decomposing $v = v_1 + v_2$ as above, and using the two-dimensional statement, any plane rotation followed by a translation is another rotation; see Section 9.) Finally, piecing these arguments together to a single proof we obtain that if U is a rotatory reflection, then so is S. (A spatial reflection and a translation commute if the translation vector is parallel to the reflection plane. Now distribute T_{v_1} and T_{v_2} to the rotation and reflection part of U.) Summarizing, every spatial isometry is either a rotation, a reflection, a rotatory reflection, a translation, a glide reflection, or a screw displacement. As a byproduct we obtain that any spatial isometry is the composition of at most four spatial reflections. (For example, a rotatory half-turn, the negative of the identity, is the composition of three reflections in mutually perpendicular planes!)

We want to consider a convex polyhedron K regular if we can carry each vertex of K to another vertex by a suitable spatial isometry in $Symm(K)$, the group of spatial isometries that leave K

invariant. In two dimensions, regularity of a polygon was certainly equivalent to this; think of the cyclic group of rotations leaving the regular n-sided polygon invariant (see Section 9). Similar symmetry conditions should be required for the edges and faces of K. Between vertices, edges, and faces there are incidence relations that essentially define K.

To incorporate all these into a single symmetry condition, we introduce the concept of a *flag* in K as a triple (p, e, f) where p is a vertex, e is an edge, f is a face of K, and $p \in e \subset f$. We now say that K is *regular* if given any two flags (p_1, e_1, f_1) and (p_2, e_2, f_2) of K, there exists a spatial isometry $S \in Symm(K)$ that carries (p_1, e_1, f_1) to (p_2, e_2, f_2); that is, $S(p_1) = p_2$, $S(e_1) = e_2$, and $S(f_1) = f_2$.

Let K be a convex polyhedron with vertices $p_1, \ldots, p_n$. Define the *centroid* of K as

$$c = \frac{p_1 + \cdots + p_n}{n}.$$

We claim that every spatial isometry in $Symm(K)$ fixes c. To do this,[2] we consider the function $f : \mathbf{R}^3 \rightarrow \mathbf{R}$ defined by $f(p) = \sum_{j=1}^{n} d(p, p_j)^2$, $p \in \mathbf{R}^3$. Let $S \in Symm(K)$. Since S permutes the vertices of K, $\{S(p_1), \ldots, S(p_n)\} = \{p_1, \ldots, p_n\}$. It follows that $f(S(p)) = f(p)$, $p \in \mathbf{R}^3$. Indeed, we compute

$$f(S(p)) = \sum_{j=1}^{n} d(S(p), p_j)^2 = \sum_{j=1}^{n} d(p, S^{-1}(p_j))^2$$

$$= \sum_{j=1}^{n} d(p, p_j)^2 = f(p).$$

To show that $S(c) = c$, we now notice that f has a global minimum

---

[2]Beautiful geometric arguments exist to prove this claim; see H.S.M. Coxeter, *An Introduction to Geometry*, Wiley, 1969. For a change, we give here an analytic proof.

at c. This follows by completing the square:[3]

$$f(p) = \sum_{j=1}^{n} d(p, p_j)^2 = \sum_{j=1}^{n} |p - p_j|^2$$

$$= \sum_{j=1}^{n} (|p|^2 - 2p \cdot p_j + |p_j|^2)$$

$$= n|p|^2 - 2np \cdot c + \sum_{j=1}^{n} |p_j|^2$$

$$= n|p - c|^2 - n|c|^2 + \sum_{j=1}^{n} |p_j|^2,$$

where we used the definition of c and the dot product. Since the last two terms do not depend on p, it is clear that $f(p)$ attains its global minimum where $|p - c|^2$ does—that is, at $p = c$. The composition $f \circ S^{-1}$ takes its global minimum at $S(c)$. But $f \circ S^{-1} = f$, so these minima must coincide. $S(c) = c$ follows.

If K is regular, its centroid has the same distance from each vertex. This is because regularity ensures that every pair of vertices can be carried into one another by a spatial isometry in $Symm\,(K)$ that fixes c. The same is true for the edges and the faces of K. In particular, a sphere that contains all the vertices can be circumscribed around K. Notice that by projecting K from the centroid to the circumscribed sphere, we obtain *spherical tessellations* (see Problem 10).

How many regular polyhedra are there? Going back to the plane is misleading; there we have infinitely many regular polygons, one for each positive integer $n \geq 3$, where n is the number of edges or vertices. As we will show in a moment, there are only five regular polyhedra in $\mathbf{R}^3$. They are called the five *Platonic solids* (Figure 17.3), since Plato gave them a prominent place in his theory of ideas.

---

[3]Did you notice how useful completing the square was? We used this to derive the quadratic formula, to find the center of a hyperbolic circle, to integrate rational functions, etc.

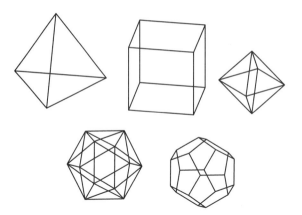

Figure 17.3

Remark.

The two halves of Bimbo's lozenge are hiding in the icosahedron with a "belt" of ten equilateral triangles separating them! The belt configuration (closed up with two regular pentagons) was called by Kepler a *pentagonal antiprism*. In a similar vein, an octahedron can be thought of as a *triangular antiprism*. (The pentagonal antiprism appeared about 100 years earlier as *octaedron elevatum* in Fra Luca Pacioli's *Da Divina Proportione*, printed in 1509. This classic is famous for its elaborate drawings of models made by Leonardo da Vinci.) Following the "icosahedral recipe," we can also insert a *square antiprism* between two square pyramids (two halves of an octahedron) and obtain a nonregular polyhedron with sixteen equilateral triangular faces (see Figure 17.4).

Figure 17.4

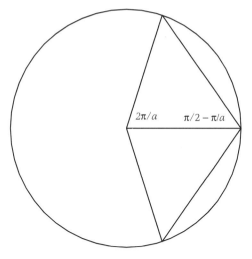

Figure 17.5

We describe a regular polyhedron by the so-called *Schläfli symbol* $\{a, b\}$, where a is the number of sides of a face and b is the number of faces meeting at a vertex. Each face is a regular a-sided polygon. The angle between two sides meeting at a vertex is therefore $\pi - 2\pi/a$ (Figure 17.5). (Split the polygon into a isosceles triangles by connecting the vertices of the polygon to the centroid.)

At a vertex of the polyhedron, b faces meet. By convexity, the sum of angles just computed for the b faces must be $< 2\pi$. We obtain

$$b(\pi - 2\pi/a) < 2\pi.$$

Dividing by π, we obtain

$$b(1 - 2/a) < 2,$$

or equivalently,

$$(a - 2)(b - 2) < 4.$$

On the other hand, $a, b > 2$ by definition. A case-by-case check gives all possibilities for the Schläfli symbol:

$$\{3, 3\}, \quad \{3, 4\}, \quad \{4, 3\}, \quad \{3, 5\}, \quad \{5, 3\}.$$

We could now go on and describe these solutions geometrically. Instead, for the moment, we take a brief look at the examples[4] (but not the last column!) in Figure 17.6.

Following Euler, we refine the somewhat crude argument above to obtain the number of possible faces, edges, and vertices of each Platonic solid. We start with an arbitrary (not necessarily regular) convex polyhedron K.

Euler's Theorem for Convex Polyhedra.[5]

Let

$$V = number\ of\ vertices\ of\ K;$$

$$E = number\ of\ edges\ of\ K;$$

$$F = number\ of\ faces\ of\ K.$$

Then we have

$$V - E + F = 2.$$

Proof.

We first associate to a convex polyhedron K a planar graph called the *Schlegel diagram* of K. To do this we use stereographic projection. Adjust K in space so that the top face is horizontal (that is, parallel to the coordinate plane spanned by the first two axes). Sit in the middle of the top face and look down to the transparent polyhedron K. The perspective image of the edges (the wireframe) gives a graph on the horizontal coordinate plane and defines a planar graph (with nonintersecting edges) if you are not too tall. Each face of K will correspond to a polygonal region in the plane bounded by the edges of the graph. We let the face you are sitting on correspond to the unbounded region that surrounds the graph. We can now analyze the last column of Schlegel diagrams of the five Platonic solids in Figure 17.6.

Under stereographic projection, edges and vertices of K correspond to edges and vertices of the Schlegel diagram, so we begin to

---

[4]See H.S.M. Coxeter, *Introduction to Geometry*, Copyright 1969 by John Wiley & Sons, Inc. Reprinted by permission of John Wiley & Sons, Inc.

[5]This was known to Descartes, and according to widespread belief, to Archimedes as well.

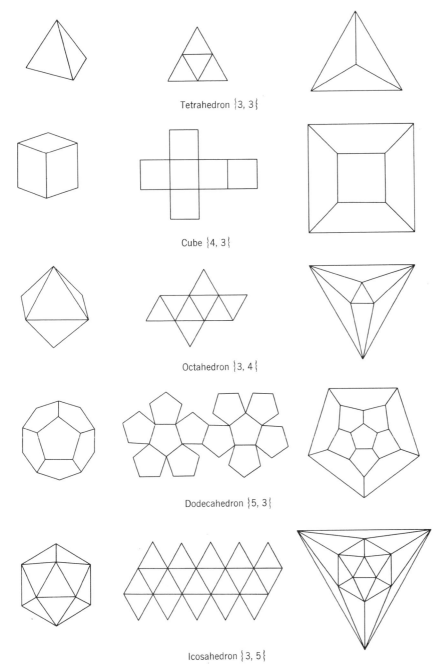

Tetrahedron {3, 3}

Cube {4, 3}

Octahedron {3, 4}

Dodecahedron {5, 3}

Icosahedron {3, 5}

Figure 17.6

suspect that Euler's theorem must be valid in general for connected planar graphs, where any graph is defined by a finite number of points of the plane (called vertices) and a finite number of non-intersecting segments (called edges) connecting the vertices. The graph is further assumed to be connected, and it decomposes the plane into nonoverlapping regions.

We now claim that Euler's theorem is true for all connected planar graphs consisting of at least one vertex. To show this, we build a planar graph step by step, starting from a single vertex graph. For this trivial case the alternating sum $V - E + F$ is 2, since $V = 1$, $E = 0$, and $F = 1$. To build the graph, at each step we apply one of the following operations:

1. A new edge is added that joins an old vertex and a new vertex;
2. A new edge is added that joins two old vertices.

In each case, we have the following changes:

1. $V + 1 \mapsto V, E + 1 \mapsto E, F \mapsto F$;
2. $V \mapsto V, E + 1 \mapsto E, F + 1 \mapsto F$.

The alternating sum remains unchanged; Euler's theorem follows!   ■

We now return to our regular polyhedra with Schläfli symbol $\{a, b\}$ and use the additional information we just gained:

$$V - E + F = 2.$$

The numbers in $\{a, b\}$ relate to V, E and F by

$$bV = 2E = aF.$$

Indeed, if we count the b edges at each vertex, we counted each edge twice. Similarly, if we count the a sides of each face, we again counted each edge twice. Combining these, we easily arrive at the following:

$$V = \frac{4a}{2a + 2b - ab}, \quad E = \frac{2ab}{2a + 2b - ab}, \quad F = \frac{4b}{2a + 2b - ab}.$$

We now look at Figure 17.6 again and see that the numbers of vertices, edges, and faces determined by these give[6] the tetrahedron, cube, octahedron, dodecahedron, and icosahedron—the five Platonic solids!

In ancient times, the existence of only five regular polyhedra called for much mysticism. In Plato's *Timaeus*, the four basic elements—air, earth, fire, and water—were mysteriously connected to the octahedron, cube, tetrahedron, and icosahedron (in this order). To the dodecahedron was associated the entire Universe. The latter is probably due to Timaeus of Locri, one of the earliest Pythagoreans. The twelve faces of the dodecahedron were believed to correspond to the twelve signs of the Zodiac. Figure 17.7 is adapted from a drawing by Kepler.

Unlike the tetrahedron, the cube and the octahedron are common, basic structures for many crystals (such as sodium sulfantimoniate, common salt, and chrome alum). The occurrence of dodecahedral and icosahedral structures are rare in nonliving nature. However, these do occur in living creatures; for example, they are found in the skeletons of some microscopic sea animals called radiolaria.[7] Moreover, a number of viruses such as the adenovirus (which causes the flu and a host of other illnesses) have icosahedral structure.

Let us now go back to mathematics and take a closer look at the Platonic solids. We have demonstrated above that there are only five regular polyhedra in space, but their actual existence was largely taken for granted. Of course, nobody doubts the existence of the regular tetrahedron, much less the cube, but how the faces of the dodecahedron, the icosahedron, and, to a lesser extent, the octahedron piece together remains to be seen. To reduce the number of cases, we now make some preparations and introduce the concept of reciprocal for regular polyhedra.

Let $\{a, b\}$, $a, b \geq 3$, be the Schläfli symbol of a Platonic solid P whose existence we now assume. Let p be a vertex of P. At p,

---

[6]It is perhaps appropriate to recall some of the Greek number prefixes here: 2 = di, 3 = tri, 4 = tetra, 5 = penta, 6 = hexa, 7 = hepta, 8 = octa, 9 = ennia, 10 = deca, 12 = dodeca, 20 = icosa. The cube does not fit in; how would you rename it?

[7]See H. Weyl, *Symmetry*, Princeton University Press, 1952.

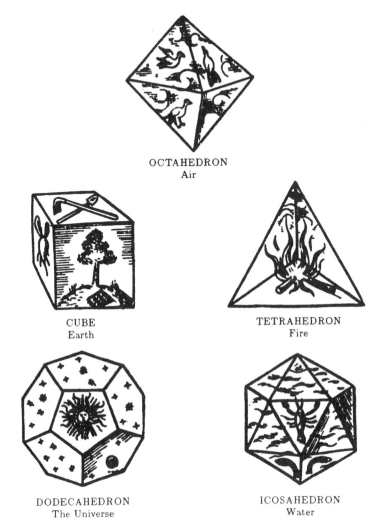

Figure 17.7.
M. Berger, *Geometry II*, 1980, 32. Reprinted by permission of Springer-Verlag New York, Inc.

exactly b edges and faces meet. We denote the edges by $e_1, \ldots, e_b$, and the faces by $f_1, \ldots, f_b$ in such a way that $e_1, e_2 \subset f_1, e_2, e_3 \subset f_2, \ldots, e_b, e_1 \subset f_b$ (Figure 17.8).

Let $S \in Symm(P)$ be an isometry that carries the flag (p, e_1, f_1) into (p, e_2, f_2). (This is the first time when regularity comes in with full force!) Like every element of $Symm(P)$, S fixes the centroid c of P, and thus it must fix the line l through c and p. (Notice that c and p must be distinct. Why?) $S(e_2) = e_3$ since $S(e_2)$ is an edge

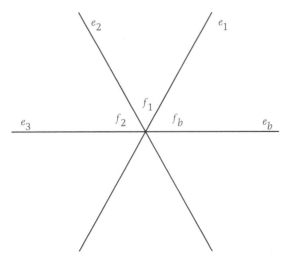

Figure 17.8

of $f_2 = S(f_1)$ other than e_2. Thus $S^2(e_1) = S(e_2) = e_3$. Iterating, we obtain $S^j(e_1) = e_{j+1}, j = 1, \ldots, b - 1$, and $S^b(e_1) = e_1$. Thus S is a rotation of order b and axis l. It is now easy to see that the cyclic group $\langle S \rangle$ generated by S is precisely the subgroup of *Symm* (P) of direct isometries that leaves p fixed.

Armed with this description of *Symm* (P) at a vertex, we are now ready to define the reciprocal of P. Let P have Schläfli symbol $\{a, b\}$. Consider a vertex p of P and denote by $e_1, \ldots, e_b$ the edges of P that meet at p. Let $m_1, \ldots, m_b$ be the midpoints of $e_1, \ldots, e_b$. Since the endpoints of $e_1, \ldots, e_b$ other than p are in the orbit of $\langle S \rangle$, so are the midpoints. Thus, $m_1, \ldots, m_b$ are the vertices of a regular b-sided polygon, and this polygon is in a plane perpendicular to the axis of rotation l of S (Figure 17.9). The polygon with vertices $m_1, \ldots, m_b$ is called the *vertex figure* of p at P. There is a vertex figure for each vertex of P, of which we have V in number. The V planes of the vertex figures enclose a polyhedron P^0 that is called the *reciprocal* of P.

What is the Schläfli symbol of P^0? Looking at the local picture of two adjacent vertices of P, we see that the edges of P^0 bisect the edges of P at right angles. Of these bisecting edges (of which we have E in number), those that bisect the a sides of a face of P all go through a vertex of P^0. Thus, exactly a edges meet at a vertex of P^0. Finally, those edges of P^0 that bisect the b edges at a vertex of

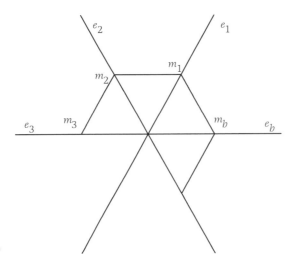

Figure 17.9

P are the edges of a face of P^0. Thus each face of P^0 is a regular b-sided polygon. We obtain that the Schläfli symbol of P^0 is $\{b, a\}$. We can now study Color Plates 5a–c, which depict the three reciprocal pairs.

Notice that the reciprocal of a tetrahedron is another tetrahedron. The reciprocal pair of tetrahedra—the *stella octangula*, as Kepler called it—is the simplest example of a *compound polyhedron*, and it occurs in nature as a crystal-twin of tetrahedrite. Historically, the first complete understanding of the relationship between a polyhedron and its reciprocal is attributed to Maurolycus (1494–1575), although inscribing various regular solids into each other appears in Book XV of the *Elements*. (Note that Book XIV was written by Hypsicles and, as the language and style suggest, Book XV is attributed to several authors.)

Reciprocity is a powerful tool in our hands, and we will use it in a variety of ways. First of all, if a regular polyhedron exists, then so does its reciprocal. Thus, the obvious existence of the cube, with Schläfli symbol $\{4, 3\}$, implies the existence of its reciprocal, the octahedron. Moreover, since the dodecahedron and icosahedron are reciprocal, it is enough to show that one of them exists! Since the existence of a tetrahedron is quite clear, our task is now reduced to showing that the dodecahedron exists.

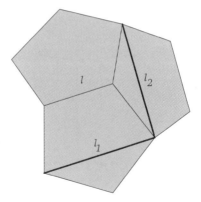

Figure 17.10

The dodecahedron has Schläfli symbol {5, 3}, and this shows that exactly three regular pentagons meet at a vertex. This configuration exists (Figure 17.10), since between two adjacent edges of a regular pentagon the angle is $\pi - 2\pi/5 = 3\pi/5$, and three of these add up to $9\pi/5 < 2\pi$.

We claim that the lines l_1, l_2 shown in Figure 17.10 are perpendicular. Indeed, l_1 is parallel to the side l, and l is certainly perpendicular to l_2. Thus we are able to pick perpendicular diagonal lines in adjacent pentagonal faces. The dodecahedron has exactly twelve faces, so we can pick twelve lines with orthogonality conditions among them. But twelve is exactly the number of edges of a cube! It is impossible to resist the temptation to pick these lines to form the edges of a cube. Figure 17.11 shows the configuration. Notice that we have not proved the existence of the dodecahedron, but have found a cube on which we want to build it. Thus, we start with a cube, pick a vertex, and arrange the three pentagonal faces to meet at this vertex (Figure 17.12). How to attach the remaining pentagonal faces? As usual, group theory helps us out. In fact, we now apply to the installed three pentagonal faces the symmetries of the cube that are spatial half-turns around the symmetry axes that go through the centroids of pairs of opposite faces (see Figure 17.13). Thus the pentagons fit together, and we created the dodecahedron! (For a different proof of the existence of the dodecahedron, cf. Problem 20.) So now we have shown that all five Platonic solids exist.

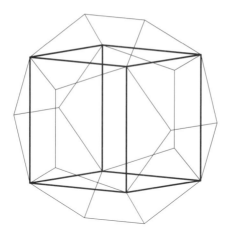

Figure 17.11

Pursuing the analogy with regular polygons, we next work out the symmetry groups of these regular polyhedra. Reciprocity reduces the cases to consider to three, since reciprocal polyhedra have the same symmetry group. This follows immediately if we recall how the reciprocal P^0 of P was constructed. (Assume that $S \in Symm(P)$, construct P^0, and realize that S carries P^0 to itself because P^0 is determined by P using data such as midpoints of edges, etc. that remain invariant under S.)

♡ Before we actually determine these three groups explicitly, we would like to see what the possibilities are for any such group. As noted above, the symmetry group of any polyhedron (regular or not) is finite, so we now take up the more ambitious task of

Figure 17.12

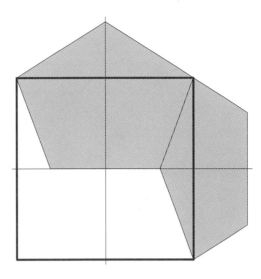

Figure 17.13

classifying all finite subgroups of *Iso* $(\mathbf{R}^3)$ (cf. Theorem 4 of Section 9 for the 2-dimensional case).

Let $G \subset$ *Iso* $(\mathbf{R}^3)$ be a finite group. First we observe the existence of a point $p_0 \in \mathbf{R}^3$ that is left fixed by every element of G. Indeed, let $p \in \mathbf{R}^3$ be arbitrary and consider the orbit

$$G(p) = \{S(p) \mid S \in G\}.$$

This is a finite set

$$G(p) = \{p_1, \ldots, p_n\}$$

since G is finite. Since the elements of G permute the p_j's, $j = 1, \ldots, n$, we can repeat the argument that was used to find the centroid of a polyhedron and conclude that

$$p_0 = \frac{p_1 + \cdots + p_n}{n}$$

is left fixed by all elements of G. It is quite remarkable that the "centroid argument" for polyhedra applies to this more general situation.

We now assume that G consists of direct isometries only. Every nontrivial element R of G is a rotation (see Theorem 10 of Section 16) around an axis l that must go through p_0, since R fixes p_0. The axis cuts the unit sphere $S^2(p_0)$ around p_0 into a pair of antipodal points. We call these points *poles* (Figure 17.14).

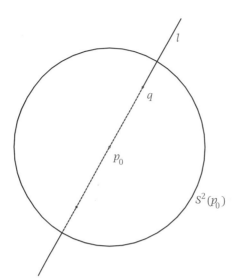

Figure 17.14

There are two poles on $S^2(p_0)$ for each rotation in G. Any rotation R in G must have finite order, since G is finite. We now let q be a pole and R be the smallest positive angle rotation with pole q. We call the order of R the *degree* of the pole q. Thus, q has degree $d \in \mathbf{N}$ if $R^d = I$, and d is the smallest positive integer with this property.

Let Q denote the (finite) set of all poles on $S^2(p_0)$. We claim that G leaves Q invariant. G certainly leaves $S^2(p_0)$ invariant, since it fixes p_0. Let $q \in Q$ as above and assume that q corresponds to the rotation $R \in G$. Let $S \in G$ be any rotation. Then $S \circ R \circ S^{-1} \in G$ is a rotation that fixes $S(q)$ so that it must be a pole. Invariance of Q under G follows. Notice that the poles q and $S(q)$ have the same degree, since $R^d = I$ iff $(S \circ R \circ S^{-1})^d = I$.

The classification of possible cases for G now depends on the successful enumeration of the elements in Q. To do this, we introduce an equivalence relation $\sim$ on Q such that $q_1 \sim q_2$, $q_1, q_2 \in Q$, if $q_2 = S(q_1)$ for some $S \in G$. That this is an equivalence follows from the fact that G is a group. The equivalence classes are actually the orbits of G on Q, and they split Q into mutually disjoint subsets. As noted above, poles in the same equivalence class have the same degree. Let $C \subset Q$ be an equivalence class and $d = d_C$

the common degree of the poles in C. We claim that

$$|C| = \frac{|G|}{d_C}.$$

To show this, let $q \in C$, and let $R \in G$ be the rotation that corresponds to the pole q as above. By assumption, R has degree d, so that the cyclic subgroup $\langle R \rangle$ consists of the distinct elements

$$I, R, R^2, \ldots, R^{d-1} \qquad (R^d = I).$$

Let $p \in S^2(p_0)$ be any point not in Q. Apply the elements in $\langle R \rangle$ to p to obtain a regular d-sided polygon with vertices

$$p, R(p), R^2(p), \ldots, R^{d-1}(p).$$

Other rotations in G transform this polygon into congruent polygons around the poles in C. We can choose p so close to q that all the transformed polygons are disjoint (Figure 17.15). The number of vertices of all these polygons is $d|C|$. On the other hand, this set of vertices is nothing but the orbit

$$G(p) = \{S(p) \mid S \in G\}.$$

We obtain that $|G(p)| = d|C|$. Finally, note that no element in G (other than the identity) fixes p, since p is not in Q. Thus, $|G(p)| = |G|$, and the claim follows.

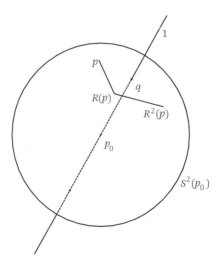

Figure 17.15

Next, we count how many nontrivial rotations have poles in C. Each axis, giving two poles in C, is the axis of $d - 1$ nontrivial rotations. Thus, the number of nontrivial rotations with poles in C is

$$\left(\frac{1}{2}\right) \frac{(d_C - 1)|G|}{d_C}, \quad d = d_C,$$

where the $(1/2)$ factor is because each rotation axis gives two poles. Thus, the total number of nontrivial rotations in G is

$$|G| - 1 = \frac{|G|}{2} \sum_C \frac{(d_C - 1)}{d_C},$$

where the summation runs through the equivalence classes of poles. Rearranging, we find

$$2 - \frac{2}{|G|} = \sum_C \left(1 - \frac{1}{d_C}\right).$$

We now see how restrictive this crucial equality is. We may assume that G consists of at least two elements. Since

$$1 \leq 2 - \frac{2}{|G|} < 2,$$

the number of equivalence classes in the summation above can only be 2 or 3. ($\neg$ If we had at least four terms $1 - 1/d_C$, they would add up to a sum $\geq 4(1 - 1/2) = 2$. $\neg$)

Assume first that we have two equivalence classes, C_1 and C_2, with $d_{C_1} = d_1$ and $d_{C_2} = d_2$. We have

$$2 - \frac{2}{|G|} = \left(1 - \frac{1}{d_1}\right) + \left(1 - \frac{1}{d_2}\right)$$

or, equivalently,

$$\frac{|G|}{d_1} + \frac{|G|}{d_2} = 2.$$

On the left-hand side the terms are positive integers, since $|C_1| = |G|/d_1$ and $|C_2| = |G|/d_2$ as proved above. Thus both terms must be equal to one. We obtain

$$d_1 = d_2 = |G|.$$

This means that G is cyclic and consists of rotations around a single axis that cuts $S_1(p_0)$ at an antipodal pair of poles.

To get something less trivial, we assume now that there are three equivalence classes:

$$2 - \frac{2}{|G|} = \left(1 - \frac{1}{d_1}\right) + \left(1 - \frac{1}{d_2}\right) + \left(1 - \frac{1}{d_3}\right)$$

(with obvious notation). We rewrite this as

$$\frac{1}{d_1} + \frac{1}{d_2} + \frac{1}{d_3} = 1 + \frac{2}{|G|}.$$

Since $1/3 + 1/3 + 1/3 = 1 < 1 + 2/|G|$, there must be at least one degree that is equal to 2. We may assume that it is d_3. Setting $d_3 = 2$, the equality above reduces to

$$\frac{1}{d_1} + \frac{1}{d_2} = \frac{1}{2} + \frac{2}{|G|}.$$

A little algebra now shows that

$$(d_1 - 2)(d_2 - 2) = 4\left(1 - \frac{d_1 d_2}{|G|}\right) < 4.$$

We obtain the same restriction as for regular polyhedra! Setting, for convenience, $d_1 \leq d_2$, we summarize the possible values of d_1, d_2, d_3, and $|G|$ in the following table:

d_1	2	3	3	3
d_2	n	3	4	5
d_3	2	2	2	2
$|G|$	$2n$	12	24	60

The first numerical column $d_1 = 2$, $d_2 = n$, $d_3 = 2$ and $|G| = 2n$ corresponds to the dihedral group D_n discussed in Section 9. There is, however, a little geometric trouble here. Recall that D_n is the symmetry group of a regular n-sided polygon and that this group includes not only rotations but reflections as well. Our symmetry

group here consists of direct isometries only. This virtual contradiction is easy to resolve. In fact, a spatial half-turn is a direct spatial
isometry, yet its restriction to a plane through its axis gives a reflection! Thus, increasing the dimension by one enables us to represent
our planar opposite isometries by spatial direct isometries. We can
now give a "geometric representation" of D_n as follows: Consider
the regular n-sided polygon $P_n \subset \mathbf{R}^2$ inscribed in the unit circle S^1
as in Section 5. Take the Cartesian product of P_n with the interval
$[-h/2, h/2] \subset \mathbf{R}$, $h > 0$. We obtain what is called a regular *prism*
$P_n \times [-h/2, h/2] \subset \mathbf{R}^2 \times \mathbf{R} = \mathbf{R}^3$ (of height h).

We now claim that $Symm^+(P_n \times [-h/2, h/2])$, the group of direct
spatial isometries of the prism, is the dihedral group D_n. (For $n = 4$,
we assume that $h \neq \sqrt{2}$ so that $P_n \times [-h/2, h/2]$ is not a cube.) Since
the centroid (the origin) of the prism must stay fixed, it is clear that
every element $S \in Symm(P_n \times [-h/2, h/2])$ leaves the middle slice
$P_n \times \{0\}$ invariant. Thus, S, restricted to the coordinate plane $\mathbf{R}^2$
spanned by the first two axes, is an element of $Symm(P_n)$.

Recall from Section 9 that the elements of $Symm(P_n)$ are reflections to the symmetry axes or rotations by angles $2k\pi/n$, $k =
0, \ldots, n - 1$. If S restricts to a planar reflection to a symmetry
axis of P_n, then S is a spatial half-turn around the same axis. If S
restricts to a planar rotation around the origin with angle $2k\pi/n$,
$k = 0, \ldots, n - 1$, then S is a spatial rotation around the vertical third coordinate axis with the same angle. Altogether, we have
$n + 1$ rotation axes, n half-turn axes in $\mathbf{R}^2$, and the third coordinate
axis (for rotations with angles $2k\pi/n$, $k = 0, \ldots, n - 1$). These give
$2(n + 1) = 2n + 2$ poles on the unit sphere S^2. The set of poles splits
into three equivalence classes. Each half-turn switches the North
and South Poles, and these form a two-element equivalence class.
On the plane of P_n, there are two intertwining equivalence classes,
with n elements in each class. Summarizing, we showed that the
dihedral group D_n of spatial rotations is the symmetry group of a
regular prism with base P_n. ♠ Conversely, if G acts on $\mathbf{R}^3$ with direct isometries and orbit structure described by the first numerical
column of the table above, then G is conjugate to D_n in $Iso(\mathbf{R}^3)$.
Indeed, G contains a degree-n rotation whose powers form a cyclic
subgroup of order n in G. The rest of G is made up by n half-turns.
The axis of the degree-n rotation is perpendicular to the axes of

the half-turns. This is because the two poles that correspond to the axis of the degree-n rotation form a single orbit, and these two poles must be interchanged by each half-turn. The $2n$ poles corresponding to the n half-turns are divided into two orbits consisting of n poles each. The degree-n rotation maps this set of poles into itself. The only way this is possible is that the angle between adjacent axes of the n half-turns is π/n. By a spatial isometry, the configuration of all the axes can be brought to that of the prism above. The same isometry conjugates G into D_n. $\heartsuit$ This completely describes the first numerical column in the table above.

The trivial case of a cyclic G with two equivalence classes, discussed above, can be geometrically represented as the symmetry group of a *regular pyramid* with base P_n.

We could go on and do the same analysis for the remaining columns of the table. Each case corresponds to a single group whose generators and defining relations can be written down explicitly. We follow here a more geometric path. We go back to our five Platonic solids and work out the three symmetry groups that arise. We will then realize (repeating the counting argument above for nearby poles p and q) that they must correspond to the last three columns of the table.

We start with the symmetry group of the regular tetrahedron $\mathcal{T}$. Looking at Figure 17.16, we see that the only spatial reflections that leave $\mathcal{T}$ invariant are those in planes that join an edge to the midpoint of the opposite edge. There are exactly six of these planes,

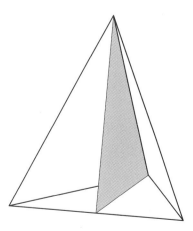

Figure 17.16

each corresponding to an edge of T. $Symm\,(T)$ thus contains six reflections. Let 1, 2, 3, 4 denote the vertices of T. Each element $S \in Symm\,(T)$ permutes these vertices; that is, S gives rise to a permutation

$$\begin{pmatrix} 1 & 2 & 3 & 4 \\ S(1) & S(2) & S(3) & S(4) \end{pmatrix}.$$

Conversely, S is uniquely determined by this permutation. (Indeed, if S_1 and S_2 in $Symm\,(T)$ give the same permutation on the vertices, then $S_2^{-1} \circ S_1$ fixes all the vertices. A spatial isometry that fixes four noncoplanar points is the identity. Thus $S_2^{-1} \circ S_1 = I$, and $S_1 = S_2$ follows.) Each reflection gives a transposition (a permutation that switches two numbers and keeps the rest of the numbers fixed). In fact, the two vertices that are on the plane of the reflection stay fixed, and the other two get switched. The six reflections thus give six transpositions. The symmetric group[8] $\mathcal{S}_4$ on four letters, consisting of all permutations of $\{1, 2, 3, 4\}$, has exactly six transpositions, so we got them all! It is a standard fact (easily verified in our case) that the transpositions generate the symmetry group. Thus, $Symm\,(T) \cong \mathcal{S}_4$.

The direct isometries in $Symm\,(T)$ are compositions of even number of reflections. They correspond to permutations that can be written as products of even number of transpositions. These are called *even* permutations, and they form a subgroup $\mathcal{A}_4 \subset \mathcal{S}_4$ called the *alternating group* on four letters. We obtain that the group of direct isometries $Symm^+(T)$ is isomorphic with $\mathcal{A}_4$. Since $|\mathcal{S}_4| = 4! = 24$, we have $|\mathcal{A}_4| = |\mathcal{S}_4|/2 = 12$. Looking at our table, we see that we recovered the second numerical column. The group $Symm^+(T) \cong \mathcal{A}_4$ is called the *tetrahedral group*. Now explore some rotations in $Symm^+(T)$ as shown in Figures 17.17 to 17.19.

Remark.
While the tetrahedral group consists only of rotations, an opposite spatial symmetry of T is not necessarily a reflection. For example,

---

[8]See "Groups" in Appendix B.

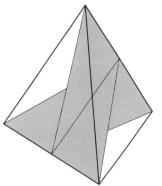

Figure 17.17

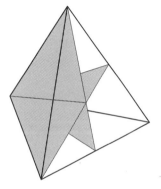

Figure 17.18

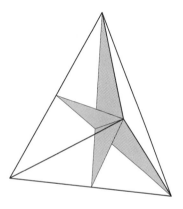

Figure 17.19

the cycle

$$\begin{pmatrix} 1 & 2 & 3 & 4 \\ 2 & 3 & 4 & 1 \end{pmatrix}$$

corresponds to a *rotatory reflection* (a rotation followed by a reflection).

Next, we have a choice between a cube and an octahedron, since they are reciprocal to each other. Let us choose the latter. We think of the regular octahedron $\mathcal{O}$ in the following way: First take a regular tetrahedron $\mathcal{T}$. At each of the four vertices, we take the vertex figure, which in this case is a triangle whose vertices are the midpoints of the three edges that meet at the given vertex of $\mathcal{T}$. We now slice off the four tetrahedra containing the vertices of $\mathcal{T}$ along the vertex figures. What is left after this truncation is a regular octahedron $\mathcal{O}$ (see Figure 17.20). You may already have noticed in Color Plate 5a that $\mathcal{O}$ is actually the intersection of $\mathcal{T}$ and its reciprocal $\mathcal{T}^0$.

Symmetries of $\mathcal{T}$ automatically become symmetries of $\mathcal{O}$:

$$Symm\,(\mathcal{T}) \subset Symm\,(\mathcal{O}).$$

There are exactly four faces of the octahedron $\mathcal{O}$ that are contained in those of $\mathcal{T}$. These four faces are nonadjacent and meet only at vertices, and any member determines the group uniquely. The

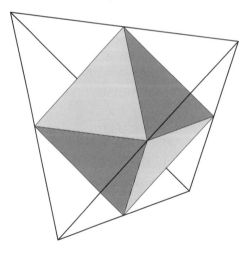

Figure 17.20

other four faces of $\mathcal{O}$ form the same configuration relative to the reciprocal tetrahedron $\mathcal{T}^0$.

Now let $S \in Symm^+(\mathcal{O})$. We claim that either $S(\mathcal{T}) = \mathcal{T}$ or $S(\mathcal{T}) = \mathcal{T}^0$, depending on whether S is in $Symm^+(\mathcal{T})$ or not. Indeed, looking at how S acts on the two groups of four nonadjacent faces above, we see that S either permutes the faces in each group separately or it interchanges the faces between the two groups. Extending the faces in each group, we obtain $\mathcal{T}$ and $\mathcal{T}^0$, and the claim follows (see Color Plate 6). Note that the second case does occur; e.g., take a spatial quarter-turn about an axis that joins two opposite vertices of $\mathcal{O}$. Considering the four possible cases of compositions of two elements of $Symm^+(\mathcal{O})$, it is now clear that there are exactly two left-cosets in $Symm^+(\mathcal{O})$ by the subgroup $Symm^+(\mathcal{T})$ (corresponding to the two cases above). Comparing $Symm^+(\mathcal{T}) \subset Symm^+(\mathcal{O})$ and $\mathcal{A}_4 \subset \mathcal{S}_4$, we see that $Symm^+(\mathcal{O})$ is isomorphic to the symmetric group $\mathcal{S}_4$. A more explicit way to obtain this isomorphism is to mark the vertices of one tetrahedron by 1, 2, 3, 4, and their antipodals by 1', 2', 3', 4' (the vertices of the reciprocal) and to consider the action of $Symm^+(\mathcal{O})$ on the "diagonals" 11', 22', 33', 44'. Since $|\mathcal{S}_4| = 24$, we recovered the third numerical column in our table! The group of direct isometries $Symm^+(\mathcal{O}) \cong \mathcal{S}_4$ is called the *octahedral group*.

Remark.

Our argument was based on the fact that a regular tetrahedron and its reciprocal intersect in an octahedron. You may be wondering what we get if we intersect the other two reciprocal pairs. Well, we obtain nonregular polyhedra! The intersection of the reciprocal pair of a cube and an octahedron gives what is called a *cuboctahedron*, a convex polygon with eight equilateral triangular faces and six square faces (Figure 17.21). The intersection of the reciprocal pair of an icosahedron and a dodecahedron is an *icosidodecahedron*. It has twenty equilateral triangular faces and twelve pentagonal faces (Figure 17.22).

Finally, we take up the task of determining the group of direct isometries of the dodecahedron. It will be easier to work with its reciprocal, the icosahedron $\mathcal{I}$. In Color Plate 7, the twenty faces of

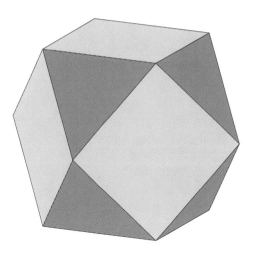

Figure 17.21

$\mathcal{I}$ have been colored with five different colors, with the property that in each color group the four faces are mutually disjoint. Figure 17.23 shows one color group in a wireframe setting.

A simple algorithm to find four faces in a color group is the following: Stand on a face F with bounding edges e_k, $k = 1, 2, 3$. For each k, step on the face F_k adjacent to F across e_k. At the vertex v_k of F_k opposite to e_k, five faces meet, three of which are not disjoint from F. The remaining two faces disjoint from F are adjacent and appear to you to the right and to the left. Now, if you are right-handed, add the right face to the color group of F, and if you are left-handed, add the left face to the color group.

The plane extensions of each of the four faces in a color group enclose a regular tetrahedron (Figure 17.24). Since each color

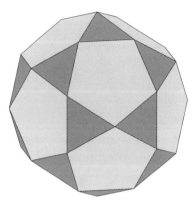

Figure 17.22

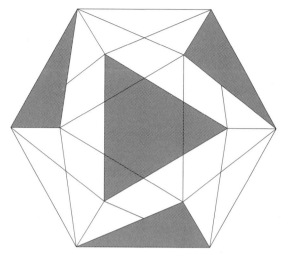

Figure 17.23

group gives one tetrahedron, altogether we have *five tetrahedra*. The five tetrahedra is one of the most beautiful compound polyhedra (Figure 17.25). Its Schläfli symbol {5, 3}[5{3, 3}]{3, 5} reflects that the twenty vertices of the five tetrahedra give the vertices of a dodecahedron (see Problem 19), and the twenty faces enclose an icosahedron (see the front cover illustration). (With this terminology, the reciprocal pair of two tetrahedra has Schläfli symbol {4, 3}[2{3, 3}]{3, 4} (see Problem 18).)

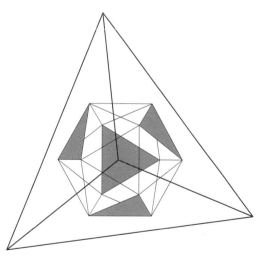

Figure 17.24

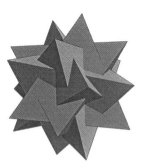

Figure 17.25

Now let $S \in Symm^+(\mathcal{I})$. Looking at the way we colored the icosahedron, we see that S acts on the set of the five tetrahedra as an even permutation, so $Symm^+(\mathcal{I})$ can be represented as a subgroup of $\mathcal{A}_5$. The twelve spatial rotations of a fixed tetrahedron act on the remaining four tetrahedra as even permutations. We obtain that the group of direct isometries $Symm^+(\mathcal{I})$ is isomorphic with the alternating group $\mathcal{A}_5$ on five letters. We arrive at the *icosahedral group*.

Remark.

♠ The icosahedral group plays a prominent role in Galois theory in connection with the problem of solving quintic (degree 5) polynomial equations in terms of radicals (in a similar way as the quadratic formula solves all quadratic equations). In fact, Galois proved that a polynomial equation is solvable by radicals iff the associated Galois group (the group of automorphisms of the splitting field) is solvable. The connection beween the symmetries of the icosahedron and unsolvable quintics is subtle[9] but it is based on the fact that a quintic with Galois group $\mathcal{A}_5$ is an example of a degree 5 equation for which no root formula exists. Any irreducible (over $\mathbf{Q}$) quintic with exactly 3 real roots (and a pair of complex conjugate roots) provides such an example.

---

[9]See F. Klein, *Lectures on the icosahedron, and the solution of equations of the fifth degree*, Trübner and Co., 1888.

Remark.

♣ Our understanding of the octahedral and icosahedral groups was based on their tetrahedral subgroups. Many different treatments of this topic exist. For example, given a Platonic solid P with Schläfli symbol $\{a, b\}$, it is clear that the axis of a rotational symmetry must go through a vertex or the midpoint of an edge or the centroid of a face. In the first case, the rotation angle is an integer multiple of $2\pi/b$; in the second, an integer multiple of π; in the third, an integer multiple of $2\pi/a$. The number of nontrivial rotations is therefore $(1/2)(V(b-1) + E + F(a-1))$, where V, E, and F are the number of vertices, edges, and faces of P. The one-half factor is present because when considering lines through vertices, midpoints of edges, and centroids of faces, we actually count the rotation axes twice. Since $bV = 2E = aF$, this number is $(1/2)(2E + 2E - 2) = 2E - 1$, where we also used Euler's theorem. Thus the order of $Symm^+(P)$ is $2E$.

We summarize our hard work in the following:[10]

Theorem 11.

The only finite groups of direct spatial isometries are the cyclic groups C_n, the dihedral groups D_n, the tetrahedral group $\mathcal{A}_4$, the octahedral group $\mathcal{S}_4$, and the icosahedral group $\mathcal{A}_5$.

The argument in the remark after the list of the 17 crystallographic groups in Section 10 can be adapted to spherical geometry to give another proof of Theorem 11. We assume that G is a finite group of rotations in $\mathbf{R}^3$ with fixed point p_0 and set of poles $Q \subset S^2(p_0)$. If Q contains only one pair of antipodal points, then G is cyclic. From now on we assume that this is not the case. Let $R_{2\alpha}(p)$ and $R_{2\beta}(q)$ be rotations in G with least positive angles 2α and 2β and distinct poles $\pm p$ and $\pm q$. Since α and β are minimal, π/α and π/β are integers (≥ 2). Following the argument in Section 10 cited above, we obtain another pole $r \in Q$ such that p, q, r are

---

[10]This result is due to Klein. Note also that, based on analogy with Euclidean plane isometries, a possible continuation of this topic would include discrete subgroups of $Iso\,(\mathbf{R}^3)$ and 3-dimensional crystallography. Alas, we are not prepared to explore this beautiful subject here.

vertices of a spherical triangle in $S^2(p_0)$ with angles α, β, γ, and we have

$$R_{2\alpha}(p) \circ R_{2\beta}(q) \circ R_{2\gamma}(r) = I.$$

In particular, if $\alpha = \beta = \pi/2$ (that is, $R_{2\alpha}(p)$ and $R_{2\beta}(q)$ are half-turns), then r is perpendicular to the great circle through $\pm p$ and $\pm q$. Then the rotation $R_{2\gamma}$ is a half-turn iff p and q are perpendicular. We see that if a noncyclic finite group G consists of half-turns only, then G must be isomorphic to the dihedral group D_2 of order 4, containing exactly three half-turns with mutually perpendicular axes. In a similar vein, if G contains exactly one pair of poles of degree greater than or equal to 3, then all the poles corresponding to the half-turns in G must be perpendicular to this, and G must be dihedral. From now on we assume that G contains at least two rotations of degrees greater than or equal to 3 and distinct axes. Using the notation above, we choose $R_{2\alpha}(p)$ and $R_{2\beta}(q)$, two rotations in G of degrees greater than or equal to 3, and distinct poles $\pm p$ and $\pm q$ such that the (spherical) distance between p and q is *minimal among all the poles in Q of degree greater than or equal to 3*. As above, we have $R_{2\gamma}(r) = R_{-2\beta}(q) \circ R_{-2\alpha}(p) \in G$. By the minimal choice of p, q, in addition to π/α and π/β being integral, π/γ must also be an integer greater than or equal to 3. The spherical excess formula (cf. Section 13) gives

$$\frac{\alpha}{\pi} + \frac{\beta}{\pi} + \frac{\gamma}{\pi} > 1.$$

In particular, at least one of the integers π/α, π/β, π/γ must be 2. Because of our initial choices, π/α, $\pi/\beta \geq 3$, we must have $\pi/\gamma = 2$. The constraint above reduces to

$$\frac{\alpha}{\pi} + \frac{\beta}{\pi} > \frac{1}{2}.$$

Rearranging, we obtain

$$(\pi/\alpha - 1)(\pi/\beta - 2) < 4, \quad \pi/\alpha, \pi/\beta \geq 3.$$

Since π/α and π/β are integers, one of them must be 3 and the other must be 3, 4, or 5. A simple analysis shows that the subgroup $G_0 \subset G$ generated by $R_{2\alpha}(p)$ and $R_{2\beta}(q)$ is the symmetry group of a tetrahedron, an octahedron/cube, or an icosahedron/dodecahedron.

More precisely, the orbit $G_0(p)$ is the set of vertices of a regular polygon with Schläfli symbol $\{\pi/\beta, \pi/\alpha\}$. In fact, the powers of $R_{2\beta}(q)$ applied to p give a (π/β)-gonal face, and the powers of $R_{2\alpha}(p)$ transform this face to the configuration of π/α faces surrounding the vertex p. The rest of G_0 installs the remaining faces of the polyhedron with vertices $G_0(p)$. In a similar vein, $G_0(q)$ is the set of vertices of the reciprocal polyhedron with Schläfli symbol $\{\pi/\alpha, \pi/\beta\}$, while $G_0(r)$ is the set of common midpoints of the edges of both polyhedra. For convenience, we now project these polyhedra to $S^2(p_0)$ radially from p_0, and obtain spherical tessellations. The spherical triangle with vertices p, q, r is *characteristic* in the sense that in a flag of the spherical polyhedron with vertices $G_0(p)$, p is a vertex, q is the midpoint of a face, and r is the midpoint of an edge in the flag. More about this in Sections 24–25. Now, it takes only a moment to realize that $G_0 = G$. Indeed, by the minimal choice of p, q, $G - G_0$ cannot contain any rotations of order greater than or equal to 3, since one of the corresponding poles of degree greater than or equal to 3 (transformed by an appropriate element in G_0) would show up in the face of the spherical polygon with centroid q and vertices in $G(p)$. Thus $G - G_0$ can contain only half-turns. If $R_\pi(s)$ were a half-turn in $G - G_0$, then $R_\pi(r) \circ R_\pi(s) \in G - G_0$ would also be a half-turn, so that r and s would be perpendicular. Thus the axis $\mathbf{R} \cdot s$ would be perpendicular to all axes $\mathbf{R} \cdot g_0(r)$, $g_0 \in G_0$, and this is impossible. Thus $G = G_0$, and Theorem 11 follows. We also realize that we have obtained the following as a byproduct: Any finite group of rotations in $\mathbf{R}^3$ is generated by one or two elements!

Remark.

♠ We now pick up the opposite isometries. Let $G \subset Iso\,(\mathbf{R}^3)$ be any finite group. Let $G^+ \subset G$ denote the subgroup consisting of the direct isometries of G. The possible choices of G^+ are listed in the theorem above. To get something new, we may assume that $G^+ \subset G$ is a proper subgroup. Since composition of two opposite isometries is direct, G^+ is of index 2 in G, that is, $|G| = 2|G^+|$. In other words, G^+ and any element in $G^- = G - G^+$ generate G. We now have to study the possible configurations of G^+ (listed above) and a single opposite isometry. We have two cases, depending on whether the antipodal map $-I : \mathbf{R}^3 \to \mathbf{R}^3$ (which is opposite in $\mathbf{R}^3$) belongs to G or not. If $-I \in G$, then $G^- = (-I) \cdot G^+$, since the right-hand side is a set of $|G^+|$ opposite isometries in G. Since $-I$ commutes with the elements of G, the cyclic subgroup $\{\pm I\} \subset G$ (isomorphic to C_2) is normal. Since index-2 subgroups (such as $G^+ \subset G$) are always normal, we obtain $G \cong G^+ \times C_2$. Using Theorem 11, we arrive at the list

$$C_n \times C_2, \qquad D_n \times C_2, \qquad \mathcal{A}_4 \times C_2, \qquad \mathcal{S}_4 \times C_2, \qquad \mathcal{A}_5 \times C_2.$$

Assume now that $-I \notin G$. We first show that G^+ is contained in a (finite) group G^* of *direct* isometries as an index-2 subgroup. To do this, we define

$$G^* = G^+ \cup (-I) \cdot G^-.$$

Notice that G^* is a group, since $-I$ commutes with the elements of G. Moreover, G^* consists of direct isometries, since $-I$ and the elements of G^- are opposite. G^+ and $(-I) \cdot G^-$ are disjoint, since $-I \notin G$. In particular, $|G^*| = 2|G^+|$, and G^+ is an index-2 subgroup in G^*. The possible inclusions $G^+ \subset G^*$ are easily listed, since all the isometries involved are direct and Theorem 11 applies. We obtain

$$C_n \subset C_{2n}, \qquad C_n \subset D_n, \qquad D_n \subset D_{2n}, \qquad \mathcal{A}_4 \subset \mathcal{S}_4.$$

Finally, G can be recovered from G^* via the formula

$$G = G^+ \cup (-I) \cdot (G^* - G^+).$$

In general, we denote by G^*G^+ the group defined by the right-hand side of this equality, when $G^+ \subset G^*$ is an inclusion of finite groups of direct isometries, and G^+ is of index 2 in G^*. We finally arrive at the list

$$C_{2n} \cdot C_n, \qquad D_n \cdot C_n, \qquad D_{2n} \cdot D_n, \qquad \mathcal{S}_4 \cdot \mathcal{A}_4.$$

For example, for the full symmetry groups of the Platonic solids, we have the following:

$$Symm\,(\mathcal{T}) = Symm^+(\mathcal{O}) \cdot Symm^+(\mathcal{T}) = \mathcal{S}_4 \cdot \mathcal{A}_4,$$

$$Symm\,(\mathcal{O}) = Symm^+(\mathcal{O}) \times C_2 = \mathcal{S}_4 \times C_2,$$

$$Symm\,(\mathcal{I}) = Symm^+(\mathcal{I}) \times C_2 = \mathcal{A}_5 \times C_2.$$

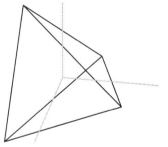

Figure 17.26

♣ Aside from constructibility, for computational purposes it will be important to realize the five Platonic solids in convenient positions in $\mathbf{R}^3$. Since every convex polyhedron is uniquely determined by its vertices, our task is to find for each Platonic solid a "symmetric" position with vertex coordinates as simple as possible.

We start with the tetrahedron $\mathcal{T}$. A natural positon of $\mathcal{T}$ in $\mathbf{R}^3$ is defined by letting the coordinate axes pass through the midpoints of three edges, as shown in Figure 17.26.

The four vertices of $\mathcal{T}$ are

$$(1,1,1), \qquad (1,-1,-1), \qquad (-1,1,-1), \qquad (-1,-1,1).$$

The tetrahedral group contains the three half-turns around the coordinate axes and the rotation around the "front vertex" $(1,1,1)$ by angle $2\pi/3$. It is easy to see that these four rotations generate the twelve rotations that make up the tetrahedral group. It would be very easy to write down these transformations in terms of orthogonal 3×3 matrices. We skip this, since the description of $Symm^+(\mathcal{T})$ is much easier using quaternions, which we will discuss in Section 23.

Truncating the tetrahedron by chopping off the four tetrahedra along the vertex figures, we arrive at the octahedron $\mathcal{O}$ with vertices

$$(\pm 1, 0, 0), \qquad (0, \pm 1, 0), \qquad (0, 0, \pm 1).$$

Taking the midpoints of the edges of $\mathcal{O}$, we obtain the midpoints of the reciprocal cube $\mathcal{C}$:

$$\left(\pm \frac{1}{2}, \pm \frac{1}{2}, \pm \frac{1}{2} \right).$$

The cube described this way has edge length 1, and the edges are parallel to the coordinate axes. The centroid of all three Platonic solids $\mathcal{T}$, $\mathcal{O}$, and $\mathcal{C}$ is the origin.

To realize the dodecahedron $\mathcal{D}$ and the icosahedron $\mathcal{I}$ in $\mathbf{R}^3$ is a less trivial task. We start with the icosahedron $\mathcal{I}$ and make the following observation. Let v be a vertex of $\mathcal{I}$. The five faces of $\mathcal{I}$ that contain v form a pyramid whose base is a regular pentagon. We call v the *vertex* of the pentagonal pyramid. Taking the edge length of $\mathcal{I}$ to be 1, the length τ of a diagonal of the regular pentagonal base is the so-called *golden section* (attributed to Eudoxus in 400 B.C.)

$$\tau = 2 \sin \left(\frac{\pi}{2} - \frac{\pi}{5} \right) = 2 \cos \left(\frac{\pi}{5} \right)$$

(Figure 17.27). Inserting two additional diagonals in the base, an interesting picture emerges (Figure 17.28). Notice that p_0, p_1, p_2, p_3 is a rhombus and the triangles $\triangle p_0 p_1 p_3$ and $\triangle p_0 p_4 p_5$ are similar. Thus, $d(p_0, p_4) = 1/\tau$. But adding 1 to this gives the diagonal again:

$$1 + 1/\tau = \tau.$$

Multiplying out, we see that τ is the positive solution of the quadratic equation

$$\tau^2 - \tau - 1 = 0.$$

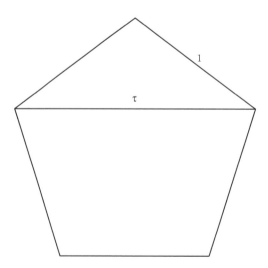

Figure 17.27

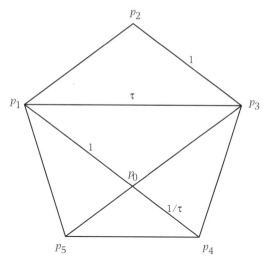

Figure 17.28

The quadratic formula gives

$$\tau = \frac{\sqrt{5}+1}{2}.$$

Remark.

Alternatively, the golden section τ can be defined as the unique ratio of side lengths of a rectangle with the property that if a square is sliced off, the remaining rectangle is similar to the original rectangle. This definition gives $\tau - 1 = 1/\tau$, which leads to the same quadratic equaion as above. Used recursively, a circular quadrant can be inserted into each sliced off square to obtain an *approximation* of an Archimedes spiral (see Figure 17.29). The most commonly known phenomenon in nature that patterns this is

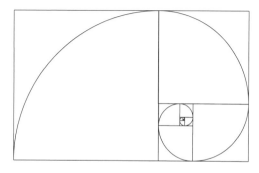

Figure 17.29

the *nautilus shell*, where the shell grows in a spiral for structural harmony in weight and strength.

Iterating $\tau = 1 + 1/\tau$, we obtain the continued fraction

$$\tau = 1 + \cfrac{1}{1 + \frac{1}{1+\cdots}},$$

and this can be used to give rational approximations of τ (cf. Problem 14). In a similar vein, iterating $\tau = \sqrt{1 + \tau}$, we arrive at the infinite radical expansion

$$\tau = \sqrt{1 + \sqrt{1 + \sqrt{1 + \cdots}}}.$$

Recall that the icosahedron $\mathcal{I}$ can be sliced into two pentagonal pyramids with opposite vertices v_1 and v_2 and a pentagonal antiprism (a belt of ten equilateral triangular faces). Consider now two opposite edges e_1 and e_2 of the pyramids emanating from v_1 and v_2. They are parallel and form the opposite sides of a rectangle whose longer sides are diagonals of two pentagonal bases! Since the side lengths of this rectangle have ratio τ : 1, it is called a *golden rectangle*. (Ancient Greeks attributed special significance to this; for example, the Parthenon in Athens (fifth century B.C.) fits perfectly in a golden rectangle.) A beautiful model of the icosahedron (due to Pacioli and shown in Figure 17.30) emerges this way; the twelve vertices of $\mathcal{I}$ are on three golden rectangles in mutually perpendicular planes! Considering now these planes as coordinate planes for our coordinate system, we see immediately that the

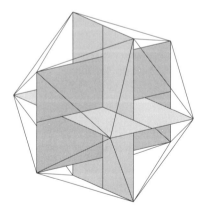

Figure 17.30

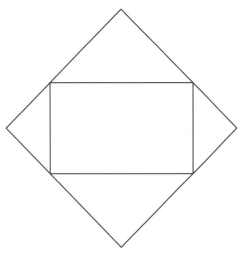

Figure 17.31

twelve vertices of the icosahedron are

$$(0, \pm\tau, \pm1), \qquad (\pm1, 0, \pm\tau), \qquad (\pm\tau, \pm1, 0).$$

A golden rectangle can be inscribed in a square such that each vertex of the golden rectangle divides a side of the square in the ratio $\tau : 1$ (Figure 17.31). Inserting these squares around the three golden rectangles that make up the icosahedron, we obtain an octahedron circumscribed about $\mathcal{I}$ (Figure 17.32). Looking at Figure 17.33, we see that the vertices of the octahedron are

$$(\pm\tau^2, 0, 0), \qquad (0, \pm\tau^2, 0), \qquad (0, 0, \pm\tau^2)$$

and this is homothetic (with ratio of magnification τ^2) to our earlier model $\mathcal{O}$.

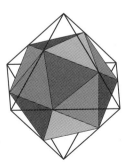

Figure 17.32

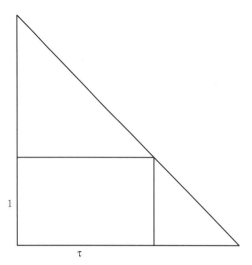

Figure 17.33

Finally, note that the dodecahedron $\mathcal{D}$ constructed on the cube with vertices $(\pm 1, \pm 1, \pm 1)$ as above has vertices

$$(0, \pm 1/\tau, \pm\tau), \qquad (\pm\tau, 0, \pm 1/\tau), \qquad (\pm 1/\tau, \pm\tau, 0).$$

That these are the correct vertices for $\mathcal{D}$ can be verified using reciprocity. An alternative approach is given in Problem 20.

Problems

1. Using the isomorphism $Symm(\mathcal{T}) \cong \mathcal{S}_4$, work out for the regular tetrahedron the rotational symmetries that correspond to all possible products of two transpositions.

2. Let $\mathcal{T}$ be a regular tetrahedron.

   (a) Given a flag (p, e, f) of $\mathcal{T}$, what is the composition of $S_1, S_2 \in Symm^+(\mathcal{T})$, where S_1 and S_2 are counterclockwise rotations by $2\pi/3$ about p and the centroid of f?

   (b) Given opposite edges e_1 and e_2, list all possible scenarios for flags (p_1, e_1, f_1) and (p_2, e_2, f_2) and symmetries of $\mathcal{T}$ that carry one flag to the other.

   (c) Let e_1 and e_2 be opposite edges. Show that the midpoints of the four complementary edges are vertices of a square. Slice $\mathcal{T}$ with the plane spanned by this square; prove that the two pieces are congruent and find a symmetry of $\mathcal{T}$ that carries one piece to the other.

3. Let $S \in Symm^+(\mathcal{O})$ be an isometry that does not leave $\mathcal{T}$ invariant. Show that

$$S \circ Symm^+(\mathcal{T}) \circ S^{-1} = Symm^+(\mathcal{T}).$$

4. Derive that $Symm^+(\mathcal{O})$ is isomorphic to $\mathcal{S}_4$ using a cube rather than an octahedron.

5. (a) Construct a golden rectangle and a regular pentagon with straightedge and compass.

   (b) Use the fact that the vertices of a regular pentagon inscribed in the unit circle are the powers ω^k, $k = 0, \ldots, 4$, of the primitive fifth root of unity $\omega = e^{2\pi i/5}$ to show that

$$\tau = \frac{\omega - \omega^4}{\omega^2 - \omega^3} = -(\omega^2 + \omega^3) = \frac{1}{\omega + \omega^4}.$$

   (c) Use (b) to derive the formula

$$\sqrt{\tau^2 + 1} = -i(\omega - \omega^4).$$

   (d) Let s and d be the side length and the diagonal length of a regular pentagon; $\tau = d/s$. Show that the side length and the diagonal length of the regular pentagon whose sides extend to the five diagonals of the original pentagon are $2s - d$ and $d - s$. Interpret this via paper folding. Conclude that τ is irrational. (Hint: Assume that $\tau = d/s$ with s and d integral, and use Fermat's method of infinite descent. Compare this with Problem 12 in Section 3.)

6. Prove that the icosahedral group $Symm^+(\mathcal{I})$, is simple[11] using the following argument: Let $N \subset Symm^+(\mathcal{I})$ be a normal subgroup.

   (a) Show that if N contains a rotation with axis through a vertex of $\mathcal{I}$ then N contains all rotations with axes through the vertices of $\mathcal{I}$.

   (b) Derive similar statements for rotations with axes through the midpoints of edges and the centroids of faces of $\mathcal{I}$.

   (c) Counting the nontrivial rotations in N, conclude from (a)–(b) that $|N| = 1 + 24a + 20b + 15c$, where a, b, c are 0 or 1.

   (d) Use the fact that $|N|$ divides 60 to show that either $a = b = c = 0$ or $a = b = c = 1$.

7. Establish an isomorphism between $Symm^+(\mathcal{I})$ and $\mathcal{A}_5$ by filling in the details for the following steps:

   (a) There are exactly five cubes inscribed in a dodecahedron (see Color Plate 8).

   (b) The icosahedral group permutes these cubes.

   (c) The map $\phi : Symm^+(\mathcal{I}) \to \mathcal{S}_5$ defined by the action in (b) is an injective homomorphism with image $\mathcal{A}_5$.

---

[11] See "Groups" in Appendix B.

8. True or false: Two cubes inscribed in a dodecahedron have a common diagonal. (The five cubes inscribed in a dodecahedron form a compound polyhedron with Schläfli symbol 2{5, 3}[5{4, 3}], where the 2 means that each vertex of the dodecahedron belongs to two of the cubes (see the back cover illustration). It can also be obtained from the compound of five tetrahedra by circumscribing a cube around each participating tetrahedron and its reciprocal.)

9. Find two flags in Bimbo's lozenge that cannot be carried into each other by a symmetry.

10. Classify spherical tilings following the argument in the remark at the end of Section 10. Observe that for a spherical triangle with angles α, β, and γ such that π/α, π/β, and π/γ are integers, the inequality $\alpha/\pi + \beta/\pi + \gamma/\pi > 1$ gives only finitely many possibilities.

11. Show that the area of a spherical n-sided polygon is the sum of its angles minus $(n - 2)\pi$.

12. Prove Euler's theorem for convex polyhedra using Problem 11 as follows: Let K be a convex polyhedron. Place K inside S^2 (by scaling if necessary) such that K contains the origin in its interior. Project the boundary of K onto S^2 from the origin. Sum up *all* angles of the projected spherical graph with V vertices, E edges, and F faces in two ways: First, counting the angles at each vertex, find that this sum is $2\pi V$. Second, count the angles for each face by converting the angle sum into spherical area (Problem 11) and use that the total area of S^2 is 4π.

13. Show that a polyhedron is regular iff its faces and vertex figures are regular.

14. ◇ Let F_n denote the nth Fibonacci[12] number defined recursively by $F_1 = 1$, $F_2 = 1$, and $F_{n+2} = F_{n+1} + F_n$, $n \in \mathbf{N}$. (F_n can be interpreted as the number of offspring generated by a pair of rabbits, assuming that the newborns mature in one iteration and no rabbits die.) (a) Let $q_n = F_{n+1}/F_n$. Show that the nth convergent of the continued fraction

$$\tau = 1 + \cfrac{1}{1 + \frac{1}{1 + \cdots}},$$

is q_n. (b) Prove that $\lim_{n \to \infty} q_n = \tau$. (This observation is due to Kepler.) Use induction with respect to n to show that the ratios q_n alternate above and below τ, that is, $q_n < \tau$ for n odd, and $q_n > \tau$ for n even. (c) Verify the recurrence relation $\tau^{n+1} = F_{n+1}\tau + F_n$. Define $F_{-n} = F_{-n+2} - F_{-n+1}$ for nonnegative integers n, so that $F_{-n} = (-1)^{n+1}F_n$, and extend the validity of the recurrence relation to all integers n. (d) Define the nth *Lucas number* L_n recursively by $L_{n+1} = L_n + L_{n-1}$, $n \in \mathbf{N}$, $L_0 = 2$, $L_1 = 1$. Show that $L_n =$

---

[12] Leonardo of Pisa (13th century) wrote under the name Fibonacci, a shortened version of filius Bonacci (son of Bonacci).

$F_{n-1} + F_{n+1}$, $n \geq 0$. Verify that $\lim_{n\to\infty} L_{n+1}/L_n = \tau$ and $\lim_{n\to\infty} L_n/F_n = \sqrt{5}$. Use (c) to show that $L_n = \tau^n + (-1)^n/\tau^n$. Derive the formula

$$F_n = \frac{\tau^n - (-1)^n/\tau^n}{\tau + 1/\tau},$$

due to Binet (1843).

15. (a) Show that the sum of vectors from the centroid to each vertex of a Platonic solid is zero.

(b) Prove that the midpoint of each edge of a Platonic solid is on a sphere.

(c) Derive an analogous statement for the centroids of the faces.

(d) Show that the centroids of the faces of a Platonic solid are the vertices of another Platonic solid. What is the relation between these two Platonic solids?

16. Show that the icosahedron can be truncated[13] such that the resulting convex polyhedron has 12 pentagonal and 20 hexagonal faces. This polyhedron is called the *buckyball* (Figure 17.34). (It has great significance in chemistry, since, besides graphite and diamond, it is a third form of pure carbon (Figure 17.35).) Show that the buckyball[14] has the following rotational symmetries:

(a) Half-turns around axes that bisect opposite pairs of edges;

(b) Rotations around axes through the centroids of opposite pairs of hexagonal faces with rotation angles that are integral multiples of $2\pi/3$;

(c) Rotations around axes through the centroids of opposite pairs of pentagonal faces, with rotation angles that are integral multiples of $2\pi/5$.

(d) Realize that (a)–(b)–(c) account for 59 nontrivial rotations, so that (adding the identity) we obtain the icosahedral group.

(e) Show that no vertex stays fixed under any nontrivial rotational symmetry of the buckyball.

Figure 17.34

---

[13]Truncating the five Platonic solids in various ways, one arrives at the thirteen Archimedean solids. It is very probable that Archimedes knew about these, but the first surviving written record is by Kepler from 1619. We already encountered two of these, the cuboctahedron and the icosidodecahedron.

[14]For an excellent article, see F. Chung and S. Sternberg, *Mathematics and the Buckyball*, American Scientist, Vol. 81 (1993).

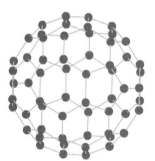

Figure 17.35

17. (a) Show that the reciprocal of the side length of a regular decagon inscribed in the unit circle is the golden section.

    (b) Inscribe an equilateral triangle into a circle. Take a segment connecting the midpoints of two sides of the triangle and extend it to a chord of the circle. Show that one midpoint splits the length of the chord in the square of the golden section.

18. The 8 vertices of a reciprocal pair of tetrahedra are those of a (circumscribed) cube (cf. Problem 4). Show that this cube is homothetic to the reciprocal cube of the octahedral intersection of the two tetrahedra. What is the ratio of magnification?

19. Show that the 20 vertices of the 5 tetrahedra circumscribed to an icosahedron are the vertices of a dodecahedron. Prove that this dodecahedron is homothetic to the dodecahedron reciprocal to the icosahedron.

20. A diagonal splits a pentagon with unit side length into an isosceles triangle and a symmetric trapezoid with a base length of the golden section. Define a convex polyhedron with a square base of side length that of the golden section, and let it have four additional faces: two isosceles triangles and two symmetric trapezoids, attached to each other along their unit-length sides in an alternating manner. We call this polyhedron a *roof*.[15] The top unit-length side of the trapezoids, which forms a single edge opposite the base, is called the *ridge*. Notice that the solid dodecahedron is the union of an inscribed cube and six roofs whose bases are the faces of the cube (Figure 17.11).

    (a) Prove the existence of the dodecahedron by showing that in a roof the dihedral angle between a triangular face and the base and the dihedral angle between a trapezoidal face and the base are complementary.

    (b) In a given roof, extend the lateral sides of a trapezoid beyond the ridge to form an isosceles triangle. Prove that the four nonbase vertices of the four

---

[15]The four planes spanned by the four faces of a tetrahedron divide $\mathbf{R}^3$ into the tetrahedron itself, four *frusta*, four *trihedra*, and six *roofs*. See M. Berger, *Geometry I–II*, Springer, 1980. Although this is unbounded, we use this classical terminology in our situation. The author's students called the argument for the existence of the dodecahedron the *roof-proof*. It is essentially contained in Book XIII of the *Elements*.

isosceles triangles obtained from four trapezoids in an opposite pair of roofs form a golden rectangle (see Color Plate 9).

(c) Show that two edges of the golden rectangle in (b) contain the ridges of another pair of opposite roofs.

(d) By (b), the three pairs of opposite roofs in a dodecahedron give three mutually perpendicular golden rectangles around which an icosahedron can be circumscribed. Prove that this icosahedron is homothetic to the reciprocal of the dodecahedron. What is the ratio of magnification?

(e) Circumscribe an octahedron to the icosahedron obtained in (d) and relate it to the reciprocal of the cube inscribed in the dodecahedron.

21. A triangular array of dots consists of rows of $1, 2, \ldots, n$ dots, $n \in \mathbf{N}$, stacked up in a triangular shape. The total number of dots in a triangular array is the nth triangular number $Tri\,(n) = 1 + 2 + \cdots + n = n(n+1)/2$ (see Problem 11 of Section 2). A tetrahedral array of dots consists of triangular arrays of $1, 3, \ldots, n(n+1)/2$ dots stacked up in a tetrahedral shape. The total number of dots $Tet\,(n) = \sum_{k=1}^{n} k(k+1)/2$ in a tetrahedral array is called the nth *tetrahedral number*. In a similar vein, define the nth *square pyramid number* $Pyr\,(n) = 1^2 + 2^2 + \cdots + n^2$ as the total number of dots in a square pyramid obtained by stacking up square arrays of $1, 4, \ldots, n^2$ dots. Finally, define the nth *octahedral number*[16] $Oct\,(n) = Pyr\,(n) + Pyr\,(n-1)$ as the total number of dots in the octahedron viewed as the union of two square pyramids.

(a) Show that $Pyr\,(n) = n(n+1)(2n+1)/6$ by fitting six square pyramids in an $n \times (n+1) \times (2n+1)$ box (cf. also Problem 8 of Section 2).

(b) Prove that $Pyr\,(n) = 2\,Tet\,(n-1) + n(n+1)/2$ by slicing the array of dots in a square pyramid into two tetrahedral arrays and a triangular array. Conclude that $Tet\,(n) = n(n+1)(n+2)/6$. (This is attributed to the Hindu mathematician Aryabhatta, c. A.D. 500.)

(c) Show that $Oct\,(n) = Tet\,(2n-1) - 4\,Tet\,(n-1) = n(2n^2+1)/3$ by considering the octahedron as a tetrahedron truncated along the vertex figures.

22. Use the 3-dimensional ball model D^3 of hyperbolic space geometry (see Problem 7 of Section 13) to prove the existence of a hyperbolic dodecahedron with right dihedral angles (cf. Problem 8 of Section 14). Show that D^3 can be tessellated by hyperbolic dodecahedra with right dihedral angles. (Observe the close analogy between dodecahedral tessellations of the hyperbolic space and tessellations of the Euclidean space by cubes.)

23. Prove that (a) the full symmetry group of the regular pyramid with base P_n is $D_n \cdot C_n$, and (b) the full symmetry group of the regular prism with base P_n is $D_n \times C_2$ for n even, and $D_{2n} \cdot D_n$ for n odd.

---

[16] For an interesting account on these and other related numbers, see J. Conway and R. Guy, *The Book of Numbers*, Springer, 1996.

Web Sites

1. www.mathsoft.com/asolve/constant/gold/gold.html

2. www.vashti.net/mceinc/golden.htm

3. www.geom.umn.edu/docs/education/institute91/handouts/node6.html

4. www.geom.umn.edu/graphics/pix/Special_Topics/Tilings /allmuseumsolids.html

5. www.frontiernet.net/~imaging/polyh.html

Film

Ch. Gunn and D. Maxwell: *Not Knot*, Geometry Center, University of Minnesota; Jones and Bartlett Publishers, Inc. (20 Park Plaza, Suite 1435, Boston, MA 02116).

18 Finite Möbius Groups

S E C T I O N

† ♣ Recall from Section 11 that a spherical reflection R_C in a great circle $C \subset S^2$ can be pulled down by the stereographic projection $h_N : S^2 \to \hat{\mathbf{C}}$ (extended to the North Pole N by $h_N(N) = \infty$) to a reflection $R_S : \hat{\mathbf{C}} \to \hat{\mathbf{C}}$ in the circle $S = h_N(C)$. In other words, $R_S = h_N R_C h_N^{-1}$. Utilizing the rich geometric setting, we were able to derive an explicit formula for R_S. A spherical rotation $R_\theta(p_0)$ around an axis $\mathbf{R} \cdot p_0$, $p_0 \in S^2$, is the composition of two spherical reflections (in great circles that meet in angle $\theta/2$ at p_0). Conjugating (this time) $R_\theta(p_0)$ by the stereographic projection, we thus obtain the composition of two reflections in circles intersecting at $h_N(p_0)$ in an angle of $\theta/2$, a direct Möbius transformation with fixed point $h_N(p_0)$. In Section 12 we learned that direct Möbius transformations are nothing but linear fractional tranformations of the extended complex plane $\hat{\mathbf{C}}$. Thus, to $R_\theta(p_0)$ there corresponds a linear fractional transformation whose coefficients depend on θ and the coordinates of p_0. It is natural to try to determine these dependencies explicitly. This is given in the folowing theorem, due to Cayley in 1879.

Cayley's Theorem.

For $p_0 = (a_0, b_0, c_0) \in S^2$ and $\theta \in \mathbf{R}$, the spherical rotation $R_\theta(p_0)$ conjugated with the stereographic projection h_N is a linear fractional transformation given by

$$(h_N \circ R_\theta(p_0) \circ h_N^{-1})(z) = \frac{\lambda z - \bar{\mu}}{\mu z + \bar{\lambda}}, \quad z \in \hat{\mathbf{C}},$$

where

$$\lambda = \cos\left(\frac{\theta}{2}\right) + \sin\left(\frac{\theta}{2}\right) c_0 i \quad \text{and} \quad \mu = \sin\left(\frac{\theta}{2}\right)(b_0 + a_0 i).$$

Remark.

♠ Under the isomorphism $M\ddot{o}b(\hat{\mathbf{C}}) = SL(2, \mathbf{C})/\{\pm I\}$, the linear fractional transformation in Cayley's theorem corresponds to the matrix

$$A = \begin{bmatrix} \lambda & -\bar{\mu} \\ \mu & \bar{\lambda} \end{bmatrix}, \quad |\lambda|^2 + |\mu|^2 = 1, \quad \lambda, \mu \in \mathbf{C}.$$

This is a *special unitary matrix*, that is,

$$A^* = \bar{A}^\top = A^{-1},$$

and A has determinant 1. The special unitary matrices form a subgroup of $SL(2, \mathbf{C})$, denoted by $SU(2)$. The importance of this subgroup will be apparent in Sections 22 to 24.

Proof.

♣ The statement is clear for $p_0 = N$, since under h_N, $R_\theta(p_0)$ corresponds to multiplication by $e^{i\theta}$. Let $\{e_1, e_2, N\} \subset \mathbf{R}^3$ denote the standard basis. We work out the composition $h_N R_\theta(p_0) h_N^{-1}$ in the particular case $p_0 = e_1 = (1, 0, 0)$. First note that the matrix of $R_\theta(e_1)$ is

$$R = \begin{bmatrix} 1 & 0 & 0 \\ 0 & \cos(\theta) & -\sin(\theta) \\ 0 & \sin(\theta) & \cos(\theta) \end{bmatrix}.$$

Using this and the explicit form of h_N given in Problem 1 of Section 7, we have

$$(h_N \circ R_\theta(p_0) \circ h_N^{-1})(z) = (h_N \circ R_\theta(p_0))\left(\frac{2z}{|z|^2 + 1}, \frac{|z|^2 - 1}{|z|^2 + 1}\right)$$

$$= h_N\left(\frac{2\Re(z)}{|z|^2 + 1}, \frac{2\Im(z)}{|z|^2 + 1}\cos(\theta) - \frac{|z|^2 - 1}{|z|^2 + 1}\sin(\theta),\right.$$

$$\left.\frac{2\Im(z)}{|z|^2 + 1}\sin(\theta) + \frac{|z|^2 - 1}{|z|^2 + 1}\cos(\theta)\right)$$

$$= \frac{2\Re(z) + 2i\Im(z)\cos(\theta) - i(|z|^2 - 1)\sin(\theta)}{|z|^2 + 1 - 2\Im(z)\sin(\theta) - (|z|^2 - 1)\cos(\theta)}$$

$$= \frac{(z\cos(\theta/2) + i\sin(\theta/2))(\bar{z}\sin(\theta/2) + i\cos(\theta/2))}{(iz\sin(\theta/2) + \cos(\theta/2))(\bar{z}\sin(\theta/2) + i\cos(\theta/2))}$$

$$= \frac{\cos(\theta/2)z + i\sin(\theta/2)}{i\sin(\theta/2)z + \cos(\theta/2)}.$$

The formula follows in this special case. Turning to the general case, we first note that $h_N R_\theta(p_0)h_N^{-1}$ is a linear fractional transformation of the form above (with $|\lambda|^2 + |\mu|^2 = 1$), since any rotation can be written as the composition of rotations with axes $\mathbf{R} \cdot N$ and $\mathbf{R} \cdot e_1$, and these cases have already been treated. Thus, it remains to show that the explicit expressions for the coefficients λ and μ are valid. To do this we use the result from Section 12 asserting that a linear fractional transformation is uniquely determined by its action on three points. For the first two points we choose the fixed points of the linear fractional transformation claimed to be equal to our conjugated spherical rotation. The fixed points are obtained by solving the quadratic equation

$$\mu z^2 - 2i\Im(\lambda)z + \bar{\mu} = 0$$

for z. We find that the fixed points are $(c_0 \pm 1)/(a_0 - b_0 i) = h_N(\pm p_0)$. For the third point we choose ∞, and verify that under h_N, $R_\theta(p_0)(N)$ corresponds to λ/μ. Indeed, since $|\lambda|^2 + |\mu|^2 = 1$, we have

$$h_N^{-1}(\lambda/\mu) = (2\lambda\bar{\mu}, |\lambda|^2 - |\mu|^2) \in S^2,$$

and this is $R_\theta(p_0)(N)$, as an easy computation shows. The theorem follows. ∎

Remark.

The first part of the proof of Cayley's theorem can be skipped if we use the fact that h_N is conformal and that any conformal transformation of $\hat{\mathbf{C}}$ (such as $h_N R_\theta(p_0) h_N^{-1}$) is a linear fractional transformation.

To simplify the terminology, we say that the linear fractional transformation with parameters λ, μ, and $R_\theta(p_0)$ *correspond to each other*. Notice that replacing the rotation angle θ by $\theta + 2\pi$ has the effect of changing λ and μ to their negatives.

Our main quest, to be fully accomplished in Section 24, is to classify all finite subgroups of the Möbius group $M\ddot{o}b\,(\hat{\mathbf{C}})$. To begin with, here we confine ourselves to giving a list of finite Möbius groups most of which arise from the geometry of Platonic solids.

An example to start with is the cyclic group C_n of order n. This group can be realized as the group of rotations

$$z \mapsto e^{2k\pi i/n} z, \quad k = 0, \ldots, n-1.$$

Notice that each rotation can be viewed as a linear fractional tranformation with $\lambda = e^{k\pi i/n}$ and $\mu = 0$, $k = 0, \ldots, n-1$. In fact, C_n corresponds to the group of rotations $R_{2k\pi/n}(N)$, $k = 0, \ldots, n-1$, with common vertical axis through the North and South Poles.

If we adjoin to C_n the linear fractional transformation $z \mapsto 1/z$ (characterized by $\lambda = 0$ and $\mu = i$) that corresponds to the half-turn $R_\pi(e_1)$, we obtain the *dihedral Möbius group* D_n of order $2n$:

$$z \mapsto e^{2k\pi i/n} z, \quad \frac{e^{-2k\pi i/n}}{z}, \quad k = 0, \ldots, n-1.$$

This group corresponds to the symmetry group of Klein's *dihedron*, the regular spherical polyhedron with two hemispherical faces, n spherical edges, and n vertices distributed equidistantly along the equator of S^2. If we fix one vertex at e_1, then the half-turn $R_\pi(e_1)$ is a symmetry of the dihedron that interchanges the two faces. We also see that the first group of linear fractional transformations in the dihedral Möbius group above corresponds to $\lambda = e^{k\pi i/n}$, $\mu = 0$,

$k = 0, \ldots, n-1$, and the second corresponds to $\lambda = 0$, $\mu = ie^{k\pi i/n}$, $k = 0, \ldots, n-1$.

As a straightforward generalization, we inscribe a Platonic solid P in S^2, apply Cayley's theorem, and obtain a finite Möbius group G isomorphic to the symmetry group of P. Since reciprocal pairs of Platonic solids have the same symmetry group, we may restrict ourselves to the tetrahedron, octahedron, and icosahedron. We call the corresponding groups *tetrahedral*, *octahedral*, and *icosahedral Möbius groups*. The dihedral Möbius group discussed above can be considered as a member of this family if we replace P with its spherical tessellation obtained by projecting P radially from the origin to S^2. In what follows we call these configurations *spherical Platonic tessellations*. Before we actually determine the Möbius groups explicitly, we should note that, again by Cayley's theorem, the Möbius groups obtained by inscribing the same Platonic solid into S^2 in two different ways are conjugate subgroups in $M\ddot{o}b\,(\hat{\mathbf{C}})$. Thus, we can choose our spherical Platonic tesselations in convenient positions in S^2.

We choose our regular tetrahedron such that its vertices are alternate vertices of the cube, and the cube is inscribed in S^2 such that its faces are orthogonal to the coordinate axes. We also agree that the first octant contains a vertex of the tetrahedron. This is a scaled version of the regular tetrahedron discussed in Section 17. (The scaling is only to circumscribe S^2 around the tetrahedron.) The vertex in the first octant must be

$$\left(\frac{1}{\sqrt{3}}, \frac{1}{\sqrt{3}}, \frac{1}{\sqrt{3}}\right).$$

The three coordinate axes go through the midpoints of the three opposite pairs of edges of the tetrahedron. The three half-turns around these axes are symmetries of the tetrahedron. Applying these half-turns to the vertex above, we obtain the remaining three vertices

$$\left(\frac{1}{\sqrt{3}}, -\frac{1}{\sqrt{3}}, -\frac{1}{\sqrt{3}}\right), \left(-\frac{1}{\sqrt{3}}, \frac{1}{\sqrt{3}}, -\frac{1}{\sqrt{3}}\right), \left(-\frac{1}{\sqrt{3}}, -\frac{1}{\sqrt{3}}, \frac{1}{\sqrt{3}}\right).$$

As noted above, the half-turn $R_\pi(e_1)$ corresponds to the linear fractional transformation $z \mapsto 1/z$. For the half-turn $R_\pi(e_2)$, we have

$\lambda = 0$ and $\mu = 1$, and the corresponding linear fractional transformation is $z \mapsto -1/z$. Finally, $R_\pi(N)$ corresponds to $z \mapsto -z$, the negative of the identity map, with $\lambda = i$ and $\mu = 0$. Adjoining the identity, we have

$$z \mapsto \pm z, \quad \pm\frac{1}{z}.$$

This is the dihedral Möbius group D_2 of order 4. We obtain that D_2 is a subgroup of the tetrahedral Möbius group T.

The four lines passing through the origin and the four vertices above intersect the opposite faces of the tetrahedron at the centroids. These are axes of symmetry rotations with angles $2\pi/3$ and $4\pi/3$. For example, for the rotation with angle $2\pi/3$ and axis through $(1/\sqrt{3}, 1/\sqrt{3}, 1/\sqrt{3})$, we obtain $\lambda = \mu = (1 + i)/2$, so that the corresponding linear fractional transformation is

$$z \mapsto \frac{(1 + i)z - (1 - i)}{(1 + i)z + (1 - i)} = \frac{z + i}{z - i}.$$

In a similar vein, the 8 linear fractional transformations corresponding to these rotations are

$$z \mapsto \pm i\frac{z + 1}{z - 1}, \quad \pm i\frac{z - 1}{z + 1}, \quad \pm\frac{z + i}{z - i}, \quad \pm\frac{z - i}{z + i}.$$

These correspond to $\lambda = (\pm 1 + i)/2, \mu = \pm(1 + i)/2$, and $\lambda = (\pm 1 - i)/2, \mu = \pm(1 - i)/2$.

Putting everything together, we arrive at the 12 elements of the *tetrahedral Möbius group* T:

$$z \mapsto \pm z, \quad \pm\frac{1}{z}, \quad \pm i\frac{z + 1}{z - 1}, \quad \pm i\frac{z - 1}{z + 1}, \quad \pm\frac{z + i}{z - i}, \quad \pm\frac{z - i}{z + i}.$$

We choose the octahedron in S^2 such that its vertices are the six intersections of the coordinate axes with S^2. This octahedron is homothetic to the intersection of the tetrahedron above and its reciprocal. As we concluded in Section 17, the octahedral group is generated by the tetrahedral group and a symmetry of the octahedron that interchanges the tetrahedron with its reciprocal. An example of the latter is the quarter-turn around the first axis. This quarter-turn corresponds to the linear fractional transformation $z \mapsto iz$ characterized by $\lambda = e^{\pi i/4}$ and $\mu = 0$. Thus, the 24 elements of the *octahedral Möbius group* O are as follows:

$$z \mapsto i^k z, \quad \frac{i^k}{z}, \quad i^k \frac{z+1}{z-1}, \quad i^k \frac{z-1}{z+1},$$

$$i^k \frac{z+i}{z-i}, \quad i^k \frac{z-i}{z+i}, \quad k = 0, 1, 2, 3.$$

Finally, we work out the icosahedral Möbius group I. We inscribe the icosahedron in S^2 such that the North and South Poles become vertices. The icosahedron can now be considered as being made up of northern and southern pentagonal pyramids separated by a pentagonal antiprism. The rotations $S^j, j = 0, \ldots, 4$, $S = R_{2\pi/5}(N)$, are symmetries of the icosahedron, and they correspond to the linear fractional transformations

$$S^j : z \mapsto \omega^j z, \quad j = 0, \ldots, 4,$$

where $\omega = e^{2\pi i/5}$ is a primitive fifth root of unity. We still have the freedom to rotate the icosahedron around the vertical axis for a convenient position. We fine-tune the position of the icosahedron by agreeing that the *second coordinate axis* must go through the midpoint of one of the cross edges of the pentagonal antiprism. The half-turn U around this axis thus becomes a symmetry of the icosahedron, and as noted above, it corresponds to the linear fractional transformation

$$U : z \mapsto -\frac{1}{z}.$$

The rotations S and U do not generate the entire symmetry group of the icosahedron because they both leave the equator invariant. We choose for another generator the half-turn V whose axis is orthogonal to the axis of U and goes through the midpoint of an edge in the base of the upper pentagonal pyramid (Figure 18.1). Since U and V are (commuting) half-turns with orthogonal rotation axes, their composition $W = UV$ is also a half-turn whose rotation axis is orthogonal to those of U and V. With the identity, U, V, and W form a dihedral subgroup D_2 of the icosahedral group I.

To see what linear fractional transformations correspond to V and W, we need to work out the coordinates of the axis of V. To do this we first claim that the vertices of the icosahedron

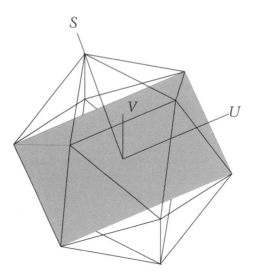

S

V

U

Figure 18.1

stereographically projected to $\hat{\mathbf{C}}$ are

$$0, \quad \infty, \quad \omega^j(\omega + \omega^4), \quad \omega^j(\omega^2 + \omega^3), \quad j = 0, \dots, 4.$$

Clearly, the poles correspond to 0 and ∞. The remaining ten vertices of the pentagonal antiprism projected to $\hat{\mathbf{C}}$ appear in two groups of five points equidistantly and alternately distributed in two concentric circles (Figure 18.2). As in Section 17, we now think of the icosahedron as being the convex hull of three mutually orthogonal golden rectangles. Consider the golden rectangle that contains the North Pole as a vertex. The two sides of this golden rectangle emanating from the North Pole can be extended to $\hat{\mathbf{C}}$ and give one projected vertex on each of the concentric circles. Considering similar triangles on the plane spanned by this golden rectangle, we see that the radii of the two concentric circles are τ and $1/\tau$ (Figure 18.3). By Problem 5 (b) in Section 17, in terms of ω, these radii are

$$\tau = -\omega^2 - \omega^3 \quad \text{and} \quad \frac{1}{\tau} = \omega + \omega^4.$$

To finish the proof of our claim, we now note that the projected vertices have a 5-fold symmetry given by multiplication by ω. By construction, one of the vertices in the outer concentric circle must

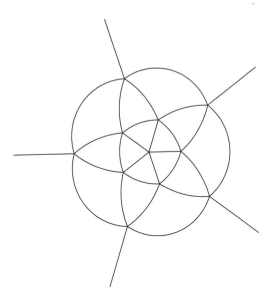

Figure 18.2

be on the negative first axis. The formula above for the projected vertices now follows by easy inspection.

Returning to the main line, we now discuss how the projected vertices can be used to obtain coordinates of the axis of rotation for the half-turn V. Since the axis of U, the second coordinate axis, is orthogonal to the axis of V, the latter goes through the midpoint of the segment connnecting $h_N^{-1}(\omega^2(\omega^2 + \omega^3))$ and $h_N^{-1}(\omega^3(\omega^2 + \omega^3))$. By the explicit form of h_N^{-1} (cf. Problem 1 (c) of Section 7), this

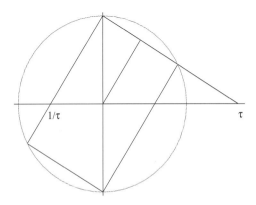

Figure 18.3

midpoint is

$$2\left(\frac{\tau^2}{\tau^2 + 1}, 0, \frac{\tau^2 - 1}{\tau^2 + 1}\right).$$

Normalizing, we obtain that $V = R_\pi(p_0)$, where

$$p_0 = \left(\frac{\tau}{\sqrt{\tau^2 + 1}}, 0, \frac{1}{\sqrt{\tau^2 + 1}}\right).$$

Using the notation of Cayley's theorem, we thus have

$$\lambda = \frac{i}{\sqrt{\tau^2 + 1}} \quad \text{and} \quad \mu = \frac{i\tau}{\sqrt{\tau^2 + 1}}.$$

We can rewrite λ and μ in terms of ω. A simple computation gives

$$\lambda = -\frac{1}{\omega - \omega^4} = \frac{\omega^2 - \omega^3}{\sqrt{5}}$$

and

$$\mu = \frac{\omega^2 + \omega^3}{\omega - \omega^4} = \frac{\omega - \omega^4}{\sqrt{5}}.$$

(cf. Problem 5 (c) of Section 17.)

Remark.
The usual argument leading to λ and μ above involves computation of the angle of the axis of the rotation of V with a coordinate axis. In our approach we adopted Schläfli's philosopy and expressed all metric properties in terms of the golden section.

The linear fractional transformation corresponding to the half-turn V finally can be written as

$$V : z \mapsto \frac{(\omega^2 - \omega^3)z + (\omega - \omega^4)}{(\omega - \omega^4)z - (\omega^2 - \omega^3)}.$$

Composing this with $U : z \mapsto -1/z$, we obtain the linear fractional transformation corresponding to $W = UV$:

$$W : z \mapsto \frac{-(\omega - \omega^4)z + (\omega^2 - \omega^3)}{(\omega^2 - \omega^3)z + (\omega - \omega^4)}.$$

As noted above, W is also a half-turn, since the axes of U and V are orthogonal. Making all possible combinations of these linear fractional transformations with those corresponding to multiplication by ω^j, $j = 0, \ldots, 4$, we finally arrive at the 60 elements of the *icosahedral Möbius group I*:

$$z \mapsto \omega^j z, \quad -\frac{1}{\omega^j z}, \quad \omega^j \frac{-(\omega - \omega^4)\omega^k z + (\omega^2 - \omega^3)}{(\omega^2 - \omega^3)\omega^k z + (\omega - \omega^4)},$$

$$\omega^j \frac{(\omega^2 - \omega^3)\omega^k z + (\omega - \omega^4)}{(\omega - \omega^4)\omega^k z - (\omega^2 - \omega^3)}, \quad j, k = 0, \ldots, 4.$$

♠ One final note. Under the aegis of Cayley's theorem we constructed a list of finite Möbius groups. Each of these groups G has a *double cover* G^* in $SU(2)$. G^* is a finite group of special unitary matrices, and $|G^*| = 2|G|$. G^* is called the *binary group associated to G*. By definition, G^* is the inverse image of G under the natural projection $SU(2) \to SU(2)/\{\pm I\}$. Thus, we can talk about the *binary dihedral group* $\mathbf{D}_n^*$, the *binary tetrahedral group* $\mathbf{T}^*$, *the binary octahedral group* $\mathbf{O}^*$, and the *binary icosahedral group* $\mathbf{I}^*$. (We left the cyclic group out. Why?) We will return to these groups in Section 23 in a more geometric setting.

19

SECTION

Detour in Topology: Euler–Poincaré Characteristic

♣ Euler's theorem for convex polyhedra states that for any convex polyhedron, the alternating sum $V - E + F$ is 2. This suggests that this quantity is a property that refers to something more general than the actual polyhedral structure. We thus venture away from convexity and try the alternating sum on nonconvex objects.

We immediately run into difficulty, since we have not defined the concept of a nonconvex polyhedron; so far, all the polyhedra we've constructed have been convex. Fortunately, this is not a serious problem. We can simply say that a general polyhedron is the union of finitely many convex polyhedra, with the property that each two in the union are either disjoint or meet at a common face. This common face is deleted from the union. Figure 19.1 shows an example. Counting, we have $V = 16$, $E = 32$, and $F = 16$, so $V - E + F = 0$! (That Euler's theorem fails for nonconvex polyhedra was first noticed by Lhuilier in 1812.) We begin to suspect that the fact that we ended up with 0 rather than 2 as before may have something to do with our new polyhedron having a "hole" in the middle. This is definitely a topological property!

The alternating sum somehow detects the presence of "holes" (rather than the actual polyhedral structure), so that it is time to say farewell (but not goodbye!) to our Platonic solids. We do this

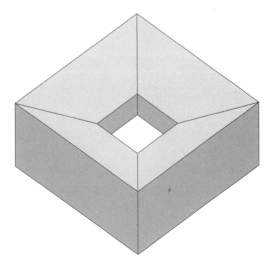

Figure 19.1

by circumscribing around each solid a sphere—a copy of S^2—and projecting the faces radially to the sphere from the centroid. The faces projected to S^2 become spherical polygons. We also realize that it does not matter whether we count $V - E + F$ on the polyhedron or on S^2. Turning the question around, we now start with S^2. Consider a spherical graph on S^2 with simply connected faces, and count $V - E + F$. Repeating the proof of Euler's theorem, we realize that we always end up with 2. The conclusion is inevitable: This magic number 2 is a property of S^2, not the actual spherical graphs!

We say that the Euler–Poincaré characteristic of S^2 is 2, and write $\chi(S^2) = 2$. One fine point: The regular tetrahedron, octahedron, and icosahedron projected to S^2 give spherical graphs whose faces are spherical triangles. When S^2 is subdivided into spherical triangles, we say that S^2 is *triangulated*. Going back to the neglected cube and dodecahedron, we see that (discarding regularity) we can split their faces into triangles without changing $V - E + F$, so that projecting these to S^2 also give triangulations of S^2. It is thus convenient to restrict ourselves to triangulations.

We have now accumulated enough information to go beyond S^2. Taking a look at Figure 19.1 again, we see that the appropriate surface to circumscribe about this polyhedron is the torus. The story is the same as before; the proof of Euler's theorem leads to the conclusion that the Euler–Poincaré characteristic of the torus T^2 is zero:

$$\chi(T^2) = 0.$$

♠ We now have great vistas ahead of us. We can take any general compact[1] surface, triangulate it, and work out the Euler–Poincaré characteristic. For example, looking at the "two-holed torus" depicted in Figure 19.2, we see that its Euler–Poincaré characteristic is −2. In general, a "p-holed torus" or, more elegantly, a closed Riemann surface M_p of genus p, has Euler–Poincaré characteristic

$$\chi(M_p) = 2 - 2p.$$

Another fine point: You might be wondering why we did not consider the cylinder and all the noncompact examples H^2/G, where G is Fuchsian, containing parabolic or hyperbolic isometries. The answer is that, at this point, we restrict ourselves to finite triangulations. Otherwise, the direct evaluation of $V - E + F$ is impossible.

Instead of going further along this line (which would lead us straight into homology theory, a branch of algebraic topology), we

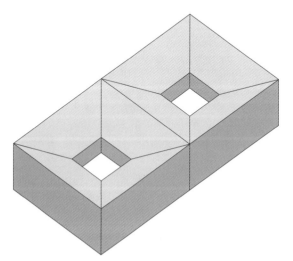

Figure 19.2

[1] See "Topology" in Appendix C.

explore some of the compact surfaces such as the Klein-bottle K^2 and the real projective plane $\mathbf{R}P^2$. Recall that K^2 is obtained from $\mathbf{R}^2$ by a discrete group generated by two glides with parallel axes. The fundamental domain again becomes a key player. In fact, if we triangulate the fundamental domain so that the side-pairing transformations map edges to edges, we obtain a triangulation of the quotient surface!

In our case, we take the fundamental domain of the Klein bottle as described in Section 16 and triangulate it as in Figure 19.3. The simplicity of this is stunning; K^2 can be obtained from two triangles by pasting their edges together appropriately! The Euler–Poincaré characteristic is

$$\chi(K^2) = 1 - 3 + 2 = 0.$$

Notice that the four vertices of the fundamental domain are identified by the side-pairing transformations so that, on K^2, we have only one vertex. Similarly, the two horizontal and the two vertical edges are identified, and the diagonal stays distinct.

We do the same for $\mathbf{R}P^2 = S^2/\{\pm I\}$ by triangulating S^2 so that the antipodal map $-I : S^2 \to S^2$ leaves the triangulation invariant; that is, it maps triangles to triangles, edges to edges, and vertices to vertices. The task is easier if we consider triangulation of the lower hemisphere only and make sure that the antipodal map restricted

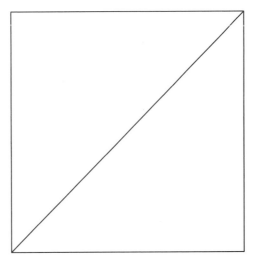

Figure 19.3

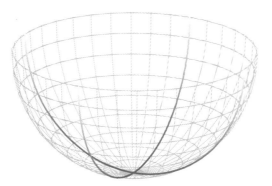

Figure 19.4

to the equatorial circle maps edges to edges and vertices to vertices. A triangulation is given in Figure 19.4. We have $V = 3$, $E = 6$, and $F = 4$. The Euler–Poincaré characteristic is

$$\chi(\mathbf{R}P^2) = 3 - 6 + 4 = 1.$$

In Section 17, when studying symmetries of the octahedron, we came very close to a polyhedral model related to $\mathbf{R}P^2$. This model is called a *heptahedron*, since it has four triangular and three square faces. For the four triangles we take four nonadjacent faces of the octahedron. The four vertices complementary to a pair of opposite vertices of the octahedron are the vertices of a square. There are exactly three opposite pairs of vertices in the octahedron whose complements give the three square faces of the heptahedron (Figure 19.5). The three square faces intersect in three diagonals (connecting the three opposite pairs of vertices), and the diagonals intersect at the centroid of the octahedron in a triple point. Although the heptahedron is self-intersecting and "singular" at the vertices, it has Euler–Poincaré characteristic of 1, since it has 6 vertices, 12 edges, and 7 faces. Furthermore, it is "nonorientable" in the sense that there is a triangle-square-triangle-square sequence of 4 adjacent faces that constitutes a polyhedral model of the Möbius band.

A natural question arises whether the heptahedron can be "deformed" into a smooth surface. Self-intersection cannot be eliminated. In fact, it is not hard to prove that any compact surface that contains a Möbius band cannot be realized as a smooth surface in $\mathbf{R}^3$ without self-intersections. As a first attempt to make

Figure 19.5

a smooth heptahedral model, we consider the quartic (degree 4) surface given in coordinates $p = (a, b, c) \in \mathbf{R}^3$ by the equation

$$a^2 b^2 + b^2 c^2 + c^2 a^2 = abc$$

(see Figure 19.6). This is called the *Roman surface*, and it was studied by Steiner. Although this surface patterns the structure and the symmetries of the heptahedron, it still contains six singular points. (Where are they?) A somewhat better model can be obtained from the hemisphere model H of $\mathbf{R}P^2$ discussed at the end of Section 16. Indeed, identifying first two pairs of equidistant antipodal pairs of points on the boundary circle of H and then pasting the remaining quarter circles together, we arrive at another model of $\mathbf{R}P^2$, algebraically given by the equation

$$(a^2 + 2b^2)(a^2 + b^2 + c^2) - 2c(a^2 + b^2) = 0.$$

In cylindrical coordinates $a = r \cos\theta, b = r \sin\theta$, and c, the equation reduces to

$$r^2 + (c - c(\theta))^2 = c(\theta)^2,$$

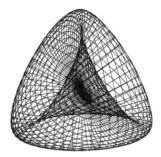

Figure 19.6

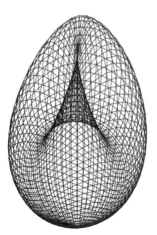

Figure 19.7

where $c(\theta) = 1/(1 + \sin^2(\theta))$. Thus the surface is swept by a rotating vertical circle with center at $(0, 0, c(\theta))$ and radius $c(\theta)$ (see Figure 19.7). This surface self-intersects in the vertical segment connecting $(0, 0, 1)$ and $(0, 0, 2)$, and at the two endpoints it is still singular! The puzzling question whether $\mathbf{R}P^2$ has a realization in $\mathbf{R}^3$ as a smooth (self-intersecting) surface without singular points has been resolved by W. Boy,[2] and the resulting surface is called the *Boy's surface*. The basic building block of the Boy's surface is a cylinder whose base curve is a leaf of the *four-leaved rose* given by the polar equation $r = \sin(2\theta)$, $0 \le \theta \le 2\pi$. In Figure 19.8, the leaf is situated in the third quadrant of the plane spanned by the second and third axes in $\mathbf{R}^3$ so that the rulings of the cylinder are parallel to the first axis. It is important to observe that the cylinder has right "dihedral" angle along the first axis. We need a finite portion of the cylinder in the negative octant of $\mathbf{R}^3$ between the origin and the negative of the arc length of the leaf.

We now rotate the cylinder around the axis $\mathbf{R} \cdot (1, 1, 1)$ by $120°$ and $240°$. The configuration of the three cylinders is shown in Figure 19.9 (a view from the positive octant) and in Figure 19.10 (a view from the negative octant). The cylinders intersect in three curves that meet in two triple intersection points. It is clear that this configuration can be made smooth along these curves, leaving

[2]D. Hilbert and S. Cohn-Vossen, *Geometry and Imagination*, Chelsea, New York, 1952, or W. Lietzmann, *Visual Topology*, Elsevier, 1969.

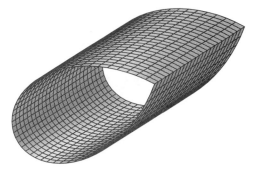

Figure 19.8

the right dihedral angles intact. For the next step we bend the first cylinder such that the original ruling on the first axis is bent exactly to the shape of a leaf in the second quadrant of the plane spanned by the first and third axes (see Figure 19.11). Notice that the right dihedral angle can be retained, and that the hole created by the bending is congruent to the initial and terminal base leaves of the bent cylinder. We cover the hole by a flat leaf and perform the same bending and covering procedure to each of the three cylinders in the configuration in Figure 19.9.

The final result is the Boy's surface depicted in Figure 19.12. Since the dihedral angles were kept 90°, this configuration is a

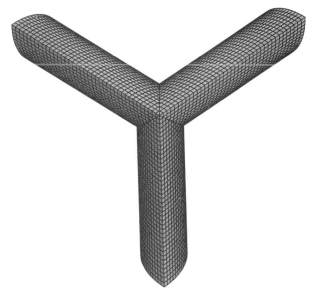

Figure 19.9

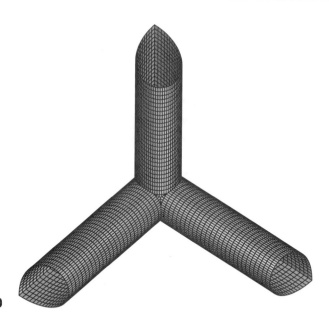

Figure 19.10

smooth self-intersecting surface with no singular points. The intersection is a closed curve consisting of three leaves forming a triple intersection point at the origin. A small strip cut from the surface along this curve gives a Möbius band, and its complement is a topological disk. It follows that the Boy's surface is a topological model of the real projective plane.

It should be clear by now that the Euler–Poincaré characteristic is a topological invariant; that is, two homeomorphic compact

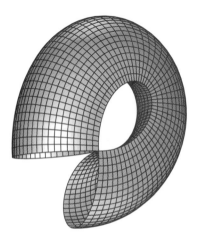

Figure 19.11

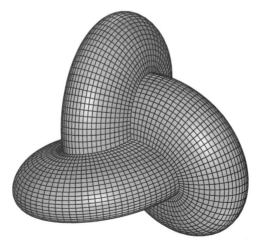

Figure 19.12

surfaces have the same Euler–Poincaré characteristic. This is because a triangulation on one surface can be carried over to the other by a homeomorphism. The big question is, of course, the converse: Given two compact surfaces with the same Euler–Poincaré characteristic, are these surfaces homeomorphic? The answer is, in general, no. For example, the torus T^2 and the Klein bottle K^2 both have vanishing Euler–Poincaré characteristic, but they are not homeomorphic, since the torus is orientable while the Klein bottle is not. It is a result of surface theory that two compact surfaces are homeomorphic iff their Euler–Poincaré characteristics are equal and they are both orientable or nonorientable. This is indeed very beautiful, since a single number, plus the knowledge of orientability, characterizes the entire surface topologically!

Many new examples of compact surfaces can be obtained by forming *connected sums*. Let M_1 and M_2 be compact surfaces. The connected sum $M_1 \# M_2$ is obtained from M_1 and M_2 by cutting out open disks $D_1 \subset M_1$ and $D_2 \subset M_2$ and then pasting $M_1 - D_1$ and $M_2 - D_2$ together along the boundary circles ∂D_1 and ∂D_2. To be precise, by a disk here we mean the inverse image of a circular disk of $\mathbf{R}^2$ in the image of a chart.

To show that we obtain a smooth surface requires some smoothing argument. It is also a technical matter (which we will not go into) that $M_1 \# M_2$ is unique up to homeomorphism; that is, its topological type does not depend on the disks chosen.

When triangulations are given on both M_1 and M_2, then D_1 and D_2 can be chosen to be the interiors of some triangles. Forming $M_1\#M_2$, we have the following changes for V_1, E_1, F_1 (for M_1) and V_2, E_2, F_2 (for M_2):

$V = V_1 + V_2 - 3$ (since 3 pairs of vertices are identified);

$E = E_1 + E_2 - 3$ (since 3 pairs of edges are identified);

$F = F_1 + F_2 - 2$ (since 2 faces are deleted).

The Euler–Poincaré charasteristic of the connected sum is therefore equal to

$$\chi(M_1 \# M_2) = \chi(M_1) + \chi(M_2) - 2.$$

Before we investigate this formula, we note that $M_1\#M_2$ is orientable iff both M_1 and M_2 are orientable.

For our first application of the formula above, we see that $\mathbf{R}P^2 \# \mathbf{R}P^2$ is homeomorphic to the Klein bottle K^2! Indeed, both are nonorientable, and $\chi(\mathbf{R}P^2 \# \mathbf{R}P^2) = 2\chi(\mathbf{R}P^2) - 2 = 0$ as for K^2. For a more direct argument, recall that $\mathbf{R}P^2$ is the Möbius band and a disk pasted together along their boundaries. To form $\mathbf{R}P^2 \# \mathbf{R}P^2$, we delete the corresponding two disks and paste the remaining Möbius bands together along their boundaries.

As a second example, we see that for any compact surface M the connected sum $M \# S^2$ is homeomorphic to M.

As a third example, the connected sum of p copies of the torus T^2 gives the p-holed torus, or a compact surface of genus p:

$$M_p = T^2 \# \cdots \# T^2 \ (p \text{ times}).$$

Here, equality means "homeomorphic." Keeping this practice, we can write the first two examples as

$$\mathbf{R}P^2 \# \mathbf{R}P^2 = K^2 \quad \text{and} \quad M \# S^2 = M.$$

(Algebraically, "twice" $\mathbf{R}P^2$ is K^2, and S^2 is the "zero" element for #. It may sound a little weird to add surfaces, but actually it is not as strange as it first sounds.)

Finally, we compile the following table:

Surface	Euler–Poincaré Characteristic
$T^2 \# \cdots \# T^2$ (p times)	$2 - 2p$
$\mathbf{R}P^2 \# \cdots \# \mathbf{R}P^2$ (p times)	$2 - p$
$\mathbf{R}P^2 \# T^2 \# \cdots \# T^2$ (p times)	$1 - 2p$
$K^2 \# T^2 \# \cdots \# T^2$ (p times)	$- 2p$

♠ Let $f : M \to N$ be a nonconstant analytic map between compact Riemann surfaces, and assume that M has genus p and N has genus q. By the table above, we have $\chi(M) = 2 - 2p$ and $\chi(N) = 2 - 2q$. As shown in complex analyis, $f : M \to N$ is an n-fold covering with finitely many branch points in M. Near a branch point on M and near its f-image (called a branch value) local coordinates z and w can be introduced that vanish at the branch point ($z = 0$) and at the branch value ($w = 0$) such that in these coordinates, f has the form $w = z^m$. We call $m - 1$ the branch number of f at the branch point (cf. Section 15). The sum of all branch numbers is called the total branch number, and it is denoted by B. A standard argument[3] shows that each point in N is assumed precisely n times on M by f, counting multiplicities. At a branch value, this means that n is equal to the sum of all m's ([branch number plus 1]'s), where the sum is over those branch points that map to the given branch value.

To relate the Euler–Poincaré characteristics of M and N, triangulate first N such that every branch value is a vertex of the triangulation. Let V, E, and F denote the number of vertices, edges, and faces of this triangulation. By Euler's theorem, we have

$$V - E + F = \chi(N) = 2 - 2q.$$

Now pull the triangulation on N back to a triangulation on M via f. Looking at what happens at a branch point reveals that the induced triangulation on M has $nV - B$ vertices, nE edges, and nF faces. Once again by the Euler's theorem, we have

$$nV - B - nE + nF = \chi(M) = 2 - 2p.$$

---

[3]See H.M. Farkas and I. Kra, *Riemann surfaces*, Springer, 1980.

Comparing this with the previous formula, we obtain the *Riemann–Hurwitz relation*

$$p = n(q - 1) + 1 + B/2.$$

There are a number of important consequences of this relation. For example, the total branching number B is always even; $p = 0$ implies $q = 0$; and if $p, q \geq 1$ and $p \neq q$, then f must have branch points. We will use the Riemann–Hurwitz relation for analytic maps $f : S^2 \to S^2$, in which case it asserts that the total branching number is given by $B = 2n - 2$.

Problems

1. Prove directly that $T^2 \# \mathbf{R}P^2$ is homeomorphic to $\mathbf{R}P^2 \# \mathbf{R}P^2 \# \mathbf{R}P^2$.

2. Check the computations leading to the table above.

Film

S. Levy, D. Maxwell and T. Munzner: *Outside In*, Geometry Center, University of Minnesota; A K Peters, Ltd. (289 Linden Street, Wellesley, MA 02181).

20

SECTION

Detour in Graph Theory: Euler, Hamilton, and the Four Color Theorem

♣ If we lived in the early 1700s in the village of Königsberg (now Kaliningrad, formerly part of East Prussia), on nice sunny Sunday afternoons we would stroll along the river Pregel and walk over its seven bridges, which connect the banks and two islands in the river (Figure 20.1). We would overhear the people who pass by say that no one has ever been able to pass over all the bridges exactly once during one stroll. Some people would even say that this is impossible. (If you plan to visit the city, you may note that two more bridges have been built since then, one serving as a railway link. Our analysis however remains the same.)

When this problem arrived at Euler's desk around 1736, graph theory was born. Euler recognized that (as far as passing the bridges is concerned) it does not matter what our exact location is at any time during the walk as long as we know which of the two banks or two islands we are on. We can thus collapse these four pieces of land to four points (called vertices), and the bridges will become edges connecting these points. We arrive at the graph shown in Figure 20.2.

The Königsberg bridge problem can thus be reformulated as follows: Does there exist a "walk" in this graph that passes along each edge exactly once? To be exact, we also have to specify whether

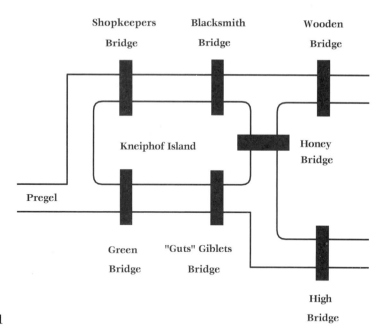

Figure 20.1

we wanted to arrive at the same spot we started at or not. In the former, we say that the walk is a "circuit"; in the latter, a "trail."

Due to the low number of vertices and edges (plus the axial symmetry), it is easy to see that there are no circuits or trails of this kind. Instead of doing a case-by-case check, we will follow Euler's simple and powerful argument. Since this applies to any graph, it is now time to give some general definitions.

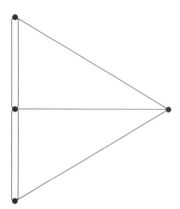

Figure 20.2

A *graph G* consists of a finite set $\mathcal{V}$ of vertices and a finite set $\mathcal{E}$ of edges. Each edge $e \in \mathcal{E}$ connects a two-element subset $\{v_1, v_2\}$ from $\mathcal{V}$. There may be multiple edges; that is, the same two vertices can be joined by more than one edge.

Given a graph G, a *walk* in G is a sequence of vertices $v_1, v_2, \ldots, v_k$, where at least one edge e connects each pair of consecutive vertices $\{v_i, v_{i+1}\}$, $i = 1, \ldots, k-1$, in the sequence. In case of multiple edges, each two-element subset in a walk specifies a chosen edge. We usually denote a walk by $v_1 \cdots v_k$. We say that the walk *joins* the vertices v_1 and v_k. The graph is *connected* if there is a walk joining any two of its vertices. A walk is *spanning* if the vertices in the walk make up the whole of $\mathcal{V}$. A walk is *closed* if $v_1 = v_k$. If no vertices are repeated in a walk, we have a *path*. If no edges are repeated, the walk is a *trail*. A closed trail is called a *circuit*. Finally, a circuit that has at least one edge, and in which the only repeated vertex is $v_1 = v_k$, is called a *cycle*.

We see that the Königsberg bridge problem is equivalent to finding a circuit that includes each edge of the graph exactly once. Such circuits are called *Eulerian*. The graph is Eulerian if it contains an Eulerian circuit. Thus our problem is to decide whether the graph of the Königsberg bridge problem is Eulerian or not. There is a very simple criterion for the Eulerian property expressed in terms of the degree of vertices of the given graph. The *degree* of a vertex v in a graph G is the number of edges having v as an endpoint.

Theorem 12.

A connected graph is Eulerian iff every vertex of the graph has even degree.

Proof.

The proof is embarrassingly simple. Assume that G is Eulerian. Given an Eulerian circuit along which we are moving, every time we traverse a vertex v we do it by leaving an "entrance" edge and getting onto an "exit" edge. Since we have to traverse each edge exactly once, we discard these two edges, which contribute 2 in the degree. As we continue discarding, the degree of each vertex goes down by 2's. Finally we run out of edges. Thus each vertex must have even degree. Conversely, assume that the degree condition holds and let us get started. At any point of our circuit-making trail, upon entering in a vertex (along an "entrance" edge), the question arises whether we have an exit edge that has not been used so far and can then be used to go on. The answer is yes, since each previous visit to v "consumed" exactly two edges, and if we were unable to go on the degree of v would be a sum of 2's plus 1 corresponding to the entrance edge—an odd number. We are done. ∎

It now takes only a second to realize that the graph in the Königsberg bridge problem has all odd-degree vertices, so it is not Eulerian! (This can be pushed a little further. We cannot traverse each bridge exactly once even if we drop the assumption that we arrive back at the same spot. For this, there must be exactly two odd-degree vertices (corresponding to departure and arrival), and all the other vertices must have even degrees.)

What has this to do with Platonic solids? To explain this, we should consider vertices and edges as some sort of reciprocal notions. In an Eulerian circuit we must traverse each edge exactly once, but we can visit the vertices as many times as needed. Switching the roles of edges and vertices, we now ask whether a given graph G has a *spanning cycle*; i.e., a circuit traversing each vertex of G exactly once. A spanning cycle is called *Hamiltonian*, and a graph possessing a Hamiltonian cycle is called Hamiltonian as well. Our first contact with the regular solids comes from Hamilton's marketing ambitions (around 1857), which involved a wooden dodecahedron with each of its 20 vertices labeled with the name of a town. The puzzle was to find a circuit along the edges of the dodecahedron which passed through each town exactly once. A solution is given in Figure 20.3. We can say elegantly that the Schlegel diagram of a dodecahedron is Hamiltonian!

Despite reciprocity of the Eulerian and Hamiltonian properties, there is no efficient algorithm to determine whether a graph is Hamiltonian. Instead of showing the subtleties of this problem, we switch to another problem closely related to regular solids. Recall that a graph was defined "abstractly" by its set of vertices and edges, and no reference was made to whether the graph can be realized (more elegantly, imbedded) in the plane such that vertices corre-

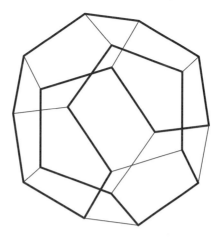

Figure 20.3

spond to points and edges to continuous curves connecting these points. No edges are allowed to intersect away from the vertices. A graph with this property is called *planar*. Figure 20.4 shows two typical examples of nonplanar graphs. (In fact, a deep result of Kuratowski asserts that a graph is planar if it does not contain any subgraphs like these.)

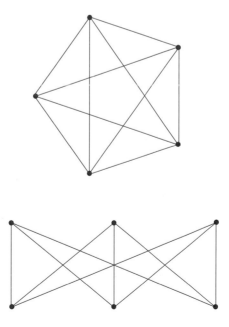

Figure 20.4

A representation of a connected planar graph on the plane is exactly what we used to prove Euler's theorem on convex polyhedra. Thus, we immediately know that

$$V - E + F = 2.$$

The structure of faces here depends on the specific representation while V and E do not. To get a result for planar graphs independent of the planar representation, we therefore try to eliminate F from this formula. We arrive at the following:

Theorem 13.

Let G be a connected planar graph with no multiple edges, and assume that $E > 1$. Then $E \leq 3V - 6$.

Proof.

The case $E = 2$ is trivial. We may therefore assume that $E > 2$. Consider a specific representation of G in the plane. Since there are no multiple edges, each face is bounded by at least three edges. When counting edges this way, going from face to face, each edge (bounding two faces) is counted at most twice. We obtain that $3F \leq 2E$. Combining this with Euler's theorem $F = 2 - V + E$, we obtain $3F = 6 - 3V + 3E \leq 2E$. ∎

In 1852, a London law student named Francis Guthrie asked the following question: "Suppose you have a map, and want to color the various countries so that any two countries which share a common border always have different colors. What is the maximum number of colors that you might need?" Note that two different countries are allowed to have the same color provided that they do not share a common border. So, for example, Canada and Mexico may have the same color, but the United States must be colored differently from both Canada and Mexico. Also note that two countries may have the same color if they touch only at a corner, so if you have sixty-four countries arranged like the squares of a checker board, then the usual red and black coloring is legal. Guthrie conjectured that every map can be colored with at most four different colors.

Guthrie was surely not the first person to guess that four colors suffice to color any map; some mapmaker must have made the same guess long before. But Guthrie seems to have been the first person to notice that this is a mathematical question. If it is really true that any map can be colored with four colors, then this ought to be a theorem with a mathematical proof. Guthrie himself was unable to find such a proof, but he was also unable to find a map that required five colors. So he told his brother Frederick about the problem. Frederick was studying mathematics at University College London with Augustus De Morgan (as Francis had done before deciding to study law) and he passed the problem on to De Morgan, who was also unable to find a proof or a counterexample. Over the years, the problem was passed around, mostly among British mathematicians, and in 1878 Arthur Cayley published the question in the Proceedings of the London Mathematical Society. The next year an amateur mathematician named Alfred Kempe published what he believed was a proof. Even Cayley was convinced by Kempe's argument, but in 1890 Percy Heawood found a subtle flaw in Kempe's reasoning. Heawood was, however, able to prove that no map requires more than five colors.

No apparent progress was made for the next 86 years, and the four color problem became one of the most famous unsolved problems in mathematics. Finally, in 1976, Wolfgang Haken and Kenneth Appel, carrying out an idea suggested a few years earlier by Heesch, proved that Guthrie was right; every map can be colored with four colors. It is not surprising that such first-class mathematicians as De Morgan and Cayley were unable to solve this problem, because the proof that Haken and Appel discovered was far longer than any previous proof in the history of mathematics. The full proof, with all the details, has never been published, or even written down. If it were written down, it would certainly occupy millions, or perhaps billions, of pages. Naturally, Haken and Appel could not check all the details themselves; instead they wrote a computer program to check the proof for them. The program ran day and night for about two months! (Longer computer-assisted proofs have been discovered since 1976. For example, in 1990 Brendan McKay proved that at any dinner party with at least 27 guests, one can always find 3 people who all know each other or 8 people who are all strangers to each other, but that this is not always true with 26 guests. McKay's proof consisted of checking every possible combination of acquaintance and non-acquaintance among the guests. Although he found an extremely efficient way to check huge numbers of combinations simultaneously, he still had to run his computer for *three years*. McKay's result is certainly a theorem, but somehow it strikes one as being of much narrower scope than the four color theorem. A handful of similar results exist in other branches of mathematics.)

The problem of coloring maps is in fact a problem in graph theory, as we can see by assigning a vertex to each country and connecting two vertices with an edge whenever the corresponding countries share a common border. In this formulation, we must color the vertices of a graph, using different colors for any two vertices connected by an edge. The minimum number of colors needed to color a graph G in this way is called the *chromatic number* of G. If a map is drawn on a plane, then the corresponding graph is planar (see Problem 17). In other words, the four color theorem is equivalent to the statement that the chromatic number of a planar graph cannot be greater than four.

Remark.

In Section 17, we described an algorithm to color the faces of the icosahedron with 5 colors such that the faces with the same color were disjoint. To obtain a 4-coloring we pick the four faces of a specific color group and recolor them using the remaining 4 colors such that no two faces with the same color meet at a common edge. (This can be done because each face has 3 adjacent faces and we have 4 colors.) This way we obtain a coloring of the faces of the icosahedron with 4 colors subject to the condition that faces with the same color can touch each other only at vertices. By reciprocity, this gives a coloring of the Schlegel diagram of the reciprocal dodecahedron with 4 colors in the sense discussed above.

The advantage of reformulating the four color problem as a question about planar graphs is that we can now apply Theorem 13. One slight obstacle remains: Theorem 13 requires that G have no multiple edges. But this is not a serious difficulty, because the chromatic number of a graph will not change if we remove all the multiple edges; all adjacent vertices will remain adjacent. (Suppose, for example, that five colors were required to color a map of Asia. The graph corresponding to this map does have multiple edges, because Russia and China share two common borders, one east and one west of Mongolia. If we were to eliminate the latter border by extending Mongolia to the west, then the map would still require five colors. In fact, conceivably it could require six, because Mongolia would now share a border with Kazakhstan. So if there existed a counterexample to the four color theorem, we could surely find it among singly connected planar graphs.)

An immediate consequence of Theorem 13 is the following:

Six Color Theorem.

The chromatic number of a planar graph cannot be greater than 6.

Proof.

First we must show that every planar graph with no multiple edges must include at least one vertex of degree 5 or less. (Before reading the rest of this proof, you may want to do Problem 8 at the end of this section.) Suppose there exists a planar graph G such that each vertex has degree 6 or more, with no multiple edges. If we add up the number of edges coming out of all the vertices, then each edge will be counted twice, because each edge connects two vertices. Therefore, $2E \geq 6V$ and $E \geq 3V$. This contradicts Theorem 13, which says that $E \leq 3V - 6$.

Now suppose there exist planar graphs with chromatic number greater than 6. Let G be such a graph with the minimum possible number of vertices (since V is positive, such a graph G must exist). Then if we remove a single vertex from G, along with its edges, the resulting graph G' will have chromatic number 6 or less. In particular, we can remove one of the vertices v_1 with five or fewer edges, and color the remaining vertices with six colors. Now restore the missing vertex v_1. Every other vertex of G has already been colored; in particular the five (or fewer) vertices which share an edge with v_1 have already been colored with five or fewer colors. Therefore, we can legally use the sixth color for v_1, so contrary to our assumption, G can be colored with six colors. This contradiction implies that G does not exist, and every planar graph can be colored with six colors.   ∎

Heawood's five color theorem[1] and Haken and Appel's four color theorem also follow from Theorem 13, but in a more roundabout way.

Five Color Theorem.

Every planar graph has a chromatic number less than or equal to 5.

Proof.

Suppose there exists a planar graph with chromatic number 6. Let G be the smallest such graph (that is, a graph with the smallest possible V). Clearly G cannot have any vertex with degree 4 or less, because we could then remove such a vertex, color the remaining (smaller) graph with five colors, and then replace the vertex, giving it a color different from that of its four neighbors. But, as we have seen, it follows from Theorem 13 that G must have a vertex of degree 5. Call one such vertex v_1. If we remove v_1, the other vertices of G can be colored with five colors. Furthermore, it must be the case that the five vertices adjacent to v_1 will use all five colors, because if only four different colors were needed for those five vertices, then the fifth color could be used for v_1.

Call the five vertices v_2, v_3, v_4, v_5, and v_6, moving clockwise in the plane around v_1. This is illustrated in Figure 20.5; note that $v_2v_3v_4v_5v_6$ is shown as a circuit with edges connecting the five vertices. There is no harm in making this assumption, because if the required edges were not originally present in G, we can surely add them without making G nonplanar or decreasing its chromatic number. Let us say that v_2, v_3, v_4, v_5, and v_6 are colored red, yellow, green, blue, and violet respectively. Now suppose we were to remove all yellow, blue, and violet vertices from our graph, along with all the edges attached to those vertices. There are two possibilities: either v_2 and v_4 are still connected (that is, there is a path from v_2 to v_4 consisting only of red and green vertices), or they are not.

---

[1]This appeared in Volume 24 of the *Quarterly Journal of Mathematics* in 1890.

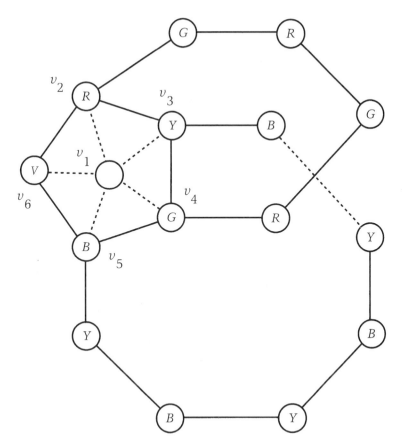

Figure 20.5

In the latter case, we can take the component of the graph connected to v_4 and color all the green vertices red and the red vertices green, while leaving the component connected to v_2 unchanged. We still have a legal coloring, because to get from one component to the other, you must pass through a region of yellow, blue, and violet vertices. But v_4 is now red, and v_2 is still red, so we can color v_1 green.

On the other hand, suppose that there is a path of red and green vertices connecting v_2 and v_4. In that case, there cannot be a path of yellow and blue vertices connecting v_3 and v_5, because the graph is planar, and the red-green path cannot cross the yellow-blue path. But if there is no such yellow-blue path, then we can swap the colors yellow and blue throughout the yellow-blue component connected to v_5, while leaving the colors alone in the yellow-blue component connected to v_3. We now have v_3 and v_5 both yellow, so we can color v_1 blue. In either case, the entire graph, including v_1, can be colored with five colors, contrary to our hypothesis. This contradiction establishes the theorem. ■

Kempe's fallacious proof of the four color theorem was in two parts. He first proved that if G is the smallest graph (in terms of vertices) with chromatic number 5, then G has no vertices of degree 4 or less (see Problem 20). This part of the proof was correct. He then continued as follows: Let v_1 be a vertex of G with degree 5 (by Theorem 13 and Problem 20, v_1 must exist), and let v_2, v_3, v_4, v_5, and v_6 be the vertices adjacent to v_1, counting clockwise, as in the proof of the five color theorem. Since G is minimal, we can color all the vertices except v_1 with four colors. Such a coloring must use all four colors for v_2, v_3, v_4, v_5, and v_6; otherwise we could use the fourth color for v_1. That means two of the five vertices must be the same color, and the other three different. The two vertices that are the same color cannot be adjacent, so we might as well assume that v_3 and v_6 are the same color (if not, just rename the vertices by moving the names around the circuit). In particular, let's suppose that v_3 and v_6 are yellow, v_2 is red, v_4 is green, and v_5 is blue (see Figure 20.6). Now suppose there is no red-green path connecting v_2 and v_4. Then we can recolor v_4 red (also recoloring the red-green component connected to v_4) while v_2 remains red; v_1 can then be green. If there is no red-blue path connecting v_2 and v_5, then we can recolor v_5 red, and color v_1 blue. The only other possibility is that a red-green path connects v_2 to v_4 and a red-blue path connects v_2 to v_5. In that case, there cannot be a yellow-blue path connecting v_3 with v_5, because, as one can see in Figure 20.6, such a yellow-blue path would have to cross the red-green path if it is to remain in the plane. Likewise, there cannot be a yellow-green path connecting v_6 and v_4, because such a path would have to cross the red-blue path. Therefore, we can swap the colors yellow and blue for all vertices inside the red-green path, and we can swap the colors yellow and green for all vertices inside the red-blue path. This changes v_3 to blue (but leaves v_5 blue) and changes v_6 to green (but leaves v_4 green). In that case, we can

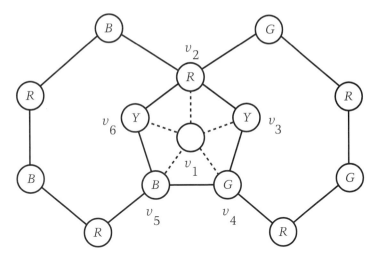

Figure 20.6

color v_1 yellow. So in all three cases, the entire graph, including v_1, can be colored with four colors, contrary to our hypothesis. This contradiction establishes the theorem.

Although Kempe's proof doesn't quite work, he was on the right track. He merely had to consider a few additional cases.

Theorem 13 tells us that the average degree of the vertices in a planar graph is always less than 6. That means every planar graph has vertices of degree 5 (or less, but Problem 20 tells us that a minimal 5-color graph cannot have vertices of degree less than 5). But Theorem 13 actually tells us a good deal more than that. For example, it rules out the possibility that every vertex of degree 5 is surrounded by five vertices of degree ≥ 7, because in such a graph, the average vertex would have degree > 6. In other words, any minimal 5-color graph must include instances where (a) two vertices of degree 5 are adjacent or (b) a vertex of degree 5 is adjacent to a vertex of degree 6. The two configurations are shown in Figure 20.7, and they are said to make up an *unavoidable set*, which we define to be any set of subgraphs with the property that at least one of them is included in a minimal 5-color graph. Of course a single vertex of degree 5 is also, by itself, an unavoidable set, but it does us no good, because we cannot rule it out by using Kempe's method of paths of alternating colors to show that a 5-color graph containing such a vertex cannot be minimal.

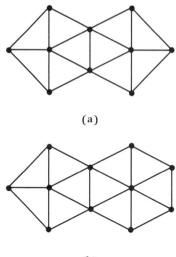

(a)

Figure 20.7   (b)

Suppose, however, it were possible to successfully carry out
Kempe's method on both the subgraphs shown in Figure 20.7. That
is, suppose that for each of these subgraphs, we could prove that
if a graph *G* containing that subgraph could be 4-colored except
for one vertex in the subgraph, then it would always be possible to
recolor *G* so that the missing vertex could also be legally colored. A
subgraph on which we successfully carry out such a proof is said to
be *reducible*. In that case, we would know that any graph containing subgraph (a) or (b) from Figure 20.7 could not be the minimal
5-color graph. But the minimal 5-color graph *must* include one of
these subgraphs. Consequently, the minimal 5-color graph could
not exist, and we would have proved the four color theorem.

Unfortunately, Kempe's method of paths of alternating colors
cannot be successfully applied to the subgraphs in Figure 20.7. But
it can be carried out with a number of slightly larger subgraphs,
such as those in Figure 20.8. So perhaps we should try to find
a somewhat larger unavoidable set of larger subgraphs. All that
we need to prove the four color theorem is an *unavoidable set of
reducible subgraphs*.

In 1969, Heesch put forward a statistical argument which suggested, but did not prove, that there exists an unavoidable set of
about 8900 reducible subgraphs, each having no more than 18

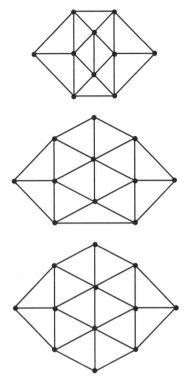

Figure 20.8

vertices around its perimeter. (In order to prove that a subgraph is reducible, you must consider every possible way to color its perimeter, so it is helpful to have the perimeter as short as possible.) He also developed methods, which could be carried out quite mechanically, for proving that a given set is unavoidable, and also for proving that a given subgraph is reducible. But Heesch had no way of finding this unavoidable set. And even if the set had somehow been handed to him (by divine revelation, say), it would have been completely impractical, given the computer hardware available in 1969, to prove that the set was unavoidable, or that each member of the set was reducible.

Haken and Appel were able to sharpen Heesch's statistical argument enough to convince themselves that an unavoidable set of reducible subgraphs could be constructed with no more than 14 vertices on the perimeter of each. This greatly decreased the time needed to prove reducibility. By 1976 computers were significantly faster than in 1969, and so, after examining about 100,000 subgraphs, Haken and Appel managed to find an unavoidable set of 1936 reducible subgraphs, finally proving the four color theorem. Later they managed to decrease the size of their set to 1476. It is, to say the least, rather difficult to check such a proof, and no one seems to have actually done so. Even if one accepts the principle that a proof can be checked by computer, one would have to rewrite the software from scratch; simply rerunning the program that Haken and Appel wrote, even on a different computer, would mean little, since the program could contain errors. However, in 1997 Neil Robertson, Daniel Sanders, Paul Seymour, and Robin Thomas were able to greatly simplify Haken and Appel's proof by finding an unavoidable set of 633 reducible subgraphs. They also found a much faster algorithm for proving unavoidability, and they claim that in principle one could check the proof by hand in a few months (needless to say, no one has actually done this!), although a computer is still needed to check that each of the 633 subgraphs is in fact reducible. For details, including pictures of all 633 subgraphs, see Web Site 1 after the problems.

Problems

1. Call a graph G *complete* if any two distinct vertices in G are connected by a single edge. Given a complete graph G, show that G is Hamiltonian and $2E = V(V - 1)$.

2. Show that if any vertex in a graph has at least degree 2, then the graph has a cycle.

3. Let G be a graph with no multiple edges. Show that if G has more than $(V - 1)(V - 2)/2$ edges, then G is connected.

4. Which Platonic solids have Hamiltonian Schlegel diagrams?

5. Define the Euler-Poincaré characteristic of a graph as the number of vertices minus the number of edges. Show that if the graph is a *tree* (connected, with no cycle), then the Euler-Poincaré characteristic is 1. (Note: The converse statement is also true.)

6. ♠ Show that the two typical examples of nonplanar graphs in Figure 20.4 cannot be imbedded into the plane.

7. Use Theorem 13 to show that a complete planar graph can have at most four vertices.

8. Use Theorem 13 to show that a connected planar graph with no multiple edges must have at least one vertex of degree ≤ 5.

9. ◇ Prove Euler's theorem for convex polyhedra using linear algebra. Assume that the connected graph G is *directed*, that is, the edges are ordered pairs of vertices of G. (Geometrically, this means that each edge has an arrow indicating its direction.) Index the vertices of G by $1, 2, \ldots, V$ and the edges by $1, 2, \ldots, E$. Define the $E \times V$ *edge-vertex matrix* A of G as follows: If $e_k = (v_i, v_j)$, $i, j = 1, \ldots, V$, $k = 1, \ldots, E$, the kth edge from the ith vertex to the jth vertex, then, in the kth row of A, define the ith entry to be -1, the jth entry $+1$, and zeros elsewhere.

   (a) Show that the kernel of $A : \mathbf{R}^V \to \mathbf{R}^E$ is one-dimensional and is spanned by $(1, 1, \ldots, 1) \in \mathbf{R}^V$.

   (b) Using (a), conclude that the rank of A is $V - 1$.

   (c) Assuming that G is planar, show that the number $F - 1$ of bounded faces is the dimension of the kernel of $A^\top : \mathbf{R}^E \to \mathbf{R}^V$, the transpose of A.

   (d) Using that A and $A^\top$ have the same rank, conclude that $E - (F - 1) = V - 1$.

10. Complete the steps in von Staudt's proof of Euler's theorem for convex polyhedra.

   (a) Let G be a connected planar graph with V vertices, E edges, and F faces. Show that G contains a spanning tree T. (Spanning means that each vertex of G is a vertex of T.)

   (b) Define a graph T' as follows: Pick a point from each region defined by G and call it a vertex of T'. Thus T' has F vertices. For each edge e of G not in T, choose a curve that avoids T and connects the two vertices of T' that are contained in the regions meeting along e. Call this curve an edge of T'. Prove that T' is a tree.

   (c) Use Problem 5 to count the edges of G by first counting those in T. Conclude that $(V - 1) + (F - 1) = E$.

11. Show that a map of the United States cannot be colored with three different colors so that no two states with a common border have the same color.

12. Find a necessary condition for a map to be colorable with three colors. Is this condition also sufficient?

13. Suppose we require that two countries have different colors even if they only touch at one point. How many colors are needed to color a checker board? What is the maximum number of colors required to color any map?

14. As in Problem 13, we require that two countries have different colors even if they only touch at one point, but this time we specify that no more than four countries may come together at one point. Make a conjecture about the maximum number of colors required to color any such map.

15. Rephrase McKay's result about the dinner party with 27 guests as a theorem about a complete graph on 27 vertices, where each edge is colored either red or green.

16. Show that the maximum number of colors required for any map drawn on the plane is the same as the maximum number of colors required for any map drawn on the sphere.

17. Prove that if a map is drawn on a plane, then the corresponding graph is planar.

18. What is the chromatic number of a complete graph on n vertices?

19. Give an example of a graph with chromatic number 4 which does not contain a complete subgraph of order 4. In other words, the four color theorem does not immediately follow from the fact that a complete planar graph cannot have five vertices.

20. Let G be the smallest graph (in terms of vertices) with chromatic number 5. Show that G has no vertices of degree 4 or less.

21. Find the flaw in Kempe's proof of the four color theorem. (This should take you much less than eleven years, even if you are not as smart as Cayley, because unlike Cayley you know that there *is* a flaw.)

22. In this problem all graphs are imbedded in the real projective plane $\mathbf{R}P^2$.
    (a) Let G be a connected graph with no multiple edges imbedded in $\mathbf{R}P^2$ and assume that $E > 0$. Using $\chi(\mathbf{R}P^2) = 1 = V - E + F$, apply the proof of Theorem 13 to conclude that $E \leq 3V - 3$.
    (b) Modifying the proof of the six color theorem, show that the chromatic number of a graph imbedded in $\mathbf{R}P^2$ cannot be greater than 6.
    (c) Show that the complete graph on six vertices can be imbedded in $\mathbf{R}P^2$. (Since the chromatic number of a complete graph on six vertices is 6 (see Problem 18), this shows that the upper bound in (b) is the best possible.)

23. Let M be a compact surface with Euler-Poincaré characteristic $\chi(M) \leq 0$. Show that the chromatic number of any connected graph G imbedded in M cannot be greater than $[(7 + \sqrt{49 - 24\chi(M)})/2]$ using the following steps: Notice first that the number inside the greatest integer function is a solution of the quadratic equation $x^2 - 7x + 6\chi(M) = 0$. Rewrite this equation in the form $6(1 - \chi(M)/x) = x - 1$ and apply the argument in the proof of Theorem 13 to conclude that there is a vertex of G with degree $\leq [x] - 1$. Finally, modify the argument in the proof of the six color theorem for the present situation. (This result is due to Heawood. With the exception of the Klein bottle, the upper bound on the chromatic number is sharp.)[2]

---

[2]See G. Ringel and W.T. Youngs, "Solution to the Heawood map-colouring problem," *Proceedings of the National Academy of Sciences (U.S.A.)*, 1968.

Web Sites

1. www.math.gatech.edu/~thomas/FC/fourcolor.html

2. www-groups.dcs.st-and.ac.uk/~history/HistTopics/

Dimension Leap

◇ The success of developing complex calculus and the beauty of Riemann surfaces come about largely because we are able to multiply complex numbers and thus can form polynomials, power series, linear fractional transformations, etc. When we view complex numbers as planar vectors, complex multiplication becomes a specific operation $F : \mathbf{R}^2 \times \mathbf{R}^2 \to \mathbf{R}^2$. To extend our development to higher dimensions, we are now motivated to find an operation

$$F : \mathbf{R}^n \times \mathbf{R}^n \to \mathbf{R}^n$$

on the n-dimensional Euclidean real number space $\mathbf{R}^n$. What conditions should F satisfy? Although opinions differ, most agree that F has to be *bilinear*, i.e., linear in both arguments:

$$F(a_1 p_1 + a_2 p_2, q) = a_1 F(p_1, q) + a_2 F(p_2, q),$$

$$F(p, a_1 q_1 + a_2 q_2) = a_1 F(p, q_1) + a_2 F(p, q_2),$$

where $p, p_1, p_2, q, q_1, q_2 \in \mathbf{R}^n$ and $a_1, a_2 \in \mathbf{R}$. (This corresponds to distributivity and homogeneity.) Instead of requiring that F be associative (which is hard to handle technically), we impose the condition that F be *normed*:

$$|F(p, q)| = |p| \cdot |q|, \quad p, q \in \mathbf{R}^n.$$

The advantage of this condition is clear. It connects algebra to geometry by simply declaring that the length of the product of two vectors must be the product of the lengths of the vectors! Our belief that this is the right condition is strengthened by our knowledge of complex multiplication, where this is a characteristic identity. A normed bilinear map $F : \mathbf{R}^n \times \mathbf{R}^n \to \mathbf{R}^n$ is called an *orthogonal multiplication*. Convinced as we are that the existence of an orthogonal multiplication is the key to developing higher dimensional analysis, our hopes are crushed by the following result of Hurwitz and Radon (c. 1898–1923).

Theorem 14.

  Orthogonal multiplications $F{:}\mathbf{R}^n \times \mathbf{R}^n \to \mathbf{R}^n$ *exist only for* $n = 1, 2, 4,$ *and* 8.

Remark 1.
One is tempted to weaken the condition that F be normed as follows: A *real division algebra structure* on $\mathbf{R}^n$ is a bilinear map $F : \mathbf{R}^n \times \mathbf{R}^n \to \mathbf{R}^n$ such that F has no zero divisors, in the sense that $p \neq 0 \neq q$ implies $F(p, q) \neq 0$. This is absolutely necessary for our purposes; otherwise, we can't form fractions. Theorem 14, however, can be generalized to the effect that real division algebra structure exists on $\mathbf{R}^n$ iff $n = 1, 2, 4,$ or 8.

Remark 2.
You might object to all this by saying that we do have a multiplication that works in every dimension—the dot product! However, the dot product is not a genuine multiplication, because the dot product of two vectors is not a vector but a number. (Also, we cannot form triple products, and we have a lot of zero divisors.) If you are somewhat more sophisticated, you may ask why we don't concentrate only on $\mathbf{R}^3$, where we have a multiplication $\times : \mathbf{R}^3 \times \mathbf{R}^3 \to \mathbf{R}^3$ given by the cross product of vectors. (Recall that, given $v_1, v_2 \in \mathbf{R}^3$, the cross product $v_1 \times v_2$ is zero iff v_1 and v_2 are linearly dependent, and if they are linearly independent, then $v_1, v_2, v_1 \times v_2$ (in this order) form a positively-oriented basis, with $|v_1 \times v_2|$ equal to the area of the parallelogram spanned by v_1 and v_2.) The cross product is

not suitable either, since $\times$ is anticommutative[1] ($v_1 \times v_2 = -v_2 \times v_1$, $v_1, v_2 \in \mathbf{R}^3$)—in particular, we cannot even form squares!

A modern proof of Theorem 14 was given by Atiyah, Bott, and Shapiro. It relies on the classification of Clifford algebras and Clifford modules, a beautiful piece of modern algebra. It would not be too difficult to reproduce their work here, but since the proof uses the concept of the tensor product of algebras, we will go only as far as the definition of Clifford algebras. This is very much in the spirit of the Glimpses, and has the further advantage that the quaternionic identities will arise naturally.

Let $F : \mathbf{R}^n \times \mathbf{R}^n \to \mathbf{R}^n$ be an orthogonal multiplication. Let $e_1 = (1, 0, \dots, 0)$, $e_2 = (0, 1, 0, \dots, 0)$, ..., $e_n = (0, \dots, 0, 1)$ denote the standard basis vectors in $\mathbf{R}^n$. We define

$$u_i^\alpha = F(e_\alpha, e_i) \in \mathbf{R}^n, \quad i, \alpha = 1, \dots, n,$$

using Greek and Latin indices to distinguish between first and second arguments. (F is not symmetric!) We claim that, for fixed α, $\{u_i^\alpha\}_{i=1}^n \subset \mathbf{R}^n$ is an orthonormal basis.

First, u_i^α is a unit vector, since

$$|u_i^\alpha| = |F(e_\alpha, e_i)| = |e_\alpha| \cdot |e_i| = 1.$$

Second, let $i \neq k$, $i, k = 1, \dots, n$. On the one hand, we have

$$|u_i^\alpha + u_k^\alpha|^2 = |u_i^\alpha|^2 + |u_k^\alpha|^2 + 2u_i^\alpha \cdot u_k^\alpha$$
$$= 2 + 2u_i^\alpha \cdot u_k^\alpha.$$

On the other hand,

$$|u_i^\alpha + u_k^\alpha|^2 = |F(e_\alpha, e_i) + F(e_\alpha, e_k)|^2$$
$$= |F(e_\alpha, e_i + e_k)|^2$$
$$= |e_\alpha|^2 |e_i + e_k|^2 = 2.$$

Combining these, we obtain

$$u_i^\alpha \cdot u_k^\alpha = 0, \quad i \neq k,$$

---

[1] This does not mean that the cross product is not useful. In fact, it is the primary example of a Lie algebra structure on $\mathbf{R}^3$.

which proves the claim. Similarly, for fixed i, $\{u_i^\alpha\}_{\alpha=1}^n \subset \mathbf{R}^n$ is also an orthonormal basis.

Next we fix α, β and consider the orthonormal bases

$$\{u_i^\alpha\}_{i=1}^n \quad \text{and} \quad \{u_k^\beta\}_{k=1}^n$$

of $\mathbf{R}^n$. Recall from linear algebra that the transfer matrix, denoted by $P^{\beta,\alpha}$, between these two orthonormal bases is an orthogonal matrix. In coordinates, $P^{\beta\alpha} = (p_{ik}^{\beta\alpha})_{i,k=1}^n$, and we have the *change of bases formula*

$$u_i^\beta = \sum_{k=1}^n p_{ik}^{\beta\alpha} u_k^\alpha.$$

Orthogonality of $P^{\beta\alpha}$ is expressed by

$$P^{\beta\alpha}(P^{\beta\alpha})^\top = I,$$

where $^\top$ stands for transpose.

We now claim that, for $\alpha \neq \beta$, $P^{\beta\alpha}$ is skew-symmetric:

$$(P^{\beta\alpha})^\top = -P^{\beta\alpha}.$$

(Notice that by assuming $\alpha \neq \beta$, we exclude the case $n = 1$.) To show this, we let $i \neq k$ and compute (using orthogonality of the u_i^α's and u_i^β's, etc.):

$$|F(e_\alpha + e_\beta, e_i + e_k)|^2 = |e_\alpha + e_\beta|^2 \cdot |e_i + e_k|^2 = 4$$
$$= |u_i^\alpha + u_k^\alpha + u_i^\beta + u_k^\beta|$$
$$= 4 + 2u_i^\alpha \cdot u_k^\beta + 2u_k^\alpha \cdot u_i^\beta.$$

We obtain the following fundamental identity:

$$u_i^\alpha \cdot u_k^\beta + u_k^\alpha \cdot u_i^\beta = 0, \quad \alpha \neq \beta.$$

This also holds for $i = k$ by orthogonality.

Substituting the change of bases formula into this, we have

$$u_i^\alpha \cdot \left(\sum_{l=1}^n p_{kl}^{\beta\alpha} u_l^\alpha\right) + u_k^\alpha \cdot \left(\sum_{l=1}^n p_{il}^{\beta\alpha} u_l^\alpha\right) = 0,$$

and orthogonality of the u_i^α's gives

$$p_{ki}^{\beta\alpha} + p_{ik}^{\beta\alpha} = 0.$$

The claimed skew-symmetry follows. Combining skew-symmetry and orthogonality of $P^{\beta,\alpha}$, we obtain

$$(P^{\beta\alpha})^2 = -I.$$

We see that $P^{\beta\alpha}$ is a *complex structure* on $\mathbf{R}^n$. In general, a complex structure on $\mathbf{R}^n$ is a linear isometry $J : \mathbf{R}^n \to \mathbf{R}^n$ (represented by an orthogonal matrix) such that $J^2 = -I$. Beyond the formal analogy with the complex identity $i^2 = -1$, a complex structure J can be thought of as a prescription for rotating vectors in $\mathbf{R}^n$ by $\pi/2$.

Let us elaborate on this. First, given $0 \neq v \in \mathbf{R}^n$, we claim that v and $J(v)$ are orthogonal and of the same length. Indeed, $|J(v)| = |v|$ since J is an isometry, and

$$v \cdot J(v) = J(v) \cdot J^2(v) = -J(v) \cdot v = -v \cdot J(v);$$

so, $v \cdot J(v)$ must be zero.

It is now clear how to rotate v by angle θ using the partial coordinate system $\{v, J(v)\}$; just define

$$R_\theta(v) = \cos\theta \cdot v + \sin\theta \cdot J(v), \quad \theta \in \mathbf{R}.$$

Let us see if we can do this inductively. Let $v_1 = v$ and denote by σ_1 the plane spanned by v and $J(v)$. Let $0 \neq v_2 \in \mathbf{R}^n$ be orthogonal to σ_1. We have

$$J(v_2) \cdot v_1 = J^2(v_2) \cdot J(v_1) = -v_2 \cdot J(v_1) = 0$$

and

$$J(v_2) \cdot J(v_1) = v_2 \cdot v_1 = 0,$$

so that $J(v_2)$ is also orthogonal to σ_1. We obtain that the plane σ_2 spanned by v_2 and $J(v_2)$ is orthogonal to σ_1. Continuing in this manner, we see that $\mathbf{R}^n$ can be decomposed into the sum of mutually orthogonal J-invariant planes:

$$\mathbf{R}^n = \sigma_1 + \sigma_2 + \cdots + \sigma_m,$$

and on each plane J acts by a quarter-turn. In particular, $n = 2m$ is even!

All this can be put into a very elegant algebraic framework. Given a complex structure J on $\mathbf{R}^n$, we can make $\mathbf{R}^n$ a *complex vector space* by defining multiplication of a vector $v \in \mathbf{R}^n$ to be given by a complex number $z = a + bi$ to be given by $z \cdot v = a \cdot v + b \cdot J(v)$.

We now return to our orthogonal multiplication $F : \mathbf{R}^n \times \mathbf{R}^n \to \mathbf{R}^n$ and investigate the fundamental identity above a little more. As before, we substitute the change of bases formula into the first term of the fundamental identity, but now we switch α and β in the change of bases formula and substitute this into the second term. We obtain

$$u_i^\alpha \cdot \left(\sum_{l=1}^n p_{kl}^{\beta\alpha} u_l^\alpha \right) + \left(\sum_{l=1}^n p_{kl}^{\alpha\beta} u_l^\beta \right) \cdot u_i^\beta = 0.$$

Using orthogonality of the u_i^α's and u_i^β's again, we arrive at

$$p_{ki}^{\beta\alpha} + p_{ki}^{\alpha\beta} = 0,$$

or equivalently

$$P^{\beta\alpha} = -P^{\alpha\beta}.$$

Finally, let α, β, and γ be distinct indices from $1, \ldots, n$ and iterate the change of bases formula twice:

$$u_i^\gamma = \sum_{k=1}^n p_{ik}^{\gamma\beta} u_k^\beta$$

$$= \sum_{l=1}^n p_{il}^{\gamma\alpha} u_l^\alpha = \sum_{l=1}^n p_{il}^{\gamma\alpha} \left(\sum_{k=1}^n p_{lk}^{\alpha\beta} u_k^\beta \right)$$

$$= \sum_{k=1}^n \left(\sum_{l=1}^n p_{il}^{\gamma\alpha} p_{lk}^{\alpha\beta} \right) u_k^\beta.$$

Equating coefficients, we obtain

$$p_{ik}^{\gamma\beta} = \sum_{l=1}^n p_{il}^{\gamma\alpha} p_{lk}^{\alpha\beta}.$$

In matrix terminology, this means that

$$P^{\gamma\beta} = P^{\gamma\alpha} P^{\alpha\beta},$$

or, using skew-symmetry in the upper indices,

$$P^{\gamma\alpha} P^{\beta\alpha} = -P^{\gamma\beta}.$$

In particular,

$$P^{\gamma\alpha}P^{\beta\alpha} = -P^{\beta\alpha}P^{\gamma\alpha},$$

so $P^{\gamma\alpha}$ and $P^{\beta\alpha}$ anticommute!

Letting $\alpha = n$ and introducing $J_\beta = P^{\beta,n}$, $\beta = 1, \ldots, n-1$, we see that

$$J_1, \ldots, J_{n-1}$$

are pairwise anticommuting complex structures on $\mathbf{R}^n$.

Summarizing, we see that the existence of an orthogonal multiplication $F : \mathbf{R}^n \times \mathbf{R}^n \to \mathbf{R}^n$ implies that there exists a family $\{J_1, \ldots, J_{n-1}\}$ of anticommuting complex structures on $\mathbf{R}^n$. $\heartsuit$ These complex structures generate (under composition by multiplication) what is called a *Clifford algebra*. Since the elements of the Clifford algebra act on $\mathbf{R}^n$ as linear transformations, the vector space $\mathbf{R}^n$ becomes a *Clifford module*.

To prove Theorem 14 we would need to show that this is possible only for $n = 2, 4$, and 8. Regrettably, this is beyond the scope of these Glimpses. Let us mollify ourselves by instead taking a closer look at what happens in four dimensions.

Problems

1. Let J_1 and J_2 be complex structures on $\mathbf{R}^2$. Show that either $J_1 = J_2$ or $J_1 = -J_2$.

2. Let $\{J_1, \ldots, J_{n-1}\}$ be a family of anticommuting complex structures on $\mathbf{R}^n$. Prove that there exists an orthogonal multiplication $F : \mathbf{R}^n \times \mathbf{R}^n \to \mathbf{R}^n$.

22

SECTION

Quaternions

◇ We consider orthogonal multiplications for the case $n = 4$, i.e., $F : \mathbf{R}^4 \times \mathbf{R}^4 \to \mathbf{R}^4$. We saw in the previous section that the existence of such F implies the existence of three complex structures J_1, J_2, J_3 on $\mathbf{R}^4$ that pairwise anticommute.

Proposition 4.

Let $\{J_1, J_2, J_3\}$ *be an anticommuting family of complex structures on* $\mathbf{R}^4$. *Then we have*

$$J_1 \circ J_2 = \pm J_3.$$

Proof.

Consider the linear isometry

$$U = J_1 \circ J_2 \circ J_3$$

of $\mathbf{R}^4$. We claim that U commutes with each complex structure J_l, $l = 1, 2, 3$, and $U^2 = I$. Both claims follow by simple computations. For the first, assuming $l = 1$, we compute

$$U \circ J_1 = J_1 \circ J_2 \circ J_3 \circ J_1 = J_1^2 \circ J_2 \circ J_3 = J_1 \circ U.$$

For the second, we have

$$U^2 = (J_1 \circ J_2 \circ J_3) \circ (J_1 \circ J_2 \circ J_3)$$

$$= J_1^2 \circ J_2 \circ J_3 \circ J_2 \circ J_3$$

$$= -J_1^2 \circ J_2^2 \circ J_3^2 = I.$$

Because U is an isometry whose square is the identity, it has only real eigenvalues, and they can only be ± 1. Moreover, the eigenspaces

$$V_+ = \{v \in \mathbf{R}^4 \mid U(v) = v\} \quad \text{and} \quad V_- = \{v \in \mathbf{R}^4 \mid U(v) = -v\}$$

are orthogonal, and together they span $\mathbf{R}^4$:

$$\mathbf{R}^4 = V_+ + V_-.$$

By definition, $U|V_+ = I$ and $U|V_- = -I$. That J_l commutes with U for each $l = 1, 2, 3$, translates into J_l leaving V_+ and V_- invariant. (In fact, if $U(v) = \pm v$, $v \in \mathbf{R}^4$, then $U(J_1(v)) = J_1(U(v)) = \pm J_1(v)$.) Thus, $\{J_1, J_2, J_3\}$ restricts to an anticommuting family of complex structures on V_+ (and on V_-). In particular, dim V_+ is even, i.e., 0, 2, or 4. The middle dimension 2 cannot occur, since on a 2-dimensional vector space a complex structure is essentially determined by the orientation. Therefore, it is impossible for three complex structures to coexist with anticommutation (cf. Problem 1 of Section 21). Thus, V_+ is either trivial or all of $\mathbf{R}^4$. The same is true in reversed order for V_-. Thus, $U = \pm I$, and we have $J_1 \circ J_2 \circ J_3 = \pm I$. Composing both sides by J_3 from the right and using $J_3^2 = -I$, we obtain $J_1 \circ J_2 = \pm J_3$ as claimed. ∎

We have the liberty of choosing a sign for $\pm J_3$ without changing the entire structure (that is, the relations). We therefore assume that J_1, J_2 and J_3 are arranged so that

$$J_1 \circ J_2 = J_3.$$

Summarizing, we obtained that the existence of an orthogonal multiplication $F : \mathbf{R}^4 \times \mathbf{R}^4 \to \mathbf{R}^4$ implies the existence of three linear isometries J_1, J_2, J_3 on $\mathbf{R}^4$ satisfying the relations:

$$J_1^2 = J_2^2 = J_3^2 = -I,$$

and

$$J_1 \circ J_2 = -J_2 \circ J_1 = J_3,$$

$$J_2 \circ J_3 = -J_3 \circ J_2 = J_1,$$

$$J_3 \circ J_1 = -J_1 \circ J_3 = J_2,$$

where the last two equalities can be derived from the first. Now look at the following analogy: $J^2 = -I$ makes $\mathbf{R}^2$ a complex vector space with complex unit $i = (0, 1)$ satisfying $i^2 = -1$. Yielding to the obvious temptation, we introduce the vectors

$$i = (0, 1, 0, 0),$$

$$j = (0, 0, 1, 0),$$

$$k = (0, 0, 0, 1),$$

in $\mathbf{R}^4$ and declare the rules for multiplication to be

$$i^2 = j^2 = k^2 = -1$$

and

$$ij = -ji = k,$$

$$jk = -kj = i,$$

$$ki = -ik = j.$$

After adding $1 = (1, 0, 0, 0)$ to $\{i, j, k\}$, each vector $q \in \mathbf{R}^4$ can be written as a linear combination of $1, i, j, k$:

$$q = a + bi + cj + dk, \quad a, b, c, d, \in \mathbf{R},$$

where we suppressed 1 from the notation. q expanded like this is called a *quaternion*. It is now clear how to multiply two quaternions using these identities. $\mathbf{R}^4$ equipped with this so-called *quaternionic multiplication* becomes a skew field that we denote by $\mathbf{H}$. ("Skew" means having noncommutative multiplication.)

Multiplication of quaternions was introduced by Hamilton in 1843. According to the story, he had struggled with the problem of defining multiplication of vectors in $\mathbf{R}^3$ since 1833, and his family took a great interest in this. Each morning at breakfast, his boys would ask, "Well, Papa, can you multiply triplets?" (meaning vectors in $\mathbf{R}^3$) and would receive the sad reply "No, I can only add and subtract them." Then, when strolling with his wife by Brougham Bridge in Dublin one day, it suddenly occurred to him that all the difficulties would disappear if he used quadruples—that is, vectors in $\mathbf{R}^4$. Overwhelmed with joy, he carved the identities above into the stonework of the bridge.

Given a quaternion $q = a + bi + cj + dk \in \mathbf{H}$, in analogy with complex numbers it is customary to define the *real part* of q as $\Re(q) = a$, and the *pure part* of q as $\mathcal{P}(q) = bi + cj + dk$. We also write $q = a + p$ with $a \in \mathbf{R}$ and $p = \mathcal{P}(q)$. The *conjugate* of $q = a + p$ is then defined as $\bar{q} = a - p$. Finally, we call a quaternion *pure* if its real part vanishes. The pure quaternions form the three-dimensional linear subspace $\mathbf{H}_0 = \{q \in \mathbf{H} \mid \bar{q} = -q\}$ of $\mathbf{H}$ spanned by i, j, k.

Quaternionic multiplication satisfies the same identities as its complex brother. A word of caution is needed, however, for the identity

$$\overline{q_1 q_2} = \bar{q}_2 \bar{q}_1, \quad q_1, q_2 \in \mathbf{H},$$

in which the factors on the right-hand side get switched!

Taking the analogy with complex arithmetic further, we now ask whether the ordinary dot product in $\mathbf{R}^4 = \mathbf{H}$ can be written in terms of quaternions. Here it is:

$$q_1 \cdot q_2 = \frac{\bar{q}_1 q_2 + \bar{q}_2 q_1}{2}, \quad q_1, q_2 \in \mathbf{H}.$$

To show this, we first note that the right-hand side is just $\Re(\bar{q}_1 q_2)$. Setting $q_1 = a_1 + b_1 i + c_1 j + d_1 k$ and $q_2 = a_2 + b_2 i + c_2 j + d_2 k$, we have

$$\Re(\bar{q}_1 q_2) = \Re((a_1 - b_1 i - c_1 j - d_1 k)(a_2 + b_2 i + c_2 j + d_2 k))$$
$$= a_1 a_2 + b_1 b_2 + c_1 c_2 + d_1 d_2,$$

since the mixed terms are all pure. The formula for the dot product follows. In particular, if $q = q_1 = q_2$, we obtain

$$|q|^2 = q \cdot q = \bar{q} q,$$

the usual Length^2-Identity. With this, the quaternionic inverse of a nonzero quaternion $q \in \mathbf{H}$ can be written as

$$q^{-1} = \frac{\bar{q}}{|q|^2}.$$

This shows that $\mathbf{H}$ is indeed a skew field.

The 3-sphere $S^3 \subset \mathbf{H}$ in quaternionic calculus is like the unit circle $S^1 \subset \mathbf{C}$ in complex calculus. In fact,

$$S^3 = \{q \in \mathbf{H} \mid |q| = 1\}$$

constitutes a group under quaternionic multiplication; this is an immediate consequence of the fact that quaternionic multiplication is normed;

$$|q_1 q_2| = |q_1| \cdot |q_2|, \quad q_1, q_2 \in \mathbf{H}.$$

To check this, we compute

$$\begin{aligned}
|q_1 q_2|^2 &= q_1 q_2 \overline{q_1 q_2} = q_1 q_2 \bar{q}_2 \bar{q}_1 \\
&= q_1 |q_2|^2 \bar{q}_1 = q_1 \bar{q}_1 |q_2|^2 \\
&= |q_1|^2 |q_2|^2,
\end{aligned}$$

where we used the fact that reals (such as $|q_2|^2$) commute with all quaternions.

What does $S^3 \subset \mathbf{H}$ look like? The answer depends on whether you want an algebraic description or a "three-dimensional vision" in $\mathbf{R}^4$! We'll use both approaches, beginning with the first.

Since we are already familiar with complex arithmetic, we just write

$$q = (a + bi) + j(c + di) = a + bi + cj - dk = z + jw,$$

$$z = a + bi, \quad w = c + di \in \mathbf{C}.$$

We obtain that a quaternion is nothing but a pair of complex numbers; $\mathbf{H} = \mathbf{C}^2$. In terms of complex variables, quaternionic multiplication can be written as

$$\begin{aligned}
q_1 q_2 &= (z_1 + jw_1)(z_2 + jw_2) \\
&= z_1 z_2 + jw_1 jw_2 + z_1 jw_2 + jw_1 z_2 \\
&= (z_1 z_2 - \bar{w}_1 w_2) + j(w_1 z_2 + \bar{z}_1 w_2),
\end{aligned}$$

so that, under the correspondence $\mathbf{H} = \mathbf{C}^2$, multiplying q_1 by q_2 corresponds to matrix multiplication

$$\begin{bmatrix} z_1 & -\bar{w}_1 \\ w_1 & \bar{z}_1 \end{bmatrix} \begin{bmatrix} z_2 \\ w_2 \end{bmatrix}.$$

Suppressing the indices, we say that multiplication by $q = z + jw$ corresponds to multiplication by the matrix

$$A = \begin{bmatrix} z & -\bar{w} \\ w & \bar{z} \end{bmatrix}.$$

Restricting q to S^3 is equivalent to assuming $|q|^2 = |z|^2 + |w|^2 = 1$, and we see that A is special unitary, that is,

$$A^* = \bar{A}^\top = A^{-1}$$

with determinant one. As observed in Section 18, these matrices constitute the important group $SU(2)$ of *special unitary 2×2 matrices.* The correspondence that associates to the quaternion $q \in S^3$ the special unitary matrix A is an isomorphism $\varphi : S^3 \to SU(2)$ (it is clearly one-to-one and onto); that is, it satisfies

$$\varphi(q_1 q_2) = \varphi(q_1) \cdot \varphi(q_2), \quad q_1, q_2 \in S^3.$$

This follows by setting $q_1 = z_1 + jw_1$ and $q_2 = z_2 + jw_2$, and comparing the first column of the product

$$\varphi(q_1)\varphi(q_2) = \begin{bmatrix} z_1 & -\bar{w}_1 \\ w_1 & \bar{z}_1 \end{bmatrix} \begin{bmatrix} z_2 & -\bar{w}_2 \\ w_2 & \bar{z}_2 \end{bmatrix}$$

with the complex expression of the product $q_1 q_2$ above.

The identification $SU(2) = S^3$ allows us to use spherical concepts such as meridians of longitude and parallels of latitude on $SU(2)$. Let

$$A(z, w) = \begin{bmatrix} z & -\bar{w} \\ w & \bar{z} \end{bmatrix}, \quad |z|^2 + |w|^2 = 1,$$

be a typical element of $SU(2)$, where we have displayed the dependence of A on z and w. The characteristic polynomial for $A(z, w)$

can be written as

$$\det(A(z,w) - tI) = (z-t)(\bar{z}-t) + w\bar{w}$$
$$= t^2 - (z+\bar{z})t + 1 = t^2 - 2\Re(z)t + 1.$$

Since the real part $\Re(z)$ of z is between -1 and $+1$, for fixed $r \in [-1,1]$, we call $\{A(z,w) \mid \Re(z) = r\}$ the parallel of latitude at r. We see that two special unitary matrices are on the same parallel of latitude iff their characteristic polynomials are the same. The parallels of latitude corresponding to $r = 1$ and $r = -1$ are the single-point sets $\{I\}$ and $\{-I\}$, which we may just as well call North and South Poles. For $-1 < r < 1$, the parallel of latitude at r is topologically a 2-sphere sitting in S^3. This is clear algebraically if we work out the equations

$$|z|^2 + |w|^2 = 1 \quad \text{and} \quad \Re(z) = r$$

in real coordinates and also geometrically, since the parellel of latitude at r is nothing but the slice cut out from S^3 by the 3-dimensional space defined by $\Re(z) = r$ in $\mathbf{C}^2$. Since $\Re(z)$ is half of the trace of $A(z,w)$, the equator $r = 0$ corresponds to traceless matrices. The meridians of longitude are great circles going through the poles. One prominent meridian of longitude is given by the diagonal matrices in $SU(2)$. For a diagonal $A(z,w)$, we have $w = 0$, so that $|z|^2 = 1$. Letting $z = e^{i\theta}$, a diagonal matrix can be written in the form

$$\begin{bmatrix} e^{i\theta} & 0 \\ 0 & e^{-i\theta} \end{bmatrix} \in SU(2), \quad \theta \in \mathbf{R}.$$

This meridian of longitude cuts the equator at

$$\begin{bmatrix} i & 0 \\ 0 & -i \end{bmatrix}$$

and its negative.

Remark.
Let $A \in SU(2)$. The conjugacy class of A is the set

$$\{BAB^{-1} \mid B \in SU(2)\}.$$

Since trace $(BAB^{-1}) =$ trace A, it is clear that each conjugacy class is contained in a parallel of latitude. ♡ A somewhat more refined analysis shows that the converse is also true, so that the parallels of latitude are exactly the conjugacy classes of matrices in $SU(2)$.

The description of $SU(2) = S^3$ in terms of parallels of latitude (as conjugacy classes), though very pleasing, does not contain anything new about the geometry of S^3. After all, the same geometric picture is valid in the one less dimension of S^2! We now give a novel insight of the subtlety of the geometry of S^3 absent in S^2.

◇ In what follows, we parameterize $S^3 \subset \mathbf{C}^2$ by two complex variables (z, w) satisfying $|z|^2 + |w|^2 = 1$. (Recall that $z + jw$ is the quaternion corresponding to $(z, w) \in \mathbf{C}^2$.) Note that z runs on the first and w on the second factor of $\mathbf{C}^2 = \mathbf{C} \times \mathbf{C}$. Consider the function $f : S^3 \to \mathbf{R}$ given by

$$f(z, w) = |z|^2 - |w|^2, \quad (z, w) \in S^3.$$

Since $|z|^2 + |w|^2 = 1$, we have $-1 \leq f \leq 1$. We now want to visualize the level sets

$$C_r = \{(z, w) \in S^3 \mid f(z, w) = r\}, \quad -1 \leq r \leq 1.$$

We have $(z, w) \in C_r$ iff

$$|z|^2 - |w|^2 = r,$$

and, since $|z|^2 + |w|^2 = 1$, adding and subtracting yields

$$|z|^2 = \frac{1 + r}{2} \quad \text{and} \quad |w|^2 = \frac{1 - r}{2}.$$

For $r = 1$, we obtain that $|z| = 1$ and $w = 0$, so that C_1 is the unit circle in the first factor of $\mathbf{C}^2 = \mathbf{C} \times \mathbf{C}$. Similarly, C_{-1} is the unit circle in the second factor of $\mathbf{C}^2 = \mathbf{C} \times \mathbf{C}$; in particular, C_1 and C_{-1} are perpendicular to each other. Now let $-1 < r < 1$ and notice that the right-hand sides of the equations above are positive. They are uncoupled; the first describes a circle around the origin with radius $\sqrt{(1 + r)/2}$ in the first factor of $\mathbf{C}^2$, and the second describes a similar circle with radius $\sqrt{(1 - r)/2}$ in the second factor of $\mathbf{C}^2$. Thus,

$$C_r = \left\{(z, w) \in \mathbf{C}^2 \mid |z|^2 = \frac{1 + r}{2}, \ |w|^2 = \frac{1 - r}{2}\right\}$$

is the Cartesian product of two circles—a torus! We see that apart from the great circles $C_{\pm 1}$, the tori C_r, $-1 < r < 1$, decompose (or *foliate*) S^3. These tori are called Clifford tori. We understand this visually as follows: Consider ourselves in S^3 moving along the great circle $C_{-1} = \{(0, e^{i\theta}) \mid \theta \in \mathbf{R}\}$. At each point we see the direction in which we are moving (given by the vector tangent to C_{-1} at our location). We see that a three-dimensional space surrounds us because we are in S^3! Thus, within S^3, we can hold a circle made of wire orthogonally to our direction of motion. If we move around C_{-1} and drag the circle along, keeping it perpendicular to our path, it will sweep a Clifford torus. By increasing the radius of the circle we carry, we get fatter and fatter tori. At the other extreme value, $r = 1$, the tori reduce to C_1. The situation is depicted in Color Plate 10.

Removing the middle torus C_0 from S^3, we see that S^3 falls into the disjoint union of two solid tori. Going backward, we reach the inevitable conclusion: The 3-sphere is obtained from two solid tori by pasting them together along their boundaries!

Going back to the group structure of S^3 leads to another interesting discovery. Consider $S^1 = \{e^{i\theta} \mid \theta \in \mathbf{R}\} \subset \mathbf{C}$ acting on S^3 by the 4-dimensional rotation $e^{i\theta}(z, w) \mapsto (e^{i\theta}z, e^{i\theta}w)$. Each orbit is a great circle and is contained in a Clifford torus. In fact, the orbit

$$S^1(z_0, w_0) = \{(e^{i\theta}z_0, e^{i\theta}w_0) \mid \theta \in \mathbf{R}\}$$

is the intersection of S^3 with the 2-dimensional linear subspace in $\mathbf{C}^2$ defined by the equation $zw_0 - wz_0 = 0$. Since

$$f(e^{i\theta}z, e^{i\theta}w) = |e^{i\theta}z|^2 - |e^{i\theta}w|^2 = |z|^2 - |w|^2 = f(z, w),$$

the second statement also follows.

What is the quotient S^3/S^1? This should be two-dimensional since in S^3 we are compressing great circles into points, so that the dimension must drop by one. We claim that S^3/S^1 and S^2 are homeomorphic. To do this, we need to understand how to associate to an orbit $S^1(z_0, w_0) = \{(e^{i\theta}z_0, e^{i\theta}w_0) \mid \theta \in \mathbf{R}\}$, $|z_0|^2 + |w_0|^2 = 1$, a unique point on the two-sphere S^2. The easiest way to do this is to identify the projection map $S^3 \to S^3/S^1$. We introduce the *Hopf map* $H : S^3 \to S^2$ given by

$$H(z, w) = (|z|^2 - |w|^2, 2z\bar{w}) \in \mathbf{R} \times \mathbf{C} = \mathbf{R}^3.$$

First, note that H maps S^3 to S^2, since

$$|H(z, w)|^2 = (|z|^2 - |w|^2)^2 + 4|z|^2|w|^2 = (|z|^2 + |w|^2)^2 = 1$$

if $(z, w) \in S^3$. Second, H is invariant under the action of S^1, since

$$H(e^{i\theta}z, e^{i\theta}w) = (|e^{i\theta}z|^2 - |e^{i\theta}w|^2, 2e^{i\theta}z\overline{e^{i\theta}w})$$
$$= (|z|^2 - |w|^2, 2z\bar{w}) = H(z, w).$$

Thus, H maps each orbit of S^1 in S^3 into a single point. To show that $S^3/S^1 = S^2$, we need to prove that the orbits are precisely the inverse images of points from S^2. In other words, we have to show that whenever $H(z_1, w_1) = H(z_2, w_2)$, the points (z_1, w_1) and (z_2, w_2) are on the same orbit under S^1. Now the fact that the Hopf images are equal translates into

$$|z_1|^2 - |w_1|^2 = |z_2|^2 - |w_2|^2$$

and

$$z_1\bar{w}_1 = z_2\bar{w}_2.$$

The first equality means that (z_1, w_1) and (z_2, w_2) are on the same Clifford torus, say C_r, so we have

$$|z_1|^2 = |z_2|^2 = \frac{1+r}{2} \quad \text{and} \quad |w_1|^2 = |w_2|^2 = \frac{1-r}{2}.$$

Letting $z_2 = e^{i\theta}z_1$ and $w_2 = e^{i\varphi}w_1$, we substitute these back to the second equation and obtain

$$z_1\bar{w}_1 = e^{i(\theta-\varphi)}z_1\bar{w}_1,$$

and so (exluding the trivial cases when $z_1 = 0$ or $w_1 = 0$, which can be handled separately)

$$e^{i(\theta-\varphi)} = 1$$

follows. By the periodicity property of the exponential function, θ and φ differ by an integer multiple of 2π. Thus $w_2 = e^{i\theta}w_1$, and we are done.

♠ We can define the *complex projective n-space* $\mathbf{C}P^n$ using the same construction as for the real projective n-space $\mathbf{R}P^n$ (Problem 5 in Section 16). $\mathbf{C}P^n$ comes equipped with the natural projection $\mathbf{C}^{n+1} - \{0\} \to \mathbf{C}P^n$ associating to a nonzero complex vector $p \in \mathbf{C}^{n+1}$

the complex line $\mathbf{C} \cdot p$ minus the origin. Restricting the projection to the unit sphere $S^{2n+1} \subset \mathbf{C}^{n+1}$ ($= \mathbf{R}^{2n+2}$), we obtain a map $S^{2n+1} \to \mathbf{C}P^n$ that associates to $p \in S^{2n+1}$ the set of multiples $e^{i\theta}p$, $\theta \in \mathbf{R}$, that span the complex projective point corresponding to p. This is better understood when we define a natural action of $S^1 \subset \mathbf{C}$ on $S^{2n+1} \subset \mathbf{C}^{n+1}$ given by multiplying complex vectors by $e^{i\theta} \in S^1$. Then $S^{2n+1} \to \mathbf{C}P^n$ is nothing but the orbit map. Note that each orbit of S^1 on S^{2n+1} is a great circle. In particular, S^{2n+1} can be thought of as composed of circles attached to every point of $\mathbf{C}P^n$! For $n = 1$, the orbit map $S^3 \to \mathbf{C}P^1$ is nothing but the Hopf map. Hence $\mathbf{C}P^1$ can be identified with S^2! This is not too surprising, though; the ordinary points $[z : w] \in \mathbf{C}P^1$ with complex homogeneous coordinates $z, w \in \mathbf{C}$ are exactly those with nonzero w. Thereby they can be made to correspond to ratios $z/w \in \mathbf{C}$, and the only ideal point is $[1 : 0]$, corresponding to ∞. We thus have $\mathbf{C}P^1 = \mathbf{C} \cup \{\infty\} = S^2$!

We close this section with an advanced remark. If we look at two orbits of the action of S^1 on S^3 (on the same Clifford torus), it is apparent that they are "linked" in S^3. This means that the Hopf map $H : S^3 \to S^2$ cannot be deformed continuously through maps into a constant map $S^3 \to S^2$ that sends the whole S^3 to a single point. We express this by saying that H is homotopically nontrivial, or, even more formally, that the third homotopy group $\pi_3(S^2)$ is nontrivial. This was a pioneering result of Hopf's during the early development of homotopy theory, since because of the apparent similarity between homotopy and homology theories, one expected to find $\pi_3(S^2)$ to be trivial since the third homology group $H_3(S^2) = 0$.

Problems

1. Consider the 3-dimensional subspace $V = \mathbf{C} \times \mathbf{R} \cdot k \subset \mathbf{H}$. Show that the map $q \mapsto kqk^{-1}$, $q \in V$, leaves V invariant. Show that this map is an opposite isometry of V and describe it geometrically using $V = \mathbf{R}^3$.

2. Let $q_1, q_2 \in \mathbf{H}_0$ be purely imaginary quaternions. Show that the quaternionic product $q_1 q_2 \in \mathbf{H}$ projected to $\mathbf{R} \cdot 1$ is the negative of the dot product of q_1 and q_2 considered as spatial vectors under the identification $\mathbf{H}_0 = \mathbf{R}^3$. Show that $q_1 q_2$ projected to $\mathbf{H}_0$ is the cross product of q_1 and q_2.

3. Use the quaternionic identity $|q_1|^2|q_2|^2 = |q_1 q_2|^2$, $q_1, q_2 \in \mathbf{H}$, to prove the four square formula:[1]

$$(a_1^2 + b_1^2 + c_1^2 + d_1^2)(a_2^2 + b_2^2 + c_2^2 + d_2^2)$$
$$= (a_1 a_2 - b_1 b_2 - c_1 c_2 - d_1 d_2)^2$$
$$+ (a_1 b_2 + b_1 a_2 + c_1 d_2 - d_1 c_2)^2$$
$$+ (a_1 c_2 - b_1 d_2 + c_1 a_2 + d_1 b_2)^2$$
$$+ (a_1 d_2 + b_1 c_2 - c_1 b_2 + d_1 a_2)^2.$$

4. Use the four square formula of Problem 3 to show that if a and b are both sums of four squares of integers, then ab is also a sum of four squares of integers. (Lagrange proved in 1772 that any positive integer can be expressed as the sum of four squares of integers (cf. Problem 7 of Section 5). This problem has a rich and complex history. In 1638, Fermat asserted that every positive number is a sum of at most three triangular numbers, four squares, five pentagonal numbers, and so on (cf. Problem 11 of Section 2). In 1796, in one of the earliest entries in his mathematical diary, Gauss recorded that he had found a proof for the triangular case. The general problem was resolved by Cauchy in 1813.)

5. Define the quaternionic Hop map $H : \mathbf{H}^2 \to \mathbf{R} \times \mathbf{H}$ by

$$H(p, q) = (|p|^2 - |q|^2, 2p\bar{q}), \quad p, q \in \mathbf{H},$$

and show that H maps the unit sphere $S^7 \subset \mathbf{H}^2(= \mathbf{R}^8)$ onto the unit sphere $S^4 \subset \mathbf{R} \times \mathbf{H}(= \mathbf{R}^5)$. Verify that the inverse image of a point in S^4 under $H : S^7 \to S^4$ is a great 3-sphere S^3 in S^7.

6. Associate to the spatial rotation with axis $\mathbf{R} \cdot p_0$, $p_0 = (a_0, b_0, c_0) \in S^2$, and angle $\theta \in \mathbf{R}$, the antipodal pair $\pm(a + bi + cj + dk) \in S^3$ of unit quaternions, where $a = \cos(\theta/2)$, $b = c_0 \sin(\theta/2)$, $c = b_0 \sin(\theta/2)$, and $d = a_0 \sin(\theta/2)$.

   (a) Describe the set of spherical rotations that correspond to parallels of latitude in S^3.

   (b) ♠ Study the correspondence between the group of spherical rotations, the quotient group $S^3/\{\pm 1\}$, and $Möb^+(\hat{\mathbf{C}})$.

Web Sites

1. www.geom.umn.edu/~banchoff/script/b3d/hypertorus.html

2. www.maths.tcd.ie/pub/HistMath/People/Hamilton/Letters /BroomeBridge.html

---

[1]This identity appears in a letter Euler wrote to Goldbach in 1705.

23

SECTION

Back to R³!

◇ After our frustration over the nonexistence of orthogonal multiplications in three dimensions, we now try to incorporate $\mathbf{R}^3$ into our skew field $\mathbf{H}$. The symmetric role of the quaternionic units i, j, and k indicates that $\mathbf{R}^3$ should sit in $\mathbf{H}$ as the linear space $\mathbf{H}_0 = \{q \in \mathbf{H} \mid \bar{q} = -q\}$ of pure quaternions. As we did in our investigations of complex arithmetic, we now want to see what kind of geometric transformations arise in $\mathbf{R}^3$ from quaternionic multiplication restricted to $\mathbf{H}_0$. As an elementary example, conjugation in $\mathbf{H}$ restricts to the antipodal map in $\mathbf{H}_0$!

Theorem 15.
Let $0 \neq q_0 \in \mathbf{H}$. Then the transformations $q \mapsto \pm q_0 q q_0^{-1}$, $q \in \mathbf{H}$, are linear isometries that leave $\mathbf{H}_0$ invariant.

(1) If $q_0 \in \mathbf{H}_0$, then the restriction of $q \mapsto -q_0 q q_0^{-1}$ to $\mathbf{H}_0$ is reflection to the plane orthogonal to q_0.

(2) If $q_0 = a_0 + p_0$, $0 \neq a_0 \in \mathbf{R}$, $0 \neq p_0 \in \mathbf{H}_0$, then the restriction of $q \mapsto q_0 q q_0^{-1}$ to $\mathbf{H}_0$ is rotation with axis $\mathbf{R} \cdot p_0$ and angle $0 < \theta < \pi$ given by $\tan(\theta/2) = |p_0|/a_0$.

Proof.

Since $|q_0qq_0^{-1}| = |q_0||q||q_0|^{-1} = |q|$, it is clear that the transformations $q \mapsto \pm q_0qq_0^{-1}$ are linear isometries of $\mathbf{H} = \mathbf{R}^4$. The invariance of $\mathbf{H}_0$ under these transformations can be proved directly. Instead, we will develop a criterion for a quaternion to belong to $\mathbf{H}_0$, which will turn out to be useful later on.

We claim that $q \in \mathbf{H}_0$ iff q^2 is a nonpositive real number. Indeed, given $q = a + p$, $a \in \mathbf{R}$, $0 \neq p \in \mathbf{H}_0$ (we may assume that p is nonzero, since otherwise the claim follows), we have

$$q^2 = a^2 + 2ap + p^2.$$

The last term p^2 is a nonpositive real number, since

$$p^2 = (bi + cj + dk)^2 = -(b^2 + c^2 + d^2)$$

(the mixed terms cancel because of anticommutativity of i, j, k). Thus, the pure part of q^2 is equal to $2a\mathcal{P}(p) = 2ap$, and this is zero iff $a = 0$.

Returning to the proof of invariance of $\mathbf{H}_0$, we need to show that $q \in \mathbf{H}_0$ implies $q_0qq_0^{-1} \in \mathbf{H}_0$. Using the criterion just proved, this is equivalent to the statement that whenever q^2 is a nonpositive real number, then so is $(q_0qq_0^{-1})^2$. We compute the latter as

$$(q_0qq_0^{-1})^2 = q_0qq_0^{-1}q_0qq_0^{-1} = q_0q^2q_0^{-1}$$
$$= q^2q_0q_0^{-1} = q^2,$$

where the last but one equality is because a real number commutes with all quaternions. Invariance of $\mathbf{H}_0$ follows. The transformations $q \to \pm q_0qq_0^{-1}$ send p_0 to $\pm p_0$, since

$$q_0p_0q_0^{-1} = q_0(q_0 - a_0)q_0^{-1} = q_0q_0q_0^{-1} - a_0q_0q_0^{-1} = q_0 - a_0 = p_0.$$

Since these are also isometries, the plane P perpendicular to p_0 remains invariant. Let $q \in P$ and compute the cosine of the angle θ between q and $q_0qq_0^{-1}$ (Figure 23.1).

Using the dot product formula, we have

$$\cos\theta = \frac{q \cdot (q_0qq_0^{-1})}{|q|^2} = \frac{1}{2|q|^2}\left(\bar{q}(q_0qq_0^{-1}) + \overline{q_0qq_0^{-1}}q\right)$$
$$= \frac{1}{2|q|^2|q_0|^2}\left(\bar{q}q_0q\bar{q}_0 + \overline{q_0q\bar{q}_0}q\right)$$

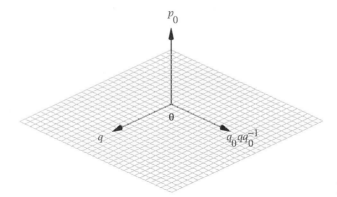

Figure 23.1

$$= \frac{1}{2|q|^2|q_0|^2}(\bar{q}q_0q\bar{q}_0 + q_0\bar{q}\bar{q}_0q)$$

$$= -\frac{1}{2|q|^2|q_0|^2}(qq_0q\bar{q}_0 + q_0q\bar{q}_0q),$$

where we used $\bar{q} = -q$. To rewrite this into a more convenient form, we now use orthogonality of q and p_0. We claim that orthogonality implies the commutation relation

$$qq_0 = \bar{q}_0q.$$

Indeed, since q and p_0 are orthogonal, $q \cdot p_0 = (\bar{q}p_0 + \bar{p}_0q)/2 = -(qp_0 + p_0q)/2 = 0$, so q and p_0 anticommute. With this, we have

$$qq_0 = q(a_0 + p_0) = qa_0 + qp_0 = a_0q - p_0q = (a_0 - p_0)q = \bar{q}_0q.$$

Returning to the main computation,

$$\cos\theta = -\frac{1}{2|q|^2|q_0|^2}(qq_0q\bar{q}_0 + q_0q\bar{q}_0q)$$

$$= -\frac{1}{2|q|^2|q_0|^2}(\bar{q}_0q^2\bar{q}_0 + q_0q^2q_0) \qquad (q^2 \text{ is real!})$$

$$= -\frac{q^2}{2|q|^2|q_0|^2}(\bar{q}_0^2 + q_0^2)$$

$$= \frac{1}{|q_0|^2}\Re(q_0^2) = \frac{a_0^2 + p_0^2}{a_0^2 - p_0^2}.$$

For case (1) of the theorem, we have $a_0 = 0$ (and $p_0 = q_0$), so that $\cos\theta = -1$ follows. Thus $\theta = \pi$ and $q \mapsto q_0qq_0^{-1}$ restricted to

P is a half-turn. Incorporating the negative sign, $q \mapsto -q_0qq_0^{-1}$ is identity on P and, as we have seen earlier, it sends q_0 to $-q_0$. Thus $q \mapsto -q_0qq_0^{-1}$ is reflection in the plane P.

For case (2) of the theorem, p_0 is left fixed, and $q \mapsto q_0qq_0^{-1}$ restricted to P is rotation with angle $0 < \theta < \pi$ where

$$\cos\theta = \frac{a_0^2 + p_0^2}{a_0^2 - p_0^2}$$

Now the trigonometric identity $\tan^2(\theta/2) = (1 - \cos\theta)/(1 + \cos\theta)$ gives $\tan(\theta/2) = |p_0|/a_0$. The theorem follows. ■

Theorem 15 means that spatial reflections and rotations can be obtained from quaternionic multiplications restricted to $\mathbf{R}^3 = \mathbf{H}_0$. These make up the group of all linear isometries of $\mathbf{R}^3$ (leaving the origin fixed), that is, they make up the *orthogonal group* $O(\mathbf{R}^3)$.

Returning to our original aim, we now place the Platonic solids in $\mathbf{H}_0$ (with centroids at the origin) and express the symmetries in terms of quaternions. In this way, we obtain a very transparent description of the symmetry groups of Platonic solids.

Before we do this, let us formalize what we just said about representing the elements of $O(\mathbf{R}^3)$ by quaternions. For simplicity, we restrict ourselves to direct linear isometries that constitute the *special orthogonal group* $SO(\mathbf{R}^3)$, a subgroup of $O(\mathbf{R}^3)$.

Theorem 16.

The map ψ that associates to each unit quaternion $q_0 \in S^3$ the transformation $q \mapsto q_0qq_0^{-1}$ restricted to $\mathbf{H}_0$ is a surjective group homomorphism

$$\psi : S^3 \to SO(\mathbf{R}^3)$$

with kernel

$$\ker\psi = \{\pm 1\}.$$

Proof.

By Theorem 15, if $q_0 \neq \pm 1$, then q_0 defines a rotation, an element of $SO(\mathbf{R}^3)$. On the other hand, $q_0 = \pm 1$ defines the identity element in $SO(\mathbf{R}^3)$ so that ψ maps into $SO(\mathbf{R}^3)$. It is clear that ψ is a homomorphism of groups, and, by what we just said, ± 1 are in the kernel of ψ. By Theorem 10 of Section 16 and Theorem 15, ψ is onto since all elements in $SO(\mathbf{R}^3)$ are rotations. It remains to show that the kernel of ψ is exactly $\{\pm 1\}$. Let $q_0 \in \ker \psi$, that is, $q_0 q q_0^{-1} = q$ for all $q \in \mathbf{H}_0$. Equivalently, q_0 commutes with all pure quaternions. Writing this condition out in terms of i, j, and k, we obtain that q_0 must be real. Since it is in S^3, it must be one of ± 1. ■

Remark.

♡ Theorem 16 implies that the group S^3 of unit quaternions modulo the normal subgroup $\{\pm 1\}$ is isomorphic with the group $SO(\mathbf{R}^3)$ of direct spatial linear isometries. The quotient group $S^3/\{\pm 1\}$ is, by definition, the group of right- (or left-) cosets of $\{\pm 1\}$. A right-coset containing $q \in S^3$ thus has the form $\{\pm 1\}q = \{\pm q\}$. Thus, topologically, $S^3/\{\pm 1\}$ can be considered as a model for the *projective space* $\mathbf{R}P^3$. The classical model of $\mathbf{R}P^3$ is the same as the model of $\mathbf{R}P^2$ discussed in Section 16; that is, a projective point is a line through the origin of $\mathbf{R}^4$, etc. By Theorem 16 above, $\mathbf{R}P^3$ can be identified by the group of direct spatial isometries $SO(\mathbf{R}^3)$!

◇ Let us explore some concrete settings. As we learned in Section 22, a quaternion q can be represented by a pair of complex numbers $(z, w) \in \mathbf{C}^2$ via $q = z + jw$. The second variable $w \in \mathbf{C}$ in this representation corresponds to $jw \in j\mathbf{C}$, and we see that $j\mathbf{C}$ is the complex plane in $\mathbf{H}_0$ spanned by the vectors j and $k(= ij)$. The complex unit $i \in \mathbf{H}_0$ is orthogonal to $j\mathbf{C}$ in $\mathbf{H}_0$.

As an example, we now describe a rotation in $\mathbf{H}_0$ with axis $\mathbf{R} \cdot i$ and angle θ, $0 < \theta < \pi$. According to Theorem 15, this rotation is described by the quaternion $q_0 = a + p_0$, $a \in \mathbf{R}$, $p_0 \in \mathbf{H}_0$, via $q \mapsto q_0 q q_0^{-1}$, $q \in \mathbf{H}_0$. Since $p_0 = i$, q_0 happens to be a complex number. The condition on the angle can be written as $\tan(\theta/2) = |i|/|a| = 1/|a|$, so that, choosing a positive, we have $q_0 = \cot(\theta/2) + i$. We normalize q_0 to a unit:

$$\frac{q_0}{|q_0|} = \frac{\cot(\theta/2) + i}{\sqrt{\cot^2(\theta/2) + 1}} = \cos\left(\frac{\theta}{2}\right) + i \sin\left(\frac{\theta}{2}\right).$$

We obtain that rotation with axis $\mathbf{R} \cdot i$ and angle θ corresponds to the complex number $e^{i\theta/2}$ viewed as a quaternion. To check that this is correct, we compute

$$q_0 i q_0^{-1} = e^{i\theta/2} i e^{-i\theta/2} = i,$$

so that i is kept fixed. Moreover, $e^{i\theta/2}$ acts on $jw \in j\mathbf{C}$ as

$$e^{i\theta/2} jw e^{-i\theta/2} = e^{i\theta} jw$$

and this is indeed rotation by angle θ on $j\mathbf{C}$! Notice now that $e^{i\theta/2}$ and its negative $-e^{i\theta/2} = e^{i(\pi + \theta/2)}$ represent the same rotation, since by Theorem 16, there is a two-to-one correspondence between unit quaternions in S^3 and linear isometries acting on $\mathbf{H}_0$.

We now consider the cone $\mathcal{C}_n$ in $\mathbf{H}_0$ with vertex hi (where $h > 0$ is the height) and base jP_n, where the regular n-sided polygon $P_n \subset \mathbf{C}$ is placed in $j\mathbf{C}$ (Figure 23.2).

We see that the cyclic group $C_{2n} = \{e^{l\pi i/n} \mid l = 0, \ldots, 2n - 1\}$ of unit quaternions is a double cover of the symmetry group $Symm^+(\mathcal{C}_n)$, in the sense that, for $l = 0, \ldots, n - 1$, $e^{l\pi i/n}$ and $-e^{l\pi i/n} = e^{(n+l)\pi i/n}$ correspond to the same rotation in $\mathbf{H}_0$.

The same argument can be used for the prism $\mathcal{P}_n$ given by

$$\left(-\frac{h}{2}, \frac{h}{2} \right) i \times jP_n.$$

Besides the elements of C_{2n}, we have half-turns at the vertices and the midpoints of edges of jP_n. They correspond to the quaternions

$$\{je^{l i\pi/n} \mid l = 0, \ldots, 2n - 1\}.$$

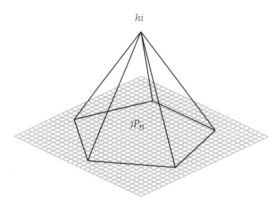

Figure 23.2

Putting these elements together, we obtain the so-called *binary dihedral group*

$$D_n^* = \{e^{l\pi i/n} \mid l = 0, \ldots, 2n-1\} \cup \{je^{l\pi i/n} \mid l = 0, \ldots, 2n-1\}$$

and this is a double cover of $Symm^+(\mathcal{P}_n)$.

We will now turn to the Platonic solids and derive double covers of the tetrahedral, octahedral, and icosahedral groups in terms of quaternions.

We begin with the tetrahedral group. We position the regular tetrahedron $\mathcal{T}$ in $\mathbf{R}^3 = \mathbf{H}_0$ as in Section 17 with vertices

$$i + j + k, \quad i - j - k, \quad -i + j - k, \quad -i - j + k.$$

$Symm^+(\mathcal{T})$ is generated by the three half-turns around the coordinate axes and by the rotation around $i + j + k$ by angle $2\pi/3$. The first three correspond to the three pairs of quaternions

$$\pm i, \quad \pm j, \quad \pm k,$$

and the latter is given by the quaternion (normalized to belong to S^3)

$$\pm \frac{1}{\sqrt{a^2 + 3}} (a + i + j + k), \quad a > 0,$$

where

$$\tan(\pi/3) = \frac{|i + j + k|}{a} = \frac{\sqrt{3}}{a}.$$

We obtain that $a = 1$, so the pair of quaternions corresponding to rotation at the front vertex $i + j + k$ is

$$\pm \frac{(1 + i + j + k)}{2}.$$

Putting these together, we see that the *binary tetrahedral group* defined by

$$\mathbf{T}^* = \{\pm 1, \pm i, \pm j, \pm k\} \cup \left\{ \frac{(\pm 1 \pm i \pm j \pm k)}{2} \right\} \subset S^3$$

is a double cover of $Symm^+(\mathcal{T})$.

Let us visualize $\mathbf{T}^* \subset S^3$ through the Clifford decomposition of S^3 discussed in Section 21. Recall that $S^3 = \bigcup_{-1 \leq r \leq 1} C_r$, where

$$C_r = \left\{ (z, w) \in \mathbf{C}^2 \mid |z|^2 = \frac{1+r}{2}, \; |w|^2 = \frac{1-r}{2} \right\}.$$

$C_{\pm 1}$ are orthogonal great circles cut out from $S^3 \subset \mathbf{R}^4$ by the coordinate planes spanned by the first two and the last two axes. For $-1 < r < 1$, C_r is a Clifford torus imbedded in S^3. Recall also that a unit quaternion is represented by a pair of complex numbers, $q = z + jw$, $z, w \in \mathbf{C}$, and $|q|^2 = 1$ corresponds to $|z|^2 + |w|^2 = 1$. With this, we see that $\{\pm 1, \pm i\} \subset \mathbf{T}^*$ correspond to the vertices of a square inscribed in C_1.

Similarly, $\{\pm j, \pm k\} \subset \mathbf{T}^*$ correspond to the same picture on C_{-1}. Finally, the elements $(\pm 1 \pm i \pm j \pm k)/2$ correspond to

$$\left(\frac{(\pm 1 \pm i)}{2}, \frac{(\pm 1 \pm i)}{2} \right) \in \mathbf{C}^2,$$

so that they are all in the middle Clifford torus C_0. We now view C_0 as $[0, 2\pi]^2$ with opposite sides identified as in Section 15. In this representation $(z, w) \in C_0$ corresponds to the point $(\arg(z), \arg(w)) \in [0, 2\pi]^2$. (Here we take the value of arg in $[0, 2\pi]$.) Working out the arguments of $(\pm 1 \pm i)/2$, we obtain all odd multiples of $\pi/4$ on $[0, 2\pi]$. Thus, on C_0, the binary tetrahedral group has sixteen points whose coordinates are odd multiples of $\pi/4$. Putting these together, we get $4 + 4 + 16 = 24 = |\mathbf{T}^*|$ points! (See Figure 23.3.)

The other examples are treated similarly. The octahedron $\mathcal{O}$ is placed in $\mathbf{R}^3 = \mathbf{H}_0$ with vertices

$$\pm i, \quad \pm j, \quad \pm k,$$

Figure 23.3

and the rotations around each vertex with angle $\pi/2$ are given by the quaternions

$$\frac{\pm 1 \pm i}{\sqrt{2}}, \quad \frac{\pm 1 \pm j}{\sqrt{2}}, \quad \frac{\pm 1 \pm k}{\sqrt{2}}.$$

Since $Symm^+(\mathcal{T}) \subset Symm^+(\mathcal{O})$ (recall that $\mathcal{O}$ can be obtained from $\mathcal{T}$ by four truncations), these and $\mathbf{T}^*$ generate the *binary octahedral group*

$$\mathbf{O}^* = \mathbf{T}^* \cup \left(\frac{1+i}{\sqrt{2}} \right) \mathbf{T}^* \subset S^3.$$

This is a double cover of $Symm^+(\mathcal{O})$. Notice that $(1+i)/\sqrt{2} = e^{i\pi/4}$ corresponds to rotation around the imaginary axis $\mathbf{R} \cdot i$ with angle $\pi/2$ and this is precisely the isometry that carries $\mathcal{T}$ into its reciprocal $\mathcal{T}^\circ$. We thus see the algebraic counterpart of the geometric argument used to determine the octahedral group in Section 17.

Multiplication by $e^{\pi i/4}$ has the effect of adding $\pi/4$ to the parameters $\arg(z)$ and $\arg(w)$ of the Clifford decomposition of S^3. Note that D_4^* becomes a subgroup of $\mathbf{O}^*$. The entire binary octahedral group is depicted in Figure 23.4.

Finally, working out all quaternions that give icosahedral rotations, we arrive at the double cover of $Symm^+(\mathcal{I})$:

$$\mathbf{I}^* = \mathbf{T}^* \cup \sigma \mathbf{T}^* \cup \sigma^2 \mathbf{T}^* \cup \sigma^3 \mathbf{T}^* \cup \sigma^4 \mathbf{T}^*,$$

where $\sigma = \frac{1}{2}(\tau + i + j/\tau)$ and τ is the golden section (see Section 17). This is called the *binary icosahedral group*. We omit the somewhat gory details (cf. Problem 2). Note again the perfect analogy between this algebraic splitting of $\mathbf{I}^*$ and the geometric description of the icosahedral group in terms of the five circumscribed tetrahedra in Section 17.

To place $\mathbf{I}^*$ in the Clifford setting needs a bit of computation. It is clear that the elements $\pm \omega^k$ and $\pm j\omega^k$, $k = 0, \ldots, 4$, make up

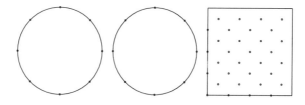

Figure 23.4

the vertices of two copies of a regular decagon, one inscribed in C_{-1}, the other in C_1. These account for 20 elements of I^*. For the remaining 100 elements, we have

$$|z|^2 - |w|^2 = \pm \frac{1}{\sqrt{5}} (|\omega - \omega^4|^2 - |\omega^2 - \omega^3|^2)$$

$$= \mp \frac{1}{\sqrt{5}} ((\omega - \omega^4)^2 - (\omega^2 - \omega^3)^2)$$

$$= \pm \frac{1}{\sqrt{5}} (\omega + \omega^4 - \omega^2 - \omega^3)$$

$$= \pm \frac{1}{\sqrt{5}} \left(\frac{1}{\tau} + \tau \right) = \pm \frac{1}{\sqrt{5}}.$$

We see that these elements (in two groups of 50) are on the two Clifford tori $C_{\pm 1/\sqrt{5}}$. Calculating the arguments is now easy, since $\omega - \omega^4$ and $\omega^2 - \omega^3$ are both purely imaginary. On $C_{1/\sqrt{5}}$, we obtain

$$\left(\frac{3\pi}{2} + \frac{2k\pi}{5}, \frac{\pi}{2} + \frac{2l\pi}{5} \right), \quad \left(\frac{\pi}{2} + \frac{2k\pi}{5}, \frac{3\pi}{2} + \frac{2l\pi}{5} \right),$$

where k, l are integers modulo 5. Similarly, on $C_{-1/\sqrt{5}}$ we have

$$\left(\frac{\pi}{2} + \frac{2k\pi}{5}, \frac{\pi}{2} + \frac{2l\pi}{5} \right), \quad \left(\frac{3\pi}{2} + \frac{2k\pi}{5}, \frac{3\pi}{2} + \frac{2l\pi}{5} \right),$$

where again k, l are integers modulo 5. The entire binary icosahedral group I^* is depicted in Figure 23.5.

Thus, quaternions can be used to describe the symmetry groups of the Platonic solids in a simple and elegant manner. Any computation involving these groups can be carried out using quaternionic arithmetic.

Figure 23.5

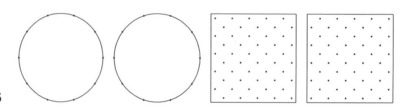

♡ One final algebraic note: In searching for quaternionic representation of the symmetries of the Platonic solids, we found the following finite subgroups of S^3:

1. $C_n = \{e^{2\pi l i/n} \mid l = 0, \ldots, n-1\}$;
2. $D_n^* = C_{2n} \cup jC_{2n}$; in particular, $D_2^* = \{\pm 1, \pm i, \pm j, \pm k\}$;
3. $\mathbf{T}^* = D_2^* \cup \{(\pm 1 \pm i \pm j \pm k)/2\}$;
4. $\mathbf{O}^* = \mathbf{T}^* \cup ((1+i)/\sqrt{2})\mathbf{T}^*$;
5. $\mathbf{I}^* = \mathbf{T}^* \cup \sigma \mathbf{T}^* \cup \sigma^2 \mathbf{T}^* \cup \sigma^3 \mathbf{T}^* \cup \sigma^4 \mathbf{T}^*$, where $\sigma = \frac{1}{2}(\tau + i + j/\tau)$.

We finish this section by showing that this is an exhaustive list of all finite subgroups of S^3.

Theorem 17.

Any finite subgroup of $S^3 = SU(2)$ is either cyclic, or conjugate to one of the binary subgroups D_n^*, T^*, O^*, or I^*.

Proof.

Let $G \subset S^3 \cong SU(2)$ be a finite subgroup with corresponding subgroup G_0 in $SU(2)/\{\pm I\}$. Let $G^* \subset S^3$ be the inverse image of G_0 under the canonical projection $SU(2) \to SU(2)/\{\pm I\}$. Clearly, $G \subset G^*$. By Theorem 16, G_0 is isomorphic to a finite subgroup of $SO(3)$. If $G = G^*$, then G^* is the double cover of the group G_0. In this case the theorem follows from the classification of finite subgroups in $SO(3)$ (Theorem 11). Thus, we need only study the case where $G \neq G^*$. In this case G is of index 2 in G^*, and G and G_0 are isomorphic. We first claim that G is of odd order. Assume, to the contrary, that G has even order. By a standard result in group theory (due to Cauchy), G must contain an element of order 2. Since the only element of order 2 in $SU(2)$ is $-I$, it must be contained in G. But $\{\pm I\}$ is the kernel of the canonical projection, and this contradicts $G \neq G^*$. Thus, G has odd order. G is isomorphic to G_0, and as noted above, the latter has an isomorphic copy in $SO(3)$. The odd order subgroups in $SO(3)$ are cyclic, as follows again from the classification of all finite subgroups in $SO(3)$. The theorem follows. ∎

♠ One truly final note: In the same way as we derived the real projective plane $\mathbf{R}P^2$ as the quotient of S^2 by the group $\{\pm I\}$, we can consider quotients of S^3 by the finite subgroups above. We obtain the *lens spaces* $L(n; 1) = S^3/C_n$, the *prism manifolds* S^3/D_n^*, the *tetrahedral manifold* $S^3/\mathbf{T}^*$, the *octahedral manifold* $S^3/\mathbf{O}^*$, and the *icosahedral manifold* $S^3/\mathbf{I}^*$.

Problems

1. Verify the following inclusions among the binary groups:

$$C_2 \subset C_4 \subset \mathbf{T}^*, \quad C_3 \subset C_6 \subset \mathbf{T}^*, \quad D_2^* \subset \mathbf{T}^*;$$

$$C_2 \subset C_4 \subset C_8 \subset \mathbf{O}^*, \quad D_2^* \subset D_4^* \subset \mathbf{O}^*, \quad D_3^* \subset \mathbf{O}^*;$$

$$C_2 \subset C_4 \subset \mathbf{I}^*, \quad C_3 \subset C_6 \subset \mathbf{I}^*, \quad C_5 \subset C_{10} \subset \mathbf{I}^*, \quad D_2^*, D_3^*, D_5^*, \subset \mathbf{I}^*.$$

2. Verify by explicit calculation that the rotation in $\mathbf{H}_0$ represented by the quaternion $\sigma = (1/2)(\tau + i + j/\tau)$ permutes the vertices of the icosahedron. (What is the rotation angle?)

24

SECTION

Invariants

† ♣ Recall from Section 18 that we set ourselves to the task of finding all finite Möbius groups. Using Cayley's theorem, we constructed a list of finite Möbius groups most of which arose from the symmetry groups of Platonic solids. To see the difficulty in our quest, we now briefly recall from Section 17 the more elementary classification of all finite groups G of spatial rotations. To pin down the structure of G, we considered the set Q of antipodal pairs of poles in S^2, the intersections of the axes of the rotations with S^2 that constituted G. We showed that Q was G-invariant and that the rotations that corresponded to the poles in a single G-orbit C had the same order. The number of rotations in G with poles in a fixed G-orbit C and of order d_C worked out to be

$$\frac{(d_C - 1)|G|}{2d_C}.$$

Adding up, we obtained the Diophantine equation

$$2|G| - 2 = |G| \sum_{C \in Q/G} \left(1 - \frac{1}{d_C}\right).$$

We finally realized that this Diophantine equation was so restrictive that we could simply list all possible scenarios.

In the more general case when G is a finite subgroup of $M\ddot{o}b\,(\hat{\mathbf{C}})$, we do not have as much Euclidean structure as for the finite rotation groups, but a simple observation gives a clue how to proceed. Assume that we can find a rational function $q : \mathbf{C} \to \mathbf{C}$ whose invariance group is G, that is, $q \circ g = q$ for $g \in M\ddot{o}b\,(\hat{\mathbf{C}})$ iff $g \in G$. Extended to $\hat{\mathbf{C}}$, q can be considered as the projection of an analytic $|G|$-fold branched covering $q : \hat{\mathbf{C}} \to \hat{\mathbf{C}}$ (denoted by the same symbol). The branch points are the fixed points of the linear fractional transformations in G, and the branch numbers correspond to the orders of the rotations (minus one) in the special case when G is defined by a rotation group via Cayley's theorem. By the Riemann–Hurwitz relation (Section 19), the total branching number is $2|G| - 2$. This will give the same Diophantine restriction for our finite Möbius group G as above!

We actually want to prove that the list of finite Möbius groups obtained in Section 18 is exhaustive, that is, any finite Möbius group G is conjugate to one of the groups in that list. To do this, we first construct, for each finite Möbius group in our list, an invariant rational function. This way we obtain a list of rational functions. We then construct an invariant rational function for a general finite Möbius group G. Finally, we compare our list of rational functions with the latter using either uniformization or counting residues. We now see the idea behind this seemingly circuitous argument: A rational function completely characterizes its invariance group, and comparing rational functions is a lot easier than comparing Möbius groups, since to accomplish the former task the entire arsenal of complex analysis is at our disposal!

Remark.

There is a quick algebraic way to find all finite Möbius groups $G \subset M\ddot{o}b\,(\hat{\mathbf{C}})$. It is based on the fact that any finite subgroup of $SL(2, \mathbf{C})$ (such as the binary cover G^* of G) is conjugate to a subgroup of $SU(2)$ (and the finite subgroups of $SU(2)$ are classified in Theorem 17). This is usually proved by averaging the standard scalar product on $\mathbf{C}^2$ over the finite subgroup of $SL(2, \mathbf{C})$. Finally, by Cayley's theorem, a finite subgroup of $SU(2)$ corresponds to a finite group of spatial rotations whose classification was accomplished in Section 17. Despite the existence of this short and elegant proof, we prefer to follow a longer path not only because of its beauty but because we will use some of the ingredients in later developments.

The geometry of the spherical Platonic tessellations (including the dihedron) can be conveniently described by the so-called *characteristic triangle*. Given a spherical Platonic tessellation, a *spherical flag* (v, e, f) consists of a vertex v, an edge e, and a face f with $v \in e \subset f(\subset S^2)$. Each spherical flag (v, e, f) contains one characteristic triangle whose vertices are $v_0 = v$, v_1, the midpoint of the edge e, and v_2, the centroid of the face f. (For spherical tessellations we use the spherical analogues of the Euclidean polyhedral concepts such as vertex, edge, face, etc., with obvious meanings.) The spherical angles of a characteristic triangle at the respective vertices are π/v_0, π/v_1, and π/v_2, where v_0, v_1, v_2 are integers greater than or equal to 2. Since v_1 is the midpoint of an edge, we always have $v_1 = 2$. For the dihedron, we have $v_0 = 2$, $v_2 = n$; for the tetrahedron, $v_0 = v_2 = 3$; for the octahedron, $v_0 = 4$, $v_2 = 3$; and, finally, for the icosahedron, $v_0 = 5$, $v_2 = 3$. (Also, v_0 is the number of faces meeting at a vertex, and v_2 is v_0 of the reciprocal; $\{v_2, v_0\}$ is then the Schläfli symbol.) ♡ Reflections in the sides of a characteristic triangle generate the symmetry group in which the group of direct symmetries of the tessellation is a subgroup of index two.

♣ Each symmetry axis of a spherical Platonic tessellation goes through a vertex, or the midpoint of an edge, or the centroid of a face. It follows that v_0, v_1, and v_2 are the orders of the rotations with axes through the respective points. In particular, the symmetry rotations around the midpoints of edges are always half-turns.

The acting symmetry group G of the tessellation has three *special orbits* on S^2. The vertices of the tessellation constitute one special orbit. Since a symmetry rotation around a vertex has order v_0, this orbit consists of $|G|/v_0$ elements. At the same time, this is the number of vertices of the tessellation. Another special orbit consists of the midpoints of the edges; the orbit consists of $|G|/v_1$ points, and this is also the number of edges. Finally, the third special orbit consists of the centroids of the faces, or equivalently, the vertices of the reciprocal tessellation. The number of faces is thus $|G|/v_2$. By Euler's theorem for convex polyhedra (Section 17), we have

$$\frac{|G|}{v_0} - \frac{|G|}{v_1} + \frac{|G|}{v_2} = 2.$$

All other orbits of G are *principal*, that is, the number of points in the orbit is the order $|G|$ of G.

Recall that our intermediate purpose is to construct a rational function $q : \mathbf{C} \to \mathbf{C}$ whose invariance group is a given finite Möbius group G in our list. It is now time to discuss our plan. We first consider the binary group $G^* \subset SU(2)$ associated to G, and study invariance of polynomials $F : \mathbf{C}^2 \to \mathbf{C}$ under G^*. We will assume that F is *homogeneous* (of degree d), that is,

$$F(tz_1, tz_2) = t^d F(z_1, z_2), \quad \text{for all } t, z_1, z_2 \in \mathbf{C}.$$

We will be able to construct two linearly independent G^*-invariant homogeneous polynomials $E, F : \mathbf{C}^2 \to \mathbf{C}$ of the same degree. (The common degree will turn out to be equal to the order $|G|$ of the group G.) Due to homogeneity, the rational quotient E/F will define our analytic function $q : \hat{\mathbf{C}} \to \hat{\mathbf{C}}$ by

$$q(z) = \frac{E(z_1, z_2)}{F(z_1, z_2)}, \quad z = \frac{z_1}{z_2}, \quad z_1, z_2 \in \mathbf{C}.$$

(The quotient E/F factors through the canonical projection $\mathbf{C}^2 - \{0\} \to \mathbf{C}P^1$, where $\mathbf{C}P^1$ is the complex projective line identified with $S^2 = \hat{\mathbf{C}}$; cf. Section 22.) This is because $q(z_1, z_2)$ depends only on the homogeneous coordinates of the projective point $[z_1 : z_2] \in \mathbf{C}P^1$. Notice, finally, that G^*-invariance can be defined up to constant multiples since common multiples, of E and F cancel in the ratio E/F.

Remark.

q is usually called the *fundamental rational function*, and the problem of inverting q (to be discussed shortly), the *fundamental problem*.

We are now ready to get started. For brevity, we call a homogeneous polynomial $F : \mathbf{C}^2 \to \mathbf{C}$ a *form*. Given any subgroup $G^* \subset SL(2, \mathbf{C})$, we say that F is G^*-invariant if there exists a *character* $\chi_F : G^* \to \mathbf{C} - \{0\}$, a homomorphism of G^* into the multiplicative group $\mathbf{C} - \{0\}$, such that

$$F \circ g = \chi_F(g) \cdot F, \quad g \in G^*.$$

Here $g \in G^*$ acts on the argument $(z_1, z_2) \in \mathbf{C}^2$ by ordinary matrix multiplication. The character χ_F is uniquely determined by F. We say that F is an *absolute invariant* of G^* if $\chi_F = 0$. In general, F is an absolute invariant of the subgroup $\ker \chi_F \subset G^*$. If G^* is finite, then χ_F maps into the unit circle $S^1 \subset \mathbf{C}^* - \{0\}$ (why?), and the quotient $G^* / \ker \chi_F$, being isomorphic to a finite subgroup of S^1, is cyclic.

According to our plan, for each finite Möbius group G in our list, with corresponding binary Möbius group $G^* \subset SU(2)$, we will exhibit two forms E, F of degree $|G|$ such that E and F are both G^*-invariant and have the *same character*. The function q defined above is the G-invariant rational function we seek.

The cyclic group $G = C_n$ does not fit in the general framework, since it is not the symmetry group of a spherical Platonic tessellation. Although it is easy to obtain the general C_n-invariant rational function $q : \mathbf{C} \to \mathbf{C}$ of degree n by inspection, it is instructive to go through the planned procedure in this simple case. We thus seek the most general C_n^*-invariant form F of degree n, where $C_n^* = C_{2n}$ is cyclic. A typical diagonal matrix in $SU(2)$ has diagonal entries $a, a^{-1} = 1/a$, where $a \in \mathbf{C} - \{0\}$. We identify this matrix with the first diagonal element a. It acts on (z_1, z_2) as $(az_1, a^{-1}z_2)$, $z_1, z_2 \in \mathbf{C}$. The character χ_F is uniquely determined by its value on the generator $\omega = e^{\pi i/n}$, a primitive $2n$th root of unity. Since $\omega^{2n} = 1$, $\chi_F(\omega)$ must be a $2n$th root of unity, that is, $\chi_F(\omega) = \omega^m$ for some $m = 0, \ldots, 2n - 1$. The condition of C_n^*-invariance for F reduces to

$$F(\omega z_1, \omega^{-1} z_2) = \omega^m F(z_1, z_2), \quad z_1, z_2 \in \mathbf{C}.$$

Being homogeneous of degree n, the typical monomial participating in F is $z_1^j z_2^{n-j}$, where $j = 0, \ldots, n$. Substituting this into the equation of invariance, we obtain $\omega^{2j} = \omega^{m+n}$. Since ω is a $2n$th root of unity, we have

$$2n \mid m + n - 2j.$$

In particular, $n|m-2j$, or equivalently, $m-2j=nl$ for some integer l. Hence, $2n|n(l+1)$, and l is odd. Inspecting the ranges of j and m, we see that $|m-2j|<2n$, so that $l=\pm 1$. We have $2j=m\pm n$, so that m and n have the same parity, and $j=(m\pm n)/2$. For $l=1$, we have $m\geq n$, and for $l=-1$, $m\leq n$. The corresponding monomials are

$$z_1^{(m\pm n)/2}z_2^{n-(m\pm n)/2}.$$

Two linearly independent C_n^*-invariant forms of degree n and the same character (the same m) exist iff $m=n$, and in this case, the general C_n^*-invariant form is a linear combination of z_1^n and z_2^n. Since $z=z_1/z_2$, the general rational function q with invariance group C_n is a quotient of two linearly independent forms. We obtain that the most general C_n-invariant rational function is a *linear fractional transformation applied to* z^n. Notice, in particular, that $z^n=z_1^n/z_2^n$ vanishes at the fixed points 0 and ∞ of the rotations that make up C_n. Analytically, the fixed points are the branch points of q considered as a self-map of $\hat{\mathbf{C}}$. (Compare this with the discussion before the proof of the FTA in Section 8!)

This last remark gives a clue how to obtain invariant forms in the case of spherical Platonic tessellations. We first discuss the dihedron with the special position given in Section 18. Recall that the vertices of a dihedron are the n points distributed uniformly along the equator of S^2 with e_1 being a vertex. On $\hat{\mathbf{C}}$, these points constitute the roots of the equation $z^n=1$. Since $z=z_1/z_2$, we see that the most general degree-n form that vanishes on these vertices is a constant multiple of $z_1^n-z_2^n$. What do we mean by a form F vanishing at a point in $\hat{\mathbf{C}}$? After all, F has two complex arguments, z_1 and z_2! If $F(z_1,z_2)=0$, then by homogeneity, we also have $F(tz_1,tz_2)=0$ for all $t\in\mathbf{C}$, so that vanishing of $F(z_1,z_2)$ is a property of the projective point $[z_1:z_2]$ rather than the property of a specific representative (z_1,z_2). Using more modern terminology, a form F factors through the canonical projection $\mathbf{C}^2-\{0\}\to\mathbf{C}P^1$ and gives a well-defined function on $\mathbf{C}P^1=\hat{\mathbf{C}}$. We set

$$\alpha(z_1,z_2)=\frac{z_1^n-z_2^n}{2}.$$

In a similar vein, the most general degree-n form that vanishes on the midpoints of the edges of the dihedron is a constant multiple of $z_1^n + z_2^n$. We define

$$\beta(z_1, z_2) = \frac{z_1^n + z_2^n}{2}.$$

Finally, the midpoints of the 2 hemispherical faces correspond to 0 and ∞, and we set

$$\gamma(z_1, z_2) = z_1 z_2.$$

We say that the forms α, β, and γ *belong to the dihedron*. As a simple computation shows, all three forms are D_n^*-invariant, with $\chi_\alpha = \chi_\beta$ and $\chi_\gamma = \pm 1$, with $+1$ corresponding to the cyclic kernel $C_n^* \subset D_n^*$. The forms α, β, and γ are algebraically dependent. In fact, we have

$$\alpha^2 - \beta^2 + \gamma^n = 0.$$

Notice that the degrees of α, β, and γ are $|D_n|/\nu_0$, $|D_n|/\nu_1$, and $|D_n|/\nu_2$, and consequently, the exponents in the equation above are ν_0, ν_1, and ν_2.

  It is worthwhile to generalize some of the properties of the forms of the dihedron derived above. In fact, in each of the remaining cases of Platonic tessellations, we will have three forms F_0, F_1, and F_2 that will be said to *belong to the tessellation* in the sense that F_0 vanishes on the projected vertices, F_1 vanishes on the projected midpoints of the edges, and F_2 vanishes on the projected centroids of the faces.

Remark.

In trying to make an up-to-date treatment of the subject, we adopted a number of changes, retaining as much classical terminology as possible. For example, our nonstandard notation for the forms F_0, F_1, and F_2 reflects the fact that our $F_j, j = 0, 1, 2$, vanishes on the centroids of the j-dimensional "cells" of the Platonic solid.

  Since the zero sets of F_0, F_1, and F_2 are the 3 special orbits of the action of G on $\hat{\mathbf{C}}$, these forms are all invariants of the corresponding binary Möbius group G^*. They have degrees $|G|/\nu_0$, $|G|/\nu_1$, and

$|G|/\nu_2$. We now claim that the forms $F_0^{\nu_0}$, $F_1^{\nu_1}$, and $F_2^{\nu_2}$ (all of degree $|G|$) are linearly dependent. To do this, recall that away from the three special orbits of G on S^2, all orbits are principal. Given a principal orbit, we can find a complex ratio $\mu_0 : \mu_1$ such that $\mu_0 F_0^{\nu_0} + \mu_1 F_1^{\nu_1}$ vanishes at one point of this orbit. By invariance, this linear combination vanishes at each point of the orbit. In a similar vein, for another ratio $\mu_1 : \mu_2$, $\mu_1 F_1^{\nu_1} + \mu_2 F_2^{\nu_2}$ vanishes on the same orbit. These two linear combinations are polynomials of degree $|G|$, and both vanish on a principal orbit containing $|G|$ points. It follows that they must be constant multiples of each other. We obtain that

$$\lambda_0 F_0^{\nu_0} + \lambda_1 F_1^{\nu_1} + \lambda_2 F_2^{\nu_2} = 0$$

for some $\lambda_0 : \lambda_1 : \lambda_2$. Notice that for the dihedron, this reduces to the algebraic relation we just derived for the dihedral forms α, β, and γ.

We finally claim that every G^*-invariant form F can be written as a polynomial in F_0, F_1, and F_2. We proceed by induction with respect to the degree of F and exhibit a polynomial factor of F in F_0, F_1, and F_2. Consider the zero set of F. By invariance of F, this set is G-invariant, the union of some G-orbits. If there is a special orbit among these, then F_0, F_1, or F_2 divides F. Otherwise, there must be a principal orbit on which F vanishes. As above, for a complex ratio $\mu_0 : \mu_1$, the linear combination $\mu_0 F_0^{\nu_0} + \mu_1 F_1^{\nu_1}$ vanishes on this principal orbit, and therefore it must be a factor of F. The claim follows.

We now consider the tetrahedron. Applying the stereographic projection to the vertices of our tetrahedron, we obtain the points

$$\pm \frac{1+i}{\sqrt{3}-1}, \quad \pm \frac{1-i}{\sqrt{3}+1}.$$

A form Φ of degree 4 that vanishes at these points can be obtained by multiplying out the linear factors:

$$\Phi(z_1, z_2) = \left(z_1^2 - \left(\frac{1+i}{\sqrt{3}-1} \right)^2 z_2^2 \right) \left(z_1^2 - \left(\frac{1-i}{\sqrt{3}+1} \right)^2 z_2^2 \right)$$

$$= z_1^4 - 2\sqrt{3}i z_1^2 z_2^2 + z_2^4.$$

The vertices of the reciprocal tetrahedron are the negatives of the original. By Problem 1 (b) in Section 7, $h_N(-p) = -1/\overline{h_N(p)}, p \in S^2$, so that these vertices projected to $\hat{\mathbf{C}}$ are

$$\mp\frac{1+i}{\sqrt{3}+1}, \quad \mp\frac{1-i}{\sqrt{3}-1}.$$

The corresponding form Ψ of degree 4 is

$$\Psi(z_1, z_2) = z_1^4 + 2\sqrt{3}iz_1^2z_2^2 + z_2^4.$$

The vertices of the tetrahedron and its reciprocal are the two alternate sets of vertices of the circumscribed cube. As a byproduct, we see that the product $\Phi\Psi$ vanishes on the vertices of this cube. Expanding, we obtain

$$\Phi\Psi(z_1, z_2) = z_1^8 + 14z_1^4z_2^4 + z_2^8.$$

The midpoints of the edges of any one of the tetrahedra are the vertices of the octahedron. Projected to $\hat{\mathbf{C}}$, these are

$$0, \quad \infty, \quad \pm 1, \quad \pm i.$$

Since z_1 vanishes at 0, and z_2 vanishes at ∞, a form Ω of degree 8 that vanishes on the vertices above takes the form

$$\Omega(z_1, z_2) = z_1z_2(z_1^2 - z_2^2)(z_1^2 + z_2^2) = z_1z_2(z_1^4 - z_2^4).$$

We say that the forms Φ, Ω, and Ψ belong to the tetrahedron. Recall that this means that $\Phi = 0$ on the vertices, $\Psi = 0$ on the midpoints of the edges, and $\Omega = 0$ on the centroids of the faces of the tetrahedron. We also see that the degrees of Φ, Ω, and Ψ are $|T|/\nu_0$, $|T|/\nu_1$, and $|T|/\nu_2$. The forms Φ and Ψ are invariants of the binary tetrahedral group T^*; $\chi_\Phi = \chi_\Psi$ with kernel D_2^*. The form Ω is an absolute invariant. By the general discussion above, Φ^3, Ψ^3, and Ω^2 are linearly dependent. Comparing some of the coefficients, we have

$$\Phi^3 - 12\sqrt{3}i\Omega^2 - \Psi^3 = 0.$$

It is again convenient to make another observation about invariant forms. Given a G^*-invariant form F, the Hessian $Hess(F)$

defined by

$$
Hess\,(F)(z_1, z_2) = \begin{vmatrix} \dfrac{\partial^2 F}{\partial z_1^2} & \dfrac{\partial^2 F}{\partial z_1 \partial z_2} \\ \dfrac{\partial^2 F}{\partial z_2 \partial z_1} & \dfrac{\partial^2 F}{\partial z_1^2} \end{vmatrix}
$$

is also G^*-invariant. Furthermore, given two G^*-invariant forms F_0 and F_1, the Jacobian $Jac\,(F_0, F_1)$ defined by

$$
Jac\,(F_0, F_1)(z_1, z_2) = \begin{vmatrix} \dfrac{\partial F_0}{\partial z_1} & \dfrac{\partial F_0}{\partial z_2} \\ \dfrac{\partial F_1}{\partial z_1} & \dfrac{\partial F_1}{\partial z_2} \end{vmatrix}
$$

is also G^*-invariant. Thus, once we have a G^*-invariant form F_0, we automatically have two additional G^*-invariant forms: the Hessian $F_1 = Hess\,(F_0)$ and then the Jacobian $F_2 = Jac\,(F_0, F_1)$.

We first try this for the tetrahedral form Φ of degree 4. Since the Hessian $Hess\,(\Phi)$ is of degree 4, it can only be a linear combination of Φ and Ψ. By an easy computation, we see that $Hess\,(\Phi) = -48\sqrt{3}i\Psi$. The Jacobian of Φ and Ψ is of degree 6, so that without any computation we see that it must be a constant multiple of Ω. Working out the leading coefficient, we see that $Jac\,(\Phi, \Psi) = 32\sqrt{3}i\Omega$.

Remark.
For the dihedral forms α, β, and γ, we have $Hess\,(\alpha) = -Hess\,(\beta) = -(n^2(n-1)^2/4)\gamma^{n-2}$, $Hess\,(\gamma) = 1$, and $Jac\,(\alpha, \beta) = \frac{n^2}{2}\gamma^{n-1}$, $Jac\,(\alpha, \gamma) = n\beta$, $Jac\,(\beta, \gamma) = n\alpha$.

Recall that the octahedral Möbius group contains the tetrahedral Möbius group, so that the same is true for the associated binary groups. Thus the O^*-invariant forms are polynomials of the tetrahedral forms Φ, Ψ, and Ω. To obtain the forms that belong to the octahedron, we first note that by construction, the octahedral form that vanishes on the vertices of the octahedron is Ω. It is equally clear that the form $\Phi\Psi$ of degree 8 vanishes on the midpoints of the faces of the octahedron, since these points are nothing but the vertices of the cubic reciprocal. Note also that $Hess\,(\Omega) = -25\Phi\Psi$. We finally have to find a form of degree 12 that vanishes on the midpoints of the edges of the octahedron. Based on analogy with the tetrahedral forms, this form of degree 12 must be the Jacobian

$$Jac\,(\Omega, \Phi\Psi) = \Phi\, Jac\,(\Omega, \Psi) + \Psi\, Jac\,(\Omega, \Phi).$$

A simple computation shows that $Jac\,(\Omega, \Phi) = -4\Psi^2$ and $Jac\,(\Omega, \Psi) = -4\Phi^2$. Normalizing, we obtain that the middle octahedral form is

$$\frac{\Phi^3 + \Psi^3}{2}.$$

An explicit expression of this form is obtained by factoring

$$\frac{\Phi^3 + \Psi^3}{2} = \frac{1}{2}(\Phi + \Psi)(\Phi^2 - \Phi\Psi + \Psi^2)$$
$$= (z_1^4 + z_2^4)(z_1^8 - 34z_1^4 z_2^4 + z_2^8)$$
$$= z_1^{12} - 33z_1^8 z_2^4 - 33z_1^4 z_2^8 + z_2^{12}.$$

Summarizing, the three forms that belong to the octahedron are Ω, $(\Phi^3 + \Psi^3)/2$, and $\Phi\Psi$. They are all absolute invariants of the binary tetrahedral group T^*. As octahedral invariants we have $\chi_\Omega = \chi_{(\Phi^3+\Psi^3)/2} = \pm 1$, while $\Phi\Psi$ is an absolute octahedral invariant. The linear relation among Ω^4, $(\Phi^3 + \Psi^3)^2/4$, and $(\Phi\Psi)^3$ can be deduced from the previous relation between Φ, Ψ, and Ω. Squaring both sides of the equation

$$\frac{\Phi^3 + \Psi^3}{2} = 6\sqrt{3}i\Omega^2,$$

we have

$$\frac{1}{4}(\Phi^3 - \Psi^3)^2 = \frac{1}{4}(\Phi^3 + \Psi^3)^2 - (\Phi\Psi)^3 = 108\Omega^4.$$

We arrive at

$$108\Omega^4 + \left(\frac{\Phi^3 + \Psi^3}{2}\right)^2 - (\Phi\Psi)^3 = 0.$$

We now turn to the final case of the icosahedron. Recall that in Section 18 we determined the vertices of the iscosahedron (in special position) projected to $\hat{\mathbf{C}}$ by the stereographic projection. An icosahedral form $\mathcal{I}$ of degree 12 that vanishes on these projected vertices is

$$
\mathcal{I}(z_1, z_2) = z_1 z_2 \prod_{j=0}^{4}(z_1 - \omega^j(\omega + \omega^4)z_2) \prod_{j=0}^{4}(z_1 - \omega^j(\omega^2 + \omega^3)z_2)
$$

$$
= z_1 z_2 (z_1^5 - (\omega + \omega^4)^5 z_2^5)(z_1^5 - (\omega^2 + \omega^3)^5 z_2^5)
$$

$$
= z_1 z_2 (z_1^5 - \tau^{-5} z_2^5)(z_1^5 + \tau^5 z_2^5)
$$

$$
= z_1 z_2 (z_1^{10} + (\tau^5 - \tau^{-5}) z_1^5 z_2^5 - z_2^{10}).
$$

Here we used the identity

$$
t^5 - 1 = \prod_{j=0}^{4}(t - \omega^j)
$$

with the substitutions $t = z/(\omega + \omega^4)$ and $t = z/(\omega^2 + \omega^3)$ (cf. Problem 5 (b) of Section 17). The coefficient $\tau^5 - 1/\tau^5$ is the Lucas number $L_5 = 11$ (cf. Problem 14 (d) of Section 17). (Another way to see this is to factor first as

$$
\tau^5 - \frac{1}{\tau^5} = \left(\tau - \frac{1}{\tau}\right)\left(\tau^4 + \tau^2 + 1 + \frac{1}{\tau^2} + \frac{1}{\tau^4}\right),
$$

and then square the defining relation $\tau - 1/\tau = 1$ to obtain $\tau^2 + 1/\tau^2 = 3$, and square again to get $\tau^4 + 1/\tau^4 = 7$. Adding up, we obtain $\tau^5 - 1/\tau^5 = 11$.)

Summarizing, we arrive at the first *icosahedral form*

$$
\mathcal{I}(z_1, z_1) = z_1 z_2 (z_1^{10} + 11 z_1^5 z_2^5 - z_2^{10}).
$$

Here $\mathcal{I}$ is an absolute invariant of I^*. In fact, all invariants of I^* are absolute, since I^* has no proper normal subgroup (cf. Problem 6 of Section 17), and thereby no nontrivial character. (The kernel of a nontrivial character would be a proper normal subgroup.) The Hessian $Hess(\mathcal{I})$ is an absolute invariant of degree 20. Thus, it must vanish on the centroids of the faces of the icosahedron, or equivalently, on the vertices of the reciprocal dodecahedron.

Normalizing, we compute

$$\mathcal{H}(z_1, z_2) = \frac{1}{121} \, Hess\,(\mathcal{I})\,(z_1, z_2)$$

$$= -(z_1^{20} + z_2^{20}) + 228(z_1^{15}z_2^5 - z_1^5z_2^{15}) - 494z_1^{10}z_2^{10}.$$

The Jacobian $Jac\,(\mathcal{I}, \mathcal{H})$ is an absolute invariant of degree 30, so it must vanish on the midpoints of the edges of the icosahedron. An easy computation shows that

$$\mathcal{J}\,(z_1, z_2) = \frac{1}{20} \, Jac\,(\mathcal{I}, \mathcal{H})(z_1, z_2)$$

$$= (z_1^{30} + z_2^{30}) + 522(z_1^{25}z_2^5 - z_1^5z_2^{25}) - 10005(z_1^{20}z_2^{10} + z_1^{10}z_2^{20}).$$

The icosahedral forms $\mathcal{I}$, $\mathcal{J}$, and $\mathcal{H}$ are algebraically dependent. A comparison of coefficients shows that

$$1728\,\mathcal{I}^5 - \mathcal{J}^2 - \mathcal{H}^3 = 0.$$

♠ Let $\mathbf{C}[z_1, z_2]$ denote the ring of polynomials in the variables z_1 and z_2. Since all icosahedral invariants are absolute, we obtain that the ring $\mathbf{C}[z, w]^{I^*}$ of icosahedral invariants is isomorphic to the polynomial ring

$$\mathbf{C}[z, w]^{I^*} = \mathbf{C}[\mathcal{I}, \mathcal{J}, \mathcal{H}]/(1728\,\mathcal{I}^5 - \mathcal{J}^2 - \mathcal{H}^3),$$

where we factor by the principal ideal generated by $1728\,\mathcal{I}^5 - \mathcal{J}^2 - \mathcal{H}^3$.

♣ We now return to the general situation. Forming the six possible quotients of $F_0^{v_0}$, $F_1^{v_1}$, and $F_2^{v_2}$, we obtain six G-invariant rational functions that solve our problem. Since we have linear relations among these forms, the six rational functions can be written in terms of each other in obvious ways. A more elegant way to express these is to use homogeneous coordinates and write

$$q : q - 1 : 1 = -\lambda_2 F_2^{v_2} : \lambda_1 F_1^{v_1} : \lambda_0 F_0^{v_0},$$

where $q = -\lambda_2 F_2^{v_2}/\lambda_0 F_0^{v_0}$. We make an exception for the dihedron and define q such that $q : q - 1 : 1 = \alpha^2 : \beta^2 : -\gamma^n$.

We summarize our results in the following tables:

Platonic solid	G	$\lvert G \rvert$	v_0	v_1	v_2	F_0	F_1	F_2
Dihedron	D_n	$2n$	2	2	n	α	β	γ
Tetrahedron	T	12	3	2	3	Φ	Ω	Ψ
Octahedron	O	24	4	2	3	Ω	$\frac{\Phi^3 + \Psi^3}{2}$	$\Phi\Psi$
Icosahedron	I	60	5	2	3	$\mathcal{I}$	$\mathcal{J}$	$\mathcal{H}$

G	$\lambda_0 F_0^{v_0} + \lambda_1 F_1^{v_1} + \lambda_2 F_2^{v_2} = 0$	$q : q - 1 : 1$
D_n	$\alpha^2 - \beta^2 + \gamma^n = 0$	$\alpha^2 : \beta^2 : -\gamma^n$
T	$\Phi^3 - 12\sqrt{3}i\Omega^2 - \Psi^3 = 0$	$\Psi^3 : -12\sqrt{3}i\Omega^2 : \Phi^3$
O	$108\Omega^4 + \left(\frac{\Phi^3 + \Psi^3}{2}\right)^2 - (\Phi\Psi)^3 = 0$	$(\Phi\Psi)^3 : \left(\frac{\Phi^3 + \Psi^3}{2}\right)^2 : 108\Omega^4$
I	$1728\mathcal{I}^5 - \mathcal{J}^2 - \mathcal{H}^3 = 0$	$\mathcal{H}^3 : -\mathcal{J}^2 : 1728\mathcal{I}^5$

We are finally ready to prove that any finite Möbius group is conjugate to one of the Möbius groups listed in Section 18. This result is due to Klein.

Theorem 18.

     Any finite Möbius group $G \subset$ Möb $(\hat{\mathbf{C}})$ is cyclic, or conjugate to D_n, T, O, or I.

Proof.

Let $G \subset$ Möb$(\hat{\mathbf{C}})$ be a finite subgroup. Let $a, b \in \mathbf{C}$ such that $g(a) \neq b$, for all $g \in G$. It follows that $g(b) \neq a$, for all $g \in G$. Consider the rational function $\tilde{q} : \mathbf{C} \to \mathbf{C}$ defined by

$$\tilde{q}(z) = \prod_{g \in G} \frac{g(z) - a}{g(z) - b}, \quad z \in \mathbf{C}.$$

The condition on a and b guarantees that $\tilde{q}$ is nonconstant. Given $w \in \mathbf{C}$, to find $z \in \mathbf{C}$ such that $\tilde{q}(z) = w$, we need to solve the equation

$$\prod_{g \in G}(g(z) - a) = w \prod_{g \in G}(g(z) - b).$$

Multiplying out the denominators in the linear fractions $g(z)$, $g \in G$, this becomes a polynomial equation of degree $|G|$ with w as a parameter. Extended to $\hat{\mathbf{C}}$, $\tilde{q}$ is an analytic map of degree $|G|$. On the other hand, $\tilde{q}$ is G-invariant. It follows that for fixed $w \in \hat{\mathbf{C}}$, the solution set of $\tilde{q}(z) = w$ is a single G-orbit, and $\tilde{q} : \hat{\mathbf{C}} \to \hat{\mathbf{C}}/G = \hat{\mathbf{C}}$ is the orbit map. The point $z \in \hat{\mathbf{C}}$ is a branch point iff the G-orbit $G(z)$ through z is not principal. In this case, $|G(z)| = |G|/\nu$, and the branch number associated to z is $\nu - 1$. Letting Π denote the set of branch points of $\tilde{q}$, the total branching number is

$$B = \sum_{\Pi/G} \frac{|G|}{\nu}(\nu - 1),$$

where the summation is over all branch values. By the Riemann–Hurwitz relation (Section 19), the total branching number is equal to $2|G| - 2$, since both the domain and the range have zero genera. We thus have

$$\sum_{\Pi/G}\left(1 - \frac{1}{\nu}\right) = 2 - \frac{2}{|G|}.$$

As noted at the beginning of this section, this is the same Diophantine restriction as the one for the classification of finite rotation groups. Adopting the analysis there in our setting, we see that the sum on the left-hand side either consists of two terms with $\nu_0 = \nu_1 = |G|$ or consists of three tems with ν_0, ν_1, and ν_2 as given in the table above. In the first case, let w_0 and w_1 be the branch values corresponding to ν_0 and ν_1. By performing a linear fractional transformation on the range, we may assume that $w_0 = 0$ and $w_1 = \infty$. For the rest of the cases, let w_0, w_1, and w_2 be the three branch values. Performing a linear fractional transformation on the range again, we may assume that these are $w_0 = 0$, $w_1 = 1$, and $w_2 = \infty$, and they correspond to ν_2, ν_1, and ν_0. (This patterns the zeros and poles of q, since

$q : q - 1 : 1 = -\lambda_2 F^{\nu_2} : \lambda_1 F_1^{\nu_1} : \lambda_0 F_0^{\nu_0}$. In the case of the dihedron ν_0 and ν_2 are switched.) Matching the branch points of $\tilde{q}$ with one of the rational functions q in our list, we arrive at a scenario in which the analytic branched coverings q and $\tilde{q}$ have the *same branch points and branch numbers*. We now apply a general uniformization theorem for branched coverings[1] and conclude that the group G is conjugate to the Möbius group that defines q and that the conjugation is a linear fractional transformation that establishes the conformal equivalence of the branched coverings q and $\tilde{q}$.   ■

Remark.
The use of the powerful uniformization therorem at the end of our proof can be dispensed with. We will give a more elementary approach to the final step in the proof above in Section 25/B.

Problem

1. Derive the following table for the absolute invariants of the finite Möbius groups:

G^*	Absolute invariants			Relation
D_n^*	$z_1^{2n} + z_2^{2n}$	$z_1 z_2(z_1^{2n} - z_2^{2n})$	$z_1^2 z_2^2$	$[z_1^{2n} + z_2^{2n}]^2 z_1^2 z_2^2 - [z_1 z_2(z_1^{2n} - z_2^{2n})]^2 = 4[z_1^2 z_2^2]^{n+1}$
T^*	Ω	$\frac{\Phi^3 + \Psi^3}{2}$	$\Phi\Psi$	$108\Omega^4 + \left(\frac{\Phi^3 + \Psi^3}{2}\right)^2 - (\Phi\Psi)^3 = 0$
O^*	Ω^2	$\Omega\frac{\Phi^3 + \Psi^3}{2}$	$\Phi\Psi$	$108[\Omega^2]^3 + \left[\Omega\frac{\Phi^3 + \Psi^3}{2}\right]^2 - \Omega^2[\Phi\Psi]^3 = 0$
I^*	$\mathcal{I}$	$\mathcal{J}$	$\mathcal{H}$	$1728\mathcal{I}^5 - \mathcal{J}^2 - \mathcal{H}^3 = 0$

---

[1] See H. Farkas and I. Kra, *Riemann surfaces*, Springer, 1980.

25

SECTION

The Icosahedron and the Unsolvable Quintic

† ♠ According to Galois theory[1] a polynomial equation is solvable by radicals iff the associated Galois group is solvable. For example, since the alternating group $\mathcal{A}_5$ is simple (cf. Problem 6 of Section 17), there is no root formula for a quintic with Galois group $\mathcal{A}_5$.

Since the symmetry group of the icosahedron is (isomorphic to) $\mathcal{A}_5$ (Section 17), the question arises naturally whether there is any connection between the icosahedron and the solutions of quintic equations. This is the subject of Klein's famous *Icosahedron Book*.[2] We devote this (admittedly long) section to sketch Klein's main result. We will treat the material here somewhat differently than Klein, and rely more on geometry. Unlike the *Icosahedron Book*,

---

[1] From now on, we will use some basic facts from Galois theory. For a quick summary, see Appendix F.

[2] The *Icosahedron Book* first appeared in German: *Vorlesungen über das Ikosaeder und die Auflösung der Gleichungen vom fünften Grade*, Teubner, Leipzig, 1884. Several English editions exist today, for example, *Lectures on the icosahedron, and the solution of equations of the fifth degree*, Kegan Paul, Trench, Trübner and Co., 1913. A new German edition with the original title (containing various comments and explanations) was published by Birkhäuser-Basel, Teubner-Leipzig in 1993. A good summary of the *Icosahedron Book* is contained in Slodowy's article, *Das Ikosaeder und die Gleichungen fünften Grades*, in Mathematischen Miniatüren, Band 3, Birkhäuser-Basel, 1986. For a recent easy-to-follow text, see J. Shurman, *Geometry of the Quintic*, John Wiley & Sons, 1997.

we will take as direct a path to the core results as possible. Due to the complexity of the exposition, we divide our treatment into subsections.

A. Polyhedral Equations

In Section 24, we defined, for each finite Möbius group G, a G-invariant rational function $q : \mathbf{C} \to \mathbf{C}$. Geometrically, the extension $q : \hat{\mathbf{C}} \to \hat{\mathbf{C}}$ is the analytic projection of a branched covering between Riemann spheres, and the branch values are $w = 0$, $w = 1$, and $w = \infty$, with branch numbers ν_2, ν_1, and ν_0 minus one. (In the case of the dihedron ν_0 and ν_2 are switched.) As in the proof of Theorem 18, for a given $w \in \mathbf{C}$, the equation $q(z) = w$ for z can be written as a degree-$|G|$ polynomial equation

$$P(z) - wQ(z) = 0,$$

where $q = P/Q$ with P and Q the polynomial numerator and denominator of q (with no common factors). We call this the *polyhedral equation associated to G*. Clearly, this polynomial equation has $|G|$ solutions (counted with multiplicity, and depending on the parameter w). We now consider this equation for each G. The case of the cyclic group C_n is obvious, since the associated equation is $z^n = w$, and the solutions are simply the nth roots of w. Using the second table in Section 24, for the equation of the dihedron we have

$$q_{D_n}(z) = -\frac{\alpha(z_1, z_2)^2}{\gamma(z_1, z_2)^n} = -\frac{(z_1^n - z_2^n)^2}{4z_1^n z_2^n} = -\frac{(z^n - 1)^2}{4z^n} = w,$$

where we have indicated the acting group by a subscript. Multiplying out, we obtain the equation of the dihedron, a quadratic equation in z^n. This can easily be solved:

$$z = q_{D_n}^{-1}(w) = \sqrt[n]{1 - 2w \pm 2\sqrt{w(w - 1)}}.$$

Inverting q_{D_n} amounts to extracting a square root followed by the extraction of an nth root.

For the tetrahedron, we have

$$q_T(z) = \frac{\Psi(z_1, z_2)^3}{\Phi(z_1, z_2)^3} = \left(\frac{z_1^4 + 2\sqrt{3}iz_1^2z_2^2 + z_2^4}{z_1^4 - 2\sqrt{3}iz_1^2z_2^2 + z_2^4} \right)^3$$

$$= \left(\frac{z^4 + 2\sqrt{3}iz^2 + 1}{z^4 - 2\sqrt{3}iz^2 + 1} \right)^3 = w.$$

Taking the cube root of both sides, we arrive at an equation that is quadratic in z^2 and can be easily solved. Inverting q_T thus amounts to extracting a cube root followed by the extraction of two square roots. Comparing the expression of q_T just obtained with that of q_{D_n} for $d = 2$, we have

$$q_T(z) = \left(\frac{q_{D_2}(z) - e^{\pi i/3}}{q_{D_2}(z) + e^{2\pi i/3}} \right)^3 .$$

As noted in Section 24, the octahedral invariants can be written as polynomials in the tetrahedral invariants. We have

$$\frac{q_O(z)}{q_O(z) - 1} = \frac{(\Phi\Psi)^3}{((\Phi^3 + \Psi^3)/2)^2} = \frac{(\Psi/\Phi)^3}{((\Psi/\Phi)^3 + 1)/2)^2} = \frac{w}{w - 1},$$

where we have omitted the arguments z_1 and z_2 for simplicity. Multiplying out, we obtain a quadratic equation in $(\Psi/\Phi)^3$. Since this is $q_T(z)$, we see that inverting q_O amounts to extracting a square root followed by the extraction of a cube root, and followed by the extraction of two square roots.

Remark.
In view of Galois theory, it is illuminating to match the sequence of roots in the root formulas for q_G^{-1} with the indices of the consecutive normal subgroups in a composition series of G, where G is one of our (solvable) groups $G = C_n, D_n, T, O$. For example, a composition series for O is $\mathcal{S}_4 \supset \mathcal{A}_4 \supset D_2 \supset C_2$, and the sequence of indices is 2, 3, 2, 2!

The situation for the icosahedron is radically different. We have

$$q_I(z) = \frac{\mathcal{H}^3}{1728 \, \mathcal{I}^5} = w.$$

As we will see later, it is *impossible* to express the solutions of the *icosahedral equation*

$$\mathcal{H}^3(z, 1) - 1728w\,\mathcal{I}^5(z, 1) = 0$$

in terms of a radical formula (depending on w). Using the explicit expressions of the forms involved, the equation of the icosahedron can be written as

$$((z^{20} + 1) - 228(z^{15} - z^5) + 494z^{10})^3 + 1728wz^5(z^{10} + 11z^5 - 1)^5 = 0.$$

B. Hypergeometric Functions

We just noted that there is no general root formula for the solutions of the icosahedral equation. The question arises naturally as to what kind of additional "transcendental" procedure is needed to express the solutions in an explicit form. In this subsection we show that any solution of a polyhedral equation can be written as the quotient of two linearly independent solutions of a homogeneous second-order linear differential equation with exactly 3 singular points, all regular. These differential equations are called *hypergeometric*.[3] Equivalently, we will show that for each spherical Platonic tessellation, the inverse q^{-1} of the rational function q is the quotient of two hypergeometric functions.

Remark.
Now some history. Bring (in 1786) and Jerrard (in 1834) independently showed that the general quintic can be reduced to $z^5 + bz + c$ (by a suitable Tschirnhaus tranformation; cf. the next subsection). This is usually called the *Bring–Jerrard form*. By scaling, the Bring–Jerrard form can be further reduced to the special quintic $z^5 + z - c = 0$. A root of this polynomial is called an *ultraradical*, and it is denoted by $\sqrt[*]{c}$. Using the defining equation, an ultraradical can be easily expanded into a convergent series. Bring and Jerrard thus showed that the general quintic can be solved by the use of radicals and ultraradicals. The relation of this special

---

[3]From now on, we use the definitions and results of Appendix E without making further references.

quintic to the so-called modular equation was used by Hermite, who pointed out that the general quintic can be solved in terms of elliptic modular functions.

As usual, we let $G \subset M\ddot{o}b\,(\hat{\mathbf{C}})$ denote the invariance group of q. Recall that q is an analytic branched covering with branch values $w_0 = 0$, $w_1 = 1$, and $w_2 = \infty$, and branch numbers v_2, v_1, and v_0 minus one. The inverse q^{-1} is multiple-valued. By G-invariance, composing q^{-1} with an element of G allows us to pass from a single-valued branch of q^{-1} to another. Since the Schwarzian is invariant under any linear fractional transformation, $\mathcal{S}(q^{-1})$ must be single-valued:

$$\mathcal{S}(q^{-1}) = s,$$

where $s : \mathbf{C} \to \mathbf{C}$ is a rational function. Following Riemann, who was a strong proponent of the principle that a rational function is best described by its poles, we determine s explicitly by taking its Laurent expansion at its poles. Near a branch value w_j, $j = 0, 1, 2$, q^{-1} can be expanded locally as

$$q^{-1}(w) - q^{-1}(w_j) = a_1(w - w_j)^{1/v_{2-j}} + a_2(w - w_j)^{2/v_{2-j}} + \cdots,$$

where $q^{-1}(w_j)$ denotes any one of the $|G|/v_{2-j}$ preimages, and $q^{-1}(w)$ is near $q^{-1}(w_j)$. As usual in complex analysis, we agree that for $w_j = \infty$, $w - w_j$ means $1/w$. Substituting this expansion into the expression for the Schwarzian $\mathcal{S}$, we obtain that the initial terms of the series for s at $w_0 = 0$, $w_1 = 1$, and $w_2 = \infty$ are

$$\frac{v_2^2 - 1}{2v_2^2 w^2}, \quad \frac{v_1^2 - 1}{2v_1^2(w-1)^2}, \quad \frac{v_0^2 - 1}{2v_0^2 w^2}.$$

(In the case of the dihedron, v_0 and v_2 are switched.) We obtain

$$s(w) = \frac{v_2^2 - 1}{2v_2^2 w^2} + \frac{A}{w} + \frac{v_1^2 - 1}{2v_1^2(w-1)^2} + \frac{B}{w-1} + C,$$

where A, B, C are complex constants. These constants are determined by the behavior of s at infinity, namely, by the requirement that s, expanded into a Laurent series at ∞, have the initial term

$(v_0^2 - 1)/2v_0^2 w^2$. We find that $A + B = 0$, $C = 0$, and

$$\frac{v_2^2 - 1}{2v_2^2} + \frac{v_1^2 - 1}{2v_1^2} + B = \frac{v_0^2 - 1}{2v_0^2}.$$

Putting all these together, we finally arrive at

$$s(w) = \frac{v_2^2 - 1}{2v_2^2 w^2} + \frac{v_1^2 - 1}{2v_1^2(w - 1)^2} + \frac{\frac{1}{v_1^2} + \frac{1}{v_2^2} - \frac{1}{v_0^2} - 1}{2w(w - 1)}.$$

Remark.

Recall that in the proof of Theorem 18 we constructed, for a given finite Möbius group G, a rational function $\tilde{q}$ that had the same branch points and branch numbers as q. Once again, passing from a single-valued branch of $\tilde{q}^{-1}$ to another amounts to the composition of $\tilde{q}^{-1}$ with a linear fractional transformation in G. We thus have $\mathcal{S}(\tilde{q}^{-1}) = \tilde{s}$, where $\tilde{s}$ is a rational function. On the other hand, since the branch points and branch numbers of q and $\tilde{q}$ are the same, the singularities of s and $\tilde{s}$ are also the same. Since a rational function is determined by its singularities, we must have $s = \tilde{s}$. We obtain $\mathcal{S}(q^{-1}) = \mathcal{S}(\tilde{q}^{-1})$, so that q and $\tilde{q}$ differ by a linear fractional transformation. This linear fractional transformation conjugates G to the finite Möbius group corresponding to q. Thus, Theorem 18 follows.

Returning to the main line, we see that q^{-1} satisfies the third-order differential equation

$$\mathcal{S}(q^{-1}) = \frac{v_2^2 - 1}{2v_2^2 w^2} + \frac{v_1^2 - 1}{2v_1^2(w - 1)^2} + \frac{\frac{1}{v_1^2} + \frac{1}{v_2^2} - \frac{1}{v_0^2} - 1}{2w(w - 1)}.$$

Although of third order, the general solution of this equation is remarkably simple and can be given in terms of solutions of the homogeneous second-order linear differential equation

$$z'' = p(w)z' + q(w)z,$$

where

$$s = p' - \frac{1}{2}p^2 - 2q.$$

We set $p(w) = 1/w$, and choose q to satisfy the equation for s. Substituting the actual expression of s into p and q, we obtain

$$z'' + \frac{z'}{w} + \frac{z}{4(w-1)^2 w^2}\left(-\frac{1}{v_2^2} + w\left(\frac{1}{v_0^2} + \frac{1}{v_2^2} - \frac{1}{v_1^2} + 1\right) - \frac{w^2}{v_0^2}\right) = 0.$$

This is a special case of the *hypergeometric differential equation*. In fact, with

$$\alpha_1 = -\alpha_2 = \frac{1}{2v_2}, \quad \beta_1 = \frac{1}{2v_1}, \quad \beta_2 = \frac{v_1^2 - 1}{2v_1}, \quad \gamma_1 = -\gamma_2 = \frac{1}{2v_0},$$

the hypergeometric differential equation reduces to our equation, since $v_1 = 2$. We have accomplished our goal and proved that the function q^{-1} can be written as the quotient of two hypergeometric functions!

C. The Tschirnhaus Transformation

In this subsection, we will reduce the general irreducible quintic

$$z^5 + a_1 z^4 + a_2 z^3 + a_3 z^2 + a_4 z + a_5 = 0$$

to a simpler form

$$z^5 + \tilde{a}_1 z^4 + \tilde{a}_2 z^3 + \tilde{a}_3 z^2 + \tilde{a}_4 z + \tilde{a}_5 = 0,$$

in which some (but not all) coefficients vanish. This reduction is made possible by the *Tschirnhaus transformation*. It is given by

$$\tilde{z} = \sum_{l=1}^{4} \lambda_l z^{(l)},$$

where

$$z^{(l)} = z^l - \frac{1}{5}\sum_{j=1}^{5} z_j^l,$$

and $z_1, \ldots, z_5$ are the roots of the original quintic. In the expression of $z^{(l)}$, the sum of powers is a symmetric polynomial in the roots, and thus, by the fundamental theorem on symmetric polynomials, it can be expressed as a polynomial in the coefficients $a_1, \ldots, a_5$. For example, since

$$\sum_{j=1}^{5} z_j = -a_1, \quad \sum_{j=1}^{5} z_j^2 = a_1^2 - 2a_2,$$

we have

$$z^{(1)} = z + \frac{a_1}{5}, \quad z^{(2)} = z^2 - \frac{1}{5}\left(a_1^2 - 2a_2\right).$$

Hence, $\tilde{z}$ is a polynomial in z of degree less than or equal to 4 with coefficients in $\mathbf{Q}[a_1, \ldots, a_5]$ depending on $\lambda_1, \ldots, \lambda_4$. The requirement on the vanishing of the prescribed coefficients in the reduced quintic amounts to polynomial relations of degree less than or equal to 4 in the coefficients $\lambda_1, \ldots, \lambda_4$. These polynomial relations are solvable by explicit root formulas (cf. Section 6).

The way the Tschirnhaus transformation $\tilde{z}$ acts on the original quintic is to transform its roots $z_1, \ldots, z_5$ to the roots $\tilde{z}_1, \ldots, \tilde{z}_5$ of the reduced quintic, where

$$\tilde{z}_j = \sum_{l=1}^{4} \lambda_l z_j^{(l)}, \quad j = 1, \ldots, 5.$$

Since $\sum_{j=1}^{5} \tilde{z}_j = \sum_{l=1}^{4} \lambda_l \sum_{j=1}^{5} z_j^{(l)} = 0$, we have $\tilde{a}_1 = 0$ for *any* Tschirnhaus transformation. The simplest Tschirnhaus transformation is

$$\tilde{z} = z^{(1)} = z + \frac{a_1}{5},$$

where we put $\lambda_1 = 1$, $\lambda_2 = \lambda_3 = \lambda_4 = 0$. (Note that the analogue of this for cubics and quartics was used in Section 6.) Next, we look for a Tschirnhaus transformation that yields $\tilde{a}_2 = 0$ in the form

$$\tilde{z} = \lambda z^{(1)} + z^{(2)} = \lambda\left(z + \frac{a_1}{5}\right) + z^2 - \frac{1}{5}\left(a_1^2 - 2a_2\right),$$

where $\lambda \in \mathbf{C}$ is a parameter to be determined. Here we set $\lambda = \lambda_1$, $\lambda_2 = 1$, and $\lambda_3 = \lambda_4 = 0$. Since $\sum_{j=1}^{5} \tilde{z}_j = 0$, by squaring we see that the vanishing of $\tilde{a}_2$ amounts to the vanishing of $\sum_{j=1}^{5} \tilde{z}_j^2$. This gives the following quadratic equation for λ:

$$\sum_{j=1}^{5} \tilde{z}_j^2 = \sum_{j=1}^{5} (\lambda z_j^{(1)} + z_j^{(2)})^2$$

$$= \lambda^2 \sum_{j=1}^{5} (z_j^{(1)})^2 + 2\lambda \sum_{j=1}^{5} z_j^{(1)} z_j^{(2)} + \sum_{j=1}^{5} (z_j^{(2)})^2 = 0.$$

Again, by the fundamental theorem on symmetric polynomials, the coefficients of this quadratic polynomial in λ depend only on $a_1, \ldots, a_5$. The corresponding quadratic equation can be solved for λ in terms of $a_1, \ldots, a_5$. The two solutions for λ involve the square root of the expression

$$4 \left(\sum_{j=1}^{5} z_j^{(1)} z_j^{(2)} \right)^2 - 4 \sum_{j=1}^{5} \left(z_j^{(1)} \right)^2 \sum_{j=1}^{5} \left(z_j^{(2)} \right)^2.$$

This is the discriminant δ multiplied by $(\sum_{j=1}^{5} (z_j^{(1)})^2)^2$. The smallest ground field over which the original quintic is defined is $k = \mathbf{Q}(a_1, \ldots, a_5)$. We have $\delta \in k$, but in general, $\sqrt{\delta} \notin k$. Thus, if we want λ to be in the ground field, $\sqrt{\delta}$ needs to be adjoined to k.

Summarizing (and adjusting the notation), the problem of solvability of the general quintic is reduced (at the expense of a quadratic extension of the ground field) to solvability of the equation

$$P(z) = z^5 + 5az^2 + 5bz + c = 0.$$

(Here we have inserted numerical factors for future convenience.) A quintic with vanishing terms of degree 3 and 4 (such as P above) is said to be *canonical*. For future reference we include here the discriminant

$$\delta = \prod_{1 \le j < l \le 5} (z_j - z_l)^2$$

of our canonical quintic as a polynomial in the coefficients a, b, c:

$$\frac{\delta}{5^5} = 108a^5c - 135a^4b^2 + 90a^2bc^2 - 320ab^3c + 256b^5 + c^4.$$

This formula is obtained by a somewhat tedious but elementary computation.[4]

Consider the roots $z_1, \ldots, z_5$ of the canonical equation as homogeneous coordinates of a point $[z_1 : \cdots : z_5]$ in the 4-dimensional complex projective space $\mathbf{C}P^4$ (cf. Section 21). (Here the trivial case $P(z) = z^5$ needs to be excluded, since it gives $z_1 = \cdots = z_5 = 0$.) The roots $z_1, \cdots, z_5$ can be reconstructed from the projective point $[z_1 : \cdots : z_5]$ by extracting a square and a cube root, so that no information is lost in the homogenization. Since ordering the roots cannot be prescribed universally, to the roots there correspond 120 projective points obtained from one another by permuting the coordinates. To incorporate this ambiguity, we consider the symmetric group $\mathcal{S}_5$ acting on $\mathbf{C}P^4$ by permuting the homogeneous coordinates. In other words, the 120 points form an orbit in $\mathbf{C}P^4$ under the action of $\mathcal{S}_5$. Notice that under the action of $\mathcal{S}_5$, the points $[z_1^{(l)} : \cdots : z_5^{(l)}]$, $l = 1, \ldots, 4$, are permuted the same way as $[z_1 : \cdots : z_5]$.

Since our equation is canonical, the roots $z_1, \ldots, z_5$, satisfy the relations

$$\sum_{j=1}^{5} z_j = 0 \quad \text{and} \quad \sum_{j=1}^{5} z_j^2 = 0.$$

Thus, $[z_1 : \cdots : z_5]$ lies in the (smooth) complex surface

$$\mathcal{Q}_0 = \left\{ [z_1 : \cdots : z_5] \in \mathbf{C}P^4 \mid \sum_{j=1}^{5} z_j = \sum_{j=1}^{5} z_j^2 = 0 \right\}.$$

Actually, $\mathcal{Q}_0$ can be identified with the so-called standard *complex projective quadric* in $\mathbf{C}P^3$, once we identify the complex projective space $\mathbf{C}P^3$ with the linear slice $\mathbf{C}P_0^3$ of $\mathbf{C}P^4$ defined by $\sum_{j=1}^{5} z_j = 0$. We thus set

$$\mathbf{C}P_0^3 = \left\{ [z_1 : \cdots : z_5] \in \mathbf{C}P^4 \mid \sum_{j=1}^{5} z_j = 0 \right\}.$$

---

[4]I must confess the advantage in using a computer algebra system such as Maple or Mathematica that reduces the computation of δ to a fraction of a second.

The geometry behind the Tschirnhaus transformation that reduces the general quintic to a canonical form is now clear. We start with the point $[z_1 : \cdots : z_5] \in \mathbf{C}P^4$ whose homogeneous coordinates are the roots of a general quintic (with no quartic term). We form the projective points $[z_1^{(1)} : \cdots : z_5^{(1)}]$ and $[z_1^{(2)} : \cdots : z_5^{(2)}]$ that both lie in $\mathbf{C}P_0^3$. We then consider one of the two intersections of the projective line through these two points with the quadric $\mathcal{Q}_0$. The Tschirnhaus transformation associates to $[z_1 : \cdots : z_5]$ this intersection point in $\mathcal{Q}_0$. As emphasized above, this process generally requires a quadratic extension of the ground field k.

D.   Quintic Resolvents of the Icosahedral Equation

Let k be our ground field. For the sake of concreteness, we assume that $k \subset \mathbf{C}$. Since k has characteristic zero, we also have $\mathbf{Q} \subset k$. Since the icosahedral equation contains an arbitrary parameter w, it is natural to consider this equation to be defined over $k(w)$, the field of rational functions in the variable w and with coefficients in k. Let z denote a solution of the icosahedral equation. Since $q(z) = w$, the field $k(z)$ contains $k(w)$. Assume from now on that $\omega = e^{2\pi i/5}$, a primitive fifth root of unity, is contained in k, so that $\mathbf{Q}(\omega) \subset k \subset \mathbf{C}$. This ensures that the linear fractional transformations in the icosahedral Möbius group $I = \mathcal{A}_5$ are defined over k. They act, by substitutions, as automorphisms of the field $k(z)$. (More precisely, $g \in I$ acts on $r^* \in k(z)$ by $g : r^* \mapsto r^* \circ g^{-1}$.) Since q is $\mathcal{A}_5$-invariant, these automorphisms fix the subfield $k(w)$. In fact, since $q : \hat{\mathbf{C}} \to \hat{\mathbf{C}}$ is the orbit map of $\mathcal{A}_5$, the fixed field $k(z)^{\mathcal{A}_5}$ is $k(w)$. We obtain that $k(z)/k(w)$ is a Galois extension with Galois group $\mathcal{A}_5$. Moreover, $k(z)$ is the splitting field of the icosahedral equation over $k(w)$; the solutions are the linear fractions that represent the transformations in the icosahedral Möbius group. Since $\mathcal{A}_5$ is transitive on the solutions, the icosahedral equation is irreducible over $k(w)$. As before, a rational function $r^* \in k(z)$ has the resolvent polynomial

$$P^*(X) = \prod_{j=1}^{n^*}(X - r_j^*),$$

with coefficients in $k(w)$, where $\mathcal{A}_5(r^*) = \{r_1^*, \ldots, r_{n^*}^*\}$ is the orbit through r^*. Since I is simple, $k(z) = k(r_1^*, \ldots, r_{n^*}^*)$.

The rest of this subsection is going to be very technical. In order to see through the details we now sketch our plan. Recall that our main objective is to establish a connection between the (irreducible) quintic in canonical form

$$P(z) = z^5 + 5az^2 + 5bz + c$$

and the solutions of the icosahedral equation

$$\mathcal{H}^3(z, 1) - 1728w\,\mathcal{I}^5(z, 1) = 0.$$

Since the latter is a polynomial equation of degree 60, based on the analogy with the cubic resolvent of quartics (cf. Section 6), we must find a suitable *quintic* resolvent of $\mathcal{H}^3 - 1728w\,\mathcal{I}^5$ *directly comparable to our canonical quintic*. Our task is actually harder than in the quartic case, since P depends on three complex coefficients.[5] Thus, we need to find a quintic resolvent in canonical form that contains, in addition to w, two extra parameters, u and v, say. We write this resolvent as

$$P^*(z) = z^5 + 5a(u, v, w)z^2 + 5b(u, v, w)z + c(u, v, w).$$

We expect the coefficients to be explicitly computable rational functions in u, v, w. Since P^* is an icosahedral resolvent, we also need to be able to express the roots of P^* in terms of the solutions of the icosahedral equation, or, what is the same, in terms of hypergeometric functions.

Once P^* is worked out explicitly, the matching with the irreducible quintic P with coefficients a, b, c amounts to solving the

---

[5]Shurman's approach in his *Geometry of the Quintic* cited above is different (see also Problem 4). Both approaches are contained in the *Icosahedron Book*.

system

$$a(u, v, w) = a,$$

$$b(u, v, w) = b,$$

$$c(u, v, w) = c.$$

This we will be able to carry out with the additional constraint

$$\sqrt{\delta(u, v, w)} = \sqrt{\delta},$$

where $\delta(u, v, w)$ is the discriminant of P^*. This matching process will establish that the roots of P and P^* coincide *as sets*. To obtain a root-by-root match we will also need to look at how the Galois group $\mathcal{A}_5$ acts on each set of roots.

The five roots of a quintic resolvent constitute an orbit of $\mathcal{A}_5$ in the extension $k(z)/k(w)$. We thus need to find five rational functions in z that are permuted among themselves by the icosahedral substitutions. During the construction of the rational function q we learned that it is much easier to construct forms first. Our task thus reduces to finding an $\mathcal{A}_5$-orbit of five forms. A further advantage in using forms is that we can derive them from geometric situations. An obvious example for five geometric objects that are permuted among themselves by the symmetries of the icosahedron are Kepler's five cubes inscribed in the reciprocal dodecahedron (cf. Section 17). We could immediately construct five degree-8 forms that vanish at the vertices of these cubes. This would give us a quintic resolvent. For technical purposes, however, it is more convenient to construct the five degree-6 forms that vanish on the vertices of the octahedral reciprocals. Thus, we stay with (Pacioli's model of) the icosahedron and notice that the 6 midpoints of the 6 shorter edges of the three mutually perpendicular golden rectangles give the 6 vertices of an octahedron. There are 5 configurations each consisting of three mutually perpendicular golden rectangles. (The 15 opposite pairs of edges of the icosahedron give 15 golden rectangles inscribed in the icosahedron. By further grouping, these 15 golden rectangles form the 5 configurations.) To each configuration we can attach an octahedron as above. Putting these together, we arrive at the compound of 5 octahedra inscribed in the icosahedron. The octahedral forms corresponding to the five octahedra will be the roots of our quintic resolvent.

To choose the first octahedron, we recall that the commuting half-turns U and V introduced in Section 18 have orthogonal axes, so that the composition $W = UV$ is also a half-turn and its axis is orthogonal to those of U and V. The linear fractional transformation that corresponds to U is $z \mapsto -1/z$, while those that correspond to V and W have been explicitly worked out in Section 18. We choose the first octahedron (projected to $\hat{\mathbf{C}}$) to have vertices as *the fixed points of these half-turns*. The fixed points of U are the solutions of $z^2 = -1$, and the form that vanishes at these points is $z_1^2 + z_2^2$, $z_1, z_2 \in \mathbf{C}$, $z = z_1/z_2$. Using the explicit form for V, for the fixed points of V we need to consider

$$\frac{(\omega^2 - \omega^3)z_1 + (\omega - \omega^4)z_2}{(\omega - \omega^4)z_1 - (\omega^2 - \omega^3)z_2} = \frac{z_1}{z_2}.$$

Multiplying out, we have

$$z_1^2 - 2(\omega + \omega^4)z_1 z_2 - z_2^2 = 0.$$

By Problem 5 (b) of Section 17, $\omega + \omega^4$ can be replaced by $1/\tau$, where τ is the golden section. We obtain that the fixed points of V are given by the zeros of the quadratic form

$$z_1^2 - \frac{2}{\tau}z_1 z_2 - z_2^2.$$

In a similar vein, the fixed points of W are given by the zeros of

$$z_1^2 + 2\tau z_1 z_2 - z_2^2.$$

We define the first octahedral form Ω_1 as the product of these three forms:

$$\Omega_1(z_1, z_2) = (z_1^2 + z_2^2)(z_1^4 + 2z_1^3 z_2 - 6z_1^2 z_2^2 - 2z_1 z_2^3 + z_2^4).$$

The remaining four octahedral forms Ω_{j+1}, $j = 1, \ldots, 4$, are obtained by applying to Ω_1 the homogeneous substitutions

$$z_1 \mapsto \pm\omega^{3j}z_1, \quad z_2 \mapsto \pm\omega^{2j}z_2,$$

which correspond to the rotations S^j, $j = 1, \ldots, 4$, introduced in Section 18. Multiplying out, we arrive at the five octahedral forms

$$\Omega_{j+1}(z_1, z_2) = \omega^{3j} z_1^6 + 2\omega^{2j} z_1^5 z_2 - 5\omega^j z_1^4 z_2^2$$

$$- 5\omega^{4j} z_1^2 z_2^4 - 2\omega^{3j} z_1 z_2^5 + \omega^{2j} z_2^6, j = 0, \ldots, 4.$$

How does $\mathcal{A}_5$ act on $\{\Omega_1, \ldots, \Omega_5\}$? It is enough to see how the icosahedral generators S and W act on these octahedral forms. By definition, S permutes Ω_j, $j = 1, \ldots, 5$, cyclically. Geometrically, the midpoints of the 5 edges that meet at the North Pole N are vertices of the 5 distinct octahedra. Since the rotation axis of S passes through N, S must permute these vertices cyclically. It is equally clear that $U, V, W(= UV)$ all fix Ω_1, since Ω_1 is defined by the requirement that it should vanish at the fixed points of U, V, and W. The simplest of these three commuting Möbius tranformations is $U : z \mapsto -1/z$. The corresponding homogeneous substitution is $z_1 \mapsto z_2, z_2 \mapsto -z_1$. Substituting this into the explicit expression of the octahedral forms above and looking at the leading coefficients, we see immediately that U acts as follows:

$$U : \Omega_1 \mapsto \Omega_1, \quad \Omega_2 \leftrightarrow \Omega_5, \quad \Omega_3 \leftrightarrow \Omega_4.$$

Recall from Section 18 that the axis of the half-turn V is perpendicular to the axis of U, and it passes through the midpoint of the base of the upper pentagonal pyramid Π of the icosahedron. The vertex of Π is N. As noted above, the midpoints of the 5 edges of Π that meet at N belong to the 5 inscribed octahedra. Since S permutes these edges cyclically, we denote them by e_1, e_2, e_3, e_4, e_5, where $S(e_j) = e_{j(\mathrm{mod}\,5)+1}, j = 1, \ldots, 5$. We adjust the index such that e_1 is an edge of a golden rectangle that defines Ω_1. Then, for all $j = 1, \ldots, 5$, e_j is an edge of a golden rectangle that defines Ω_j. Due to this arrangement the fixed point of V is the midpoint of the base of the triangular face with sides e_3 and e_4. Now a careful look at how V acts on e_2, e_3, e_4, e_5 reveals that $V(e_3)$ and e_5 are edges of two golden rectangles in the same configuration. Similarly, $V(e_4)$ and e_2 are edges of two golden rectangles in another configration. Since V is a half-turn, we obtain that it acts on the five octahedra as follows:

$$V : \Omega_1 \mapsto \Omega_1, \quad \Omega_2 \leftrightarrow \Omega_4, \quad \Omega_3 \leftrightarrow \Omega_5.$$

Finally, the way the permutation W acts follows, since $W = UV$. For future reference, we summarize the actions of S and W as follows:

$$S : \Omega_j \mapsto \Omega_{j(\mathrm{mod}\,5)+1}, \quad j = 1, \ldots, 5,$$

$$W : \Omega_1 \mapsto \Omega_1, \quad \Omega_2 \leftrightarrow \Omega_3, \quad \Omega_4 \leftrightarrow \Omega_5.$$

Despite their simplicity, these octahedral forms will not be suitable for our purposes, since $\sum_{j=1}^5 \Omega_j^2 \neq 0$, so that the quintic icosahedral resolvent constructed from these forms will not be canonical. (Nevertheless, this resolvent is of great interest, since $\sum_{j=1}^5 \Omega_j = 0$ and $\sum_{j=1}^5 \Omega_j^3 = 0$; see Problem 4.) This temporary setback is easy to fix. Consider the Hessians $\Xi_j = Hess(\Omega_j)$, $j = 1, \ldots, 5$, that are forms of degree 8, and also constitute an $\mathcal{A}_5$-orbit. They clearly satisfy the relations

$$\sum_{j=1}^5 \Xi_j = 0, \quad \sum_{j=1}^5 \Xi_j^2 = 0,$$

since there are no icosahedral forms in degrees 8 and 16. (Any icosahedral form is a polynomial in $\mathcal{I}$, $\mathcal{J}$, and $\mathcal{H}$; cf. Section 24.) Thus, the quintic icosahedral resolvent constructed from these forms will be canonical! For future reference we include here the explicit form

$$\begin{aligned}
\Xi_{j+1}(z_1, z_2) &= -\omega^{4j} z_1^8 + \omega^{3j} z_1^7 z_2 - 7\omega^{2j} z_1^6 z_2^2 - 7\omega^j z_1^5 z_2^3 \\
&\quad + 7\omega^{4j} z_1^3 z_2^5 - 7\omega^{3j} z_1^2 z_2^6 - \omega^{2j} z_1 z_2^7 - \omega^j z_2^8 \\
&= (\omega^{4j} z_1 - \omega^{3j} z_2)(-z_1^7 + 7z_1^2 z_2^5) \\
&\quad + (\omega^{2j} z_1 - \omega^j z_2)(-7z_1^5 z_2^2 - z_2^7), \quad j = 0, \ldots, 4.
\end{aligned}$$

Recall that we need a two-parameter family of these forms. Notice that we have

$$\sum_{j=1}^5 \Omega_j \Xi_j = 0, \quad \sum_{j=1}^5 \Omega_j \Xi_j^2 = 0, \quad \sum_{j=1}^5 (\Omega_j \Xi_j)^2 = 0,$$

since left-hand sides are icosahedral invariants of degrees 14, 22, and 28, respectively, and (once again) there are no icosahedral invariants in these degrees. For future reference, we include here the explicit forms

$$\Omega_{j+1}(z_1, z_2)\, \Xi_{j+1}(z_1, z_2)$$

$$= (\omega^{4j}z_1 - \omega^{3j}z_2)(-26z_1^{10}z_2^3 + 39z_1^5z_2^8 + z_2^{13})$$

$$+ (\omega^{2j}z_1 - \omega^{j}z_2)(-z_1^{13} + 39z_1^8z_2^5 + 26z_1^3z_2^{10}), \quad j = 0, \ldots, 4.$$

For greater generality, we seeek a resolvent for the linear combinations

$$\Upsilon_j = E\, \Xi_j + F\, \Omega_j \Xi_j, \quad j = 1, \ldots, 5,$$

where by homogeneity, the coefficients E and F are forms of degree 30 and 24. In fact, we will put

$$E = 12u\, \mathcal{J}, \quad F = 144v\, \mathcal{I}^2,$$

where u and v are complex parameters, but for the time being, we keep E and F arbitrary. It remains to work out the resolvent explicitly. Staying in the realm of forms for a while, we write the resolvent polynomial as

$$\prod_{j=1}^{5}(\mathcal{X} - \Upsilon_j) = \mathcal{X}^5 + b_1\mathcal{X}^4 + b_2\mathcal{X}^3 + b_3\mathcal{X}^2 + b_4\mathcal{X} + b_5,$$

where we used $\mathcal{X}$ as variable. By what we said above, the relations above imply that $b_1 = b_2 = 0$. Expanding, and using the fact that the coefficients must be (absolute) invariants of the icosahedral group, after somewhat tedious computations, we arrive at the following:

$$\mathcal{X}^5 + 5\mathcal{X}^2(8E^3\mathcal{I}^2 + E^2F\mathcal{J} + 72EF^2\mathcal{I}^3 + F^3\mathcal{I}\mathcal{J})$$

$$+ 5\mathcal{X}(-E^4\mathcal{I}\mathcal{H} + 18E^2F^2\mathcal{I}^2\mathcal{H} + EF^3\mathcal{H}\mathcal{J} + 27F^4\mathcal{I}^3\mathcal{H})$$

$$+ (E^5\mathcal{H}^2 - 10E^3F^2\mathcal{I}\mathcal{H}^2 + 45EF^4\mathcal{I}^2\mathcal{H}^2 + F^5\mathcal{J}\mathcal{H}^2).$$

We introduce the new variable

$$X = \frac{\mathcal{I}}{\mathcal{J}\mathcal{H}}\mathcal{X}$$

along with

$$r_j = 12\frac{\mathcal{I}^2}{\mathcal{J}}\Omega_j, \quad s_j = 12\frac{\mathcal{I}}{\mathcal{H}}\Xi_j, \quad j = 1, \ldots, 5.$$

Comparing the degrees of the forms involved we see that r_j and s_j are rational functions in $z = z_1/z_2$. Rewriting our variables in terms of these, we obtain

$$\Upsilon_j = \frac{\mathcal{J}\mathcal{H}}{\mathcal{I}} t_j, \quad j = 1, \ldots, 5,$$

where

$$t_j = us_j + vr_j s_j, \quad j = 1, \ldots, 5.$$

In terms of these new variables, our resolvent polynomial becomes

$$P^*(X) = \prod_{j=1}^{5}(X - t_j)$$

$$= X^5 + \frac{5X^2}{w}\left(8u^3 + 12u^2v + \frac{6uv^2 + v^3}{1 - w}\right)$$

$$+ \frac{15X}{w}\left(-4u^4 + \frac{6u^2v^2 + 4uv^3}{1 - w} + \frac{3v^4}{4(1 - w)^2}\right)$$

$$+ \frac{3}{w}\left(48u^5 - \frac{40u^3v^2}{1 - w} + \frac{15uv^4 + 4v^5}{(1 - w)^2}\right).$$

This is called the *canonical resolvent polynomial* of the icosahedral equation. By homogeneity, *the roots* $t_j = t_j(u, v, z) = us_j(z) + vr_j(z)s_j(z), j = 1, \ldots, 5,$ *of the resolvent* P^* *are rational functions in* $z = z_1/z_2$ *and linear functions in* u *and* v. As in the case of the octahedral forms, the icosahedral generators S and W induce the following permutations on the roots:

$$S : t_j \mapsto t_{j(\bmod 5)+1}, \quad j = 1, \ldots, 5,$$

$$W : t_1 \mapsto t_1, \quad t_2 \leftrightarrow t_3, \quad t_4 \leftrightarrow t_5.$$

As in the case of the canonical quintic, the projective points corresponding to the roots of the canonical resolvent lie in the complex projective quadric

$$\mathcal{Q}_0 = \left\{[t_1 : \ldots : t_5] \in \mathbf{C}P^4 \mid \sum_{j=1}^{5} t_j = \sum_{j=1}^{5} t_j^2 = 0\right\}.$$

Finally, since the solution set contains two parameters, it is reasonable to expect that the projective points corresponding to the roots of the canonical resolvent fill $\mathcal{Q}_0$. This will follow as a byproduct of stronger results in Subsection G.

Summarizing, we have found that the complex projective quadric $\mathcal{Q}_0$ in $\mathbf{C}P_0^3(\subset \mathbf{C}P^4)$ serves as a simple "parameter space" for the solutions of the canonical resolvent of the icosahedral equation. Since the solutions of the canonical resolvent depend rationally on the solutions of the icosahedral equation, our remaining task is to find a matching parametrization of the solutions of the quintic in canonical form. This we will accomplish in the remaining three subsections.

E. Solvability of the Quintic à la Klein

In view of the fact that $z^n - w = 0$ is the "polyhedral equation" for the cyclic group C_n, it is natural to look for a noncommutative analogue of the theorem in Appendix F, in which the field extension K/k is generated by any solution of the equation $z^n - w = 0$, and the Galois group of the field extension is C_n. In order that the linear fractional transformations that make up the icosahedral Möbius group I become k-automorphisms of the splitting field of the icosahedral equation, we need to assume that $\omega \in k$, where $\omega = e^{2\pi i/5}$.

Theorem 19.

Let k be a field satisfying $\mathbf{Q}(\omega) \subset k \subset \mathbf{C}$, and let $K \subset \mathbf{C}$ be a Galois extension of k with Galois group $\mathcal{A}_5$. Then, replacing k by a suitable quadratic extension, there exists $w^* \in k$ such that K is generated by any solution z^* of the icosahedral equation with parameter $w^* = q(z^*)$. Moreover, each solution z^* gives rise to an isomorphism $\varphi:\mathcal{A}_5 \to I$ of the Galois group $\mathcal{A}_5$ to the icosahedral Möbius group I such that if $\sigma \in \mathcal{A}_5$ is a k-automorphism of K that is mapped under this isomorphism to

$$\varphi(\sigma) = \pm \begin{bmatrix} a(\sigma) & b(\sigma) \\ c(\sigma) & d(\sigma) \end{bmatrix},$$

then

$$\sigma^{-1}(z^*) = \varphi(\sigma)(z^*) = \frac{a(\sigma)z^* + b(\sigma)}{c(\sigma)z^* + d(\sigma)}.$$

Remark.

Theorem 19, the so called "Normalformsatz," is the cornerstone of Klein's theory of the icosahedron. In 1861, Kronecker showed that the "suitable quadratic extension" k'/k ("akzessorische Irrationalität" as Klein called it) in the Normalformsatz cannot, in general, be dispensed with. As shown in the text, this extension appears in reducing the general quintic to a canonical form (called "Hauptgleichung" by Klein) by a Tschirnhaus transformation.

Remark.

Quadratic extensions do not change the setting in Theorem 19. In fact, if k' is a quadratic extension of the ground field k, then k' is not contained in K, since the Galois group $\mathcal{A}_5$ of the extension K/k cannot contain any subgroup of index 2. We thus have $G(K \cdot k'/k') = G(K/k) = \mathcal{A}_5$.

Every irreducible quintic

$$P(z) = z^5 + a_1 z^4 + a_2 z^3 + a_3 z^2 + a_4 z + a_5, \quad a_1, \ldots, a_5 \in \mathbf{C},$$

over k, $\mathbf{Q}(\omega) \subset k \subset \mathbf{C}$, with Galois group $\mathcal{A}_5$ has a splitting field K as in Theorem 19. Conversely, given K/k as in Theorem 19, there exists an irreducible quintic over k whose splitting field is K, and whose Galois group is $\mathcal{A}_5$. Indeed, consider a subgroup of $\mathcal{A}_5$ isomorphic to $\mathcal{A}_4$. By abuse of notation, we denote this subgroup by $\mathcal{A}_4$. The field k is properly contained in the fixed field $K^{\mathcal{A}_4}$, so that there exists $z_1 \in K^{\mathcal{A}_4} - k$. The $\mathcal{A}_5$-orbit of z_1 consists of 5 elements $z_1, \ldots, z_5$, since $\mathcal{A}_4$ is maximal in $\mathcal{A}_5$. Let $P(z) = \prod_{j=1}^{5}(z - z_j)$ be the quintic resolvent associated to z_1. Then P is irreducible over k, K is the splitting field of P over k, and the Galois group $\mathcal{A}_5$ can be identified with the group of even permutations of the roots $z_1, \ldots, z_5$.

To prove Theorem 19, we will view K as the splitting field of an irreducible quintic (with Galois group $\mathcal{A}_5$). The "suitable quadratic extension" $k(\sqrt{\delta})$ of the ground field k is due to the reduction of the quintic to canonical form by a Tschirnhaus transformation as discussed in Subsection C.

F. Geometry of the Canonical Equation: General Considerations

We saw in Subsections D and E that the complex projective quadric $\mathcal{Q}_0$ parametrizes the points that correspond to the roots of the canonical resolvent of the icosahedral equation, and to the roots of the irreducible quintic in canonical form (obtained from the general quintic by a Tschirnhaus transformation). In the two subsections that follow, by "matching" these two parametrizations we obtain a constructive proof of Theorem 19 as well as the explicit formulas that solve the canonical quintics. In this subsection we will follow the main argument with few technical details. In the last subsection we will work out all formulas in detail.

The geometry of the projective quadric $\mathcal{Q}_0 \subset \mathbf{C}P_0^3$ as a "doubly ruled surface" is well known. In fact, the so-called Lagrange substitution, a linear equivalence between $\mathbf{C}P_0^3$ and $\mathbf{C}P^3$, transforms the defining equations $\sum_{j=1}^{5} z_j = \sum_{j=1}^{5} z_j^2 = 0$ of $\mathcal{Q}_0$ into the single equation

$$\xi_1\xi_4 + \xi_2\xi_3 = 0, \qquad [\xi_1 : \cdots : \xi_4] \in \mathbf{C}P^3.$$

(The Lagrange substitution will be given explicitly in Subsection G; cf. also the analogous case in Section 6 for solving cubic equations.) This equation defines the complex surface

$$\mathcal{Q} = \{[\xi_1 : \cdots : \xi_4] \in \mathbf{C}P^3 \mid \xi_1\xi_4 + \xi_2\xi_3 = 0\}$$

in $\mathbf{C}P^3$. In what follows, we will identify $\mathcal{Q}_0$ with $\mathcal{Q}$ under this linear equivalence. For each value of a parameter $c^* \in \hat{\mathbf{C}}$, the equations

$$-\frac{\xi_1}{\xi_2} = \frac{\xi_3}{\xi_4} = c^*$$

define a complex projective line in $\mathcal{Q}$; we call this a *generating line of the first kind* (with parameter c^*). In a similar vein, for $c^{**} \in \hat{\mathbf{C}}$, the equations

$$\frac{\xi_1}{\xi_3} = -\frac{\xi_2}{\xi_4} = c^{**}$$

define in $\mathcal{Q}$ a *generating line of the second kind* (with parameter c^{**}). As can be seen from the defining formulas, the two families of generating lines satisfy the following properties: (1) Each point of the quadric is the intersection of two generating lines of different kinds; (2) any two generating lines of different kinds intersect at exactly one point; (3) any two distinct generating lines are disjoint. (For a (more than) good analogy, consider a hyperboloid of one sheet as a doubly ruled surface with the two rulers corresponding to the generating lines of the first and second kind.)

Due to the linear equivalence between $\mathcal{Q}$ and $\mathcal{Q}_0$, the entire construction in $\mathcal{Q}$ can be carried over to our initial quadric $\mathcal{Q}_0$. The parameters c^* and c^{**} become quotients of linear froms in the variables $z_1, \ldots, z_5$ subject to $\sum_{j=1}^{5} z_j = \sum_{j=1}^{5} z_j^2 = 0$.

If we fix a point $o \in \mathcal{Q}_0$ as the origin, then the generating lines $\mathbf{C}P^*$ and $\mathbf{C}P^{**}$ of the first and second kind through o can be viewed as axes of a "rectilinear" coordinate system for $\mathcal{Q}_0$. With respect to this coordinate system, any point in $\mathcal{Q}_0$ can be uniquely represented by a pair of complex coordinates $(c^*, c^{**}) \in \mathbf{C}P^* \times \mathbf{C}P^{**}$ in an obvious manner. This gives a conformal equivalence

$$\mathcal{Q}_0 = \mathbf{C}P^* \times \mathbf{C}P^{**}.$$

Recall that the symmetric group $\mathcal{S}_5$ acts on $\mathbf{C}P^4$ by permuting the homogeneous coordinates. In view of the symmetries, this action leaves $\mathcal{Q}_0 \subset \mathbf{C}P_0^3$ invariant. Each permutation in $\mathcal{S}_5$ maps generating lines to generating lines. This follows from well-known projective geometric considerations, or by the explicit formulas given in Subsection G. In fact, the elements of $\mathcal{S}_5$ act on $\mathbf{C}P^4$ as projective transformations, since they act on $\mathbf{C}^5$ as permutation matrices. In particular, restricted to $\mathcal{Q}_0$, the elements of $\mathcal{S}_5$ act as projective collineations; i.e., they take projective lines to projective lines. By continuity, each collineation in $\mathcal{S}_5$ either maps the generating lines within a family to generating lines in the same family, or interchanges the generating lines between the two families. Let $\mathcal{G} \subset \mathcal{S}_5$ be the subgroup that preserves the generating lines in each family. We claim that $\mathcal{G} = \mathcal{A}_5$. It is clear that the index of $\mathcal{G}$ in $\mathcal{S}_5$ is at most 2. Since the alternating group $\mathcal{A}_5$ is the only index 2 subgroup in $\mathcal{S}_5$, it follows that $\mathcal{A}_5 \subset \mathcal{G}$. For the reverse inclusion, notice that $\mathcal{G}$ acts on $\mathbf{C}P^*$ by complex automorphisms. This realizes $\mathcal{G}$ as a subgroup of $Aut\,(\mathbf{C}P^*)$, where the latter is the group of all complex automorphisms of the projective line $\mathbf{C}P^*$. The choice of an inhomogeneous coordinate in $\mathbf{C}P^*$ identifies $\mathbf{C}P^*$ with $\hat{\mathbf{C}}$, and $Aut\,(\mathbf{C}P^*)$ with the Möbius group $M\ddot{o}b\,(\hat{\mathbf{C}})$. This identification makes $\mathcal{G}$ a finite subgroup of $M\ddot{o}b\,(\hat{\mathbf{C}})$. On the other hand, by Theorem 18, the largest finite subgroup of $M\ddot{o}b\,(\hat{\mathbf{C}})$ is $\mathcal{A}_5$. Thus $\mathcal{G} = \mathcal{A}_5$ follows. In particular, we also obtain that the collineations that correspond to the odd permutations in $\mathcal{S}_5$ do interchange the two families of generating lines.

By the very definition of the conformal equivalence above, the action of $\mathcal{A}_5$ on the two families of generating lines induces an action of $\mathcal{A}_5$ on both $\mathbf{C}P^*$ and $\mathbf{C}P^{**}$ such that the conformal equivalence is $\mathcal{A}_5$-equivariant, where $\mathcal{A}_5$ is considered to act on the product $\mathbf{C}P^* \times \mathbf{C}P^{**}$ diagonally.

As we saw above, the action of $\mathcal{A}_5$ on generating lines of the first and second kind realizes $\mathcal{A}_5$ as a subgroup in $Aut\,(\mathbf{C}P^*)$ and also in $Aut\,(\mathbf{C}P^{**})$. In a similar vein, an odd permutation in $\mathcal{S}_5$ gives rise to a conformal equivalence of $\mathbf{C}P^*$ and $\mathbf{C}P^{**}$. Conjugating $\mathcal{A}_5$ by this odd permutation (within $\mathcal{S}_5$) then carries the action of $\mathcal{A}_5$ on $\mathbf{C}P^{**}$ into an action of $\mathcal{A}_5$ on $\mathbf{C}P^*$, and this latter action is equivalent to the original action of $\mathcal{A}_5$ on $\mathbf{C}P^*$ via complex automorphisms of $\mathbf{C}P^*$.

We now consider the projections $\pi^* : \mathcal{Q}_0 \to \mathbf{C}P^*$ and $\pi^{**} : \mathcal{Q}_0 \to \mathbf{C}P^{**}$. Let

$$\tilde{\mathcal{Q}}_0 = \left\{ (z_1, \ldots, z_5) \in \mathbf{C}^5 - \{0\} \mid \sum_{j=1}^{5} z_j = \sum_{j=1}^{5} z_j^2 = 0 \right\},$$

and define $\pi : \tilde{\mathcal{Q}}_0 \to \mathcal{Q}_0$ to be the restriction of the canonical projection $\mathbf{C}^5 - \{0\} \to \mathbf{C}P^4$. We set

$$z^* = \pi^* \circ \pi \quad \text{and} \quad z^{**} = \pi^{**} \circ \pi.$$

We take a closer look at z^*. An inhomogeneous coordinate on $\mathbf{C}P^*$ identifies $\mathbf{C}P^*$ with $\hat{\mathbf{C}}$, and z^* can be viewed as the composition

$$(\mathbf{C}^5 - \{0\} \supset) \tilde{\mathcal{Q}}_0 \overset{\pi}{\to} \mathcal{Q}_0 \overset{\pi^*}{\to} \mathbf{C}P^* = \hat{\mathbf{C}}(\supset \mathbf{C}).$$

In view of the linear equivalence of $\mathcal{Q}_0$ and $\mathcal{Q}$, we see that z^* is a rational function in the variables $z_1, \ldots, z_5$ subject to $\sum_{j=1}^{5} z_j = \sum_{j=1}^{5} z_j^2 = 0$. By construction, $\mathcal{A}_5$ acts on these variables by even permutations, and this induces an action of $\mathcal{A}_5$ on $\mathbf{C}P^*$ by complex automorphisms. With a choice of an inhomogeneous coordinate on $\mathbf{C}P^*$, this latter action is by linear fractional transformations. This identifies $\mathcal{A}_5$ with a subgroup of $M\ddot{o}b\,(\hat{\mathbf{C}})$. Different choices of inhomogeneous coordinates on $\mathbf{C}P^*$ give rise to conjugate subgroups in $M\ddot{o}b\,(\hat{\mathbf{C}})$. By Theorem 18, there is an inhomogeneous coordinate on $\mathbf{C}P^*$ with respect to which $\mathcal{A}_5$ is identified with the icosahedral Möbius group I. From now on we assume that this choice has been made, and we let $\varphi : \mathcal{A}_5 \to I$ denote the corresponding isomorphism. Summarizing, we see that $z^* : \tilde{\mathcal{Q}}_0 \to \hat{\mathbf{C}}$ is φ-equivariant, where $\mathcal{A}_5$ acts on $\tilde{\mathcal{Q}}_0$ by permuting the coordinates, and I acts on $\hat{\mathbf{C}}$ as the icosahedral Möbius group.

When K is considered as the splitting field of a canonical quintic $P(z) = z^5 + az^2 + bz + c$ with roots $z_1, \ldots, z_5$ and Galois group $\mathcal{A}_5$, then $z^*(z_1, \ldots, z_5)$ becomes an element of $K = k(z_1, \ldots, z_5)$, since z^* depends rationally on $z_1, \ldots, z_5$. Since z^* is φ-equivariant, the 60 roots of the resolvent polynomial are nothing but the icosahedral linear fractional tranformations applied to $z^*(z_1, \ldots, z_5)$. Summarizing, we have shown that z^* satisfies the icosahedral equation

$$q(z^*(z_1, \ldots, z_5)) = w^*(a, b, c, \sqrt{\delta}),$$

where $z_1, \ldots, z_5$ are subject to $\sum_{j=1}^{5} z_j = \sum_{j=1}^{5} z_j^2$, and the parameter w^* on the right-hand side depends on $a, b, c, \sqrt{\delta}$ rationally. (The presence of $\sqrt{\delta}$ is due to the fact that the Galois group is $\mathcal{A}_5$ not $\mathcal{S}_5$.) To complete the proof of Theorem 19, we need to show that z^* generates K over k. (Along with the claimed technical details, we will do this in the next subsection.) The situation is completely analogous for the generating lines of the second kind. We obtain an element $z^{**}(z_1, \ldots, z_5) \in K$ that satisfies the icosahedral equation

$$q(z^{**}(z_1, \ldots, z_5)) = w^{**}(a, b, c, \sqrt{\delta}).$$

The parameters a, b, c are invariant under the entire symmetric group $\mathcal{S}_5$, while $\sqrt{\delta}$ changes its sign when the roots are subjected to odd permutations. Since the two actions of $\mathcal{A}_5$ on $\mathbf{CP}^*$ and $\mathbf{CP}^{**}$ are conjugate under the odd permutations in $\mathcal{S}_5$, we obtain

$$w^*(a, b, c, -\sqrt{\delta}) = w^{**}(a, b, c, \sqrt{\delta}).$$

G. Geometry of the Canonical Equation: Explicit Formulas

To exhibit the stated linear equivalence between $\mathcal{Q}_0$ and $\mathcal{Q}$, we first define $\iota : \mathbf{CP}^3 \to \mathbf{CP}^4$ by

$$\iota([\xi_1 : \cdots : \xi_4]) = [z_1 : \cdots : z_5],$$

where

$$z_j = \sum_{l=1}^{4} \omega^{-(j-1)l} \xi_l, \quad j = 1, \ldots, 5.$$

Since $1 + \omega + \omega^2 + \omega^3 + \omega^4 = 0$, we have

$$\sum_{j=1}^{5} z_j = \sum_{l=1}^{4} \left(\sum_{j=1}^{5} \omega^{-(j-1)l} \right) \xi_l = 0.$$

Thus, the linear map ι sends $\mathbf{CP}^3$ into the linear slice $\mathbf{CP}_0^3 \subset \mathbf{CP}^4$. Actually, ι is a linear isomorphism between $\mathbf{CP}^3$ and $\mathbf{CP}_0^3$. As simple computation shows, the inverse $\iota^{-1} : \mathbf{CP}_0^3 \to \mathbf{CP}^3$ is given by

$$\xi_l = \frac{1}{5} \sum_{j=1}^{5} \omega^{(j-1)l} z_j, \quad l = 1, \ldots, 4.$$

To translate the defining equation $\sum_{j=1}^{5} z_j^2 = 0$ of $\mathcal{Q}_0$ in terms of the ξ_l's, we compute

$$\sum_{j=1}^{5} z_j^2 = \sum_{j=1}^{5} \left(\sum_{l=1}^{4} \omega^{-(j-1)l} \xi_l \right)^2$$

$$= \sum_{l,l'=1}^{4} \left(\sum_{j=1}^{5} \omega^{-(j-1)(l+l')} \right) \xi_l \xi_{l'}$$

$$= 10(\xi_1 \xi_4 + \xi_2 \xi_3).$$

The last equality holds because $\sum_{j=1}^{5} \omega^{-(j-1)(l+l')} = 5$ iff $l + l' = 5$, and zero otherwise. This shows that the quadrics $\mathcal{Q}$ and $\mathcal{Q}_0$ correspond to each other.

Given that the variables ξ_l, $l = 1, \ldots, 4$, are linear forms in z_1, z_2, z_3, z_4, z_5, the equations for the generating lines give explicit rational dependence of z^* and z^{**} on z_1, z_2, z_3, z_4, z_5. Due to our explicit formulas, we can determine the linear fractional transformations that z^* and z^{**} undergo when the variables z_1, z_2, z_3, z_4, z_5 are subjected to even permutations. It is enough to work this out for π^*, since the homogeneous and inhomogeneous coordinates are permuted in the same way. (The case of π^{**} can be treated analogously.) We will give explicit formulas only for the generators S and W of the icosahedral group. We claim that S (multiplication by ω) corresponds to the cyclic permutation

$$S : z_j \mapsto z_{j(\bmod 5)+1}, \quad j = 1, \ldots, 5.$$

Indeed, applying the latter, we have

$$c^* = -\frac{\xi_1}{\xi_2} = -\frac{\sum_{j=1}^{5} \omega^{j-1} z_j}{\sum_{j=1}^{5} \omega^{2(j-1)} z_j} \mapsto -\frac{\sum_{j=1}^{5} \omega^{(j-1)} z_{j(\bmod 5)+1}}{\sum_{j=1}^{5} \omega^{2(j-1)} z_{j(\bmod 5)+1}}$$

$$= -\frac{\sum_{j=1}^{5} \omega^{j-2} z_j}{\sum_{j=1}^{5} \omega^{2(j-2)} z_j} = \omega c^*,$$

and the claim follows. In a similar vein, W corresponds to the permutation

$$W : z_1 \mapsto z_1, \quad z_2 \leftrightarrow z_3, \quad z_4 \leftrightarrow z_5.$$

Since this is tedious, we give a proof. Using the explicit form of W derived in Section 18, and Problem 5(b) of Section 17, we have

$$Wc^* = -\frac{(\omega - \omega^4)\xi_1 + (\omega^2 - \omega^3)\xi_2}{(\omega^2 - \omega^3)\xi_1 - (\omega - \omega^4)\xi_2} = -\frac{\tau\xi_1 + \xi_2}{\xi_1 - \tau\xi_2},$$

where τ is the golden section. On the other hand, permuting the z_j's according to the recipe above, and rewriting the corresponding quotient in terms of the ξ_i's, we have

$$-\frac{z_1 + \omega^2 z_2 + \omega z_3 + \omega^4 z_4 + \omega^3 z_5}{z_1 + \omega^4 z_2 + \omega^2 z_3 + \omega^3 z_4 + \omega z_5}$$

$$= -\frac{(1 + 2\omega + 2\omega^4)\xi_1 + (3 + \omega^2 + \omega^3)\xi_2 + (3 + \omega + \omega^4)\xi_3 + (1 + 2\omega^2 + 2\omega^3)\xi_4}{(3 + \omega^2 + \omega^3)\xi_1 + (1 + 2\omega^2 + 2\omega^3)\xi_2 + (1 + 2\omega + 2\omega^4)\xi_3 + (3 + \omega + \omega^4)\xi_4}$$

$$= -\frac{(1 + 2/\tau)\xi_1 + (3 - \tau)\xi_2 + (3 + 1/\tau)\xi_3 + (1 - 2\tau)\xi_4}{(3 - \tau)\xi_1 + (1 - 2\tau)\xi_2 + (1 + 2/\tau)\xi_3 + (3 + 1/\tau)\xi_4}$$

$$= -\frac{\xi_1 + \xi_2/\tau + \xi_3\tau - \xi_4}{\xi_1/\tau - \xi_2 + \xi_3 + \xi_4\tau}$$

$$= -\frac{(\tau\xi_1 + \xi_2)(1/\tau + \xi_3/\xi_1)}{(\xi_1 - \tau\xi_2)(1/\tau + \xi_3/\xi_1)} = -\frac{\tau\xi_1 + \xi_2}{\xi_1 - \tau\xi_2}.$$

Here we used $1 + 2/\tau = \sqrt{5}$, $3 - \tau = \sqrt{5}/\tau$, $3 + 1/\tau = \sqrt{5}\tau$, $1 - 2\tau = -\sqrt{5}$, and $\xi_1\xi_4 + \xi_2\xi_3 = 0$. The permutation rule for W follows.

Comparing how S and W transform the z_j's and the t_j's, we see that when z^* is subjected to the linear fractional transformations of the icosahedral group I, then this action can be realized as even permutations on its variables *in exactly the same manner* as $\mathcal{A}_5$ acts on the roots $t_1, \ldots, t_5$ of the icosahedral resolvent. This means that an $\mathcal{A}_5$-equivariant one-to-one correspondence $z_j \leftrightarrow t_j, j = 1, \ldots, 5$, can be established between the two sets $\{z_1, \ldots, z_5\}$ and $\{t_1, \ldots, t_5\}$. (z_1 and t_1 are the unique fixed points of W, and $z_{j+1} = S^j(z_1)$ corresponds to $t_{j+1} = S^j(t_1), j = 1, \ldots, 4$.)

Remark.
The odd permutation

$$z_1 \mapsto z_1, \quad z_2 \mapsto z_4, \quad z_3 \mapsto z_2, \quad z_4 \mapsto z_5, \quad z_5 \mapsto z_3$$

has the effect of replacing ω by ω^2 and interchanging c^* and c^{**}. Thus, the formulas for the generating lines of the second kind can be derived from those of the first kind by the substitution $\omega \mapsto \omega^2$.

Remark.

The Lagrange substitutions above that establish the linear equivalence between the quadrics Q and Q_0, as well as make the structure of the generating lines transparent, are *derived* in the *Icosahedron Book* based on the requirement that the transformation rules for S and W should match with those of the roots of the quintic icosahedral resolvent. Here we followed a somewhat ad hoc but quicker approach.

To get a closer look at the correspondence above, we now work out the generating lines *in terms of the roots* $\{t_1, \ldots, t_5\}$ *of the icosahedral resolvent*. In perfect analogy with the Lagrange substitutions, we put

$$t_j = \sum_{l=1}^{4} \omega^{-(j-1)l} X_l, \quad j = 1, \ldots, 5,$$

and

$$X_l = \frac{1}{5} \sum_{j=1}^{5} \omega^{(j-1)l} t_j, \quad l = 1, \ldots, 4.$$

To work out X_l, we use the explicit forms of Ξ_j and $\Omega_j\Xi_j$ derived in Subsection C. Substituting them into the expression for t_j above, we obtain

$$t_j = (\omega^{4(j-1)}z_1 - \omega^{3(j-1)}z_2)A + (\omega^{2(j-1)}z_1 + \omega^{j-1}z_2)B,$$

where A, B are linear in u and v. (Here z_1 and z_2 are the complex arguments of our forms with $z = z_1/z_2$. Do not confuse them with the first two roots of the canonical quintic! In any case, z_1 and z_2 appear here only briefly.) With this, the inverse of the Lagrange substitution becomes

$$X_l = \frac{1}{5} \sum_{j=1}^{5} \omega^{(j-1)l} (\omega^{4(j-1)} z_1 - \omega^{3(j-1)} z_2) A$$

$$+ \frac{1}{5} \sum_{j=1}^{5} \omega^{(j-1)l} (\omega^{2(j-1)} z_1 - \omega^{j-1} z_2) B$$

$$= (\delta_{1l} z_1 - \delta_{2l} z_2) A + (\delta_{3l} z_1 + \delta_{4l} z_2) B,$$

where we used the Kronecker delta function δ_{jl} ($= 1$ iff $j = l$ and zero otherwise). Writing the cases out, we have

$$X_1 = z_1 A, \quad X_2 = -z_2 A, \quad X_3 = z_1 B, \quad X_4 = z_2 B.$$

For the parameters C^* and C^{**} of the generating lines defined by

$$-\frac{X_1}{X_2} = \frac{X_3}{X_4} = C^*, \quad \frac{X_1}{X_3} = -\frac{X_2}{X_4} = C^{**},$$

we obtain

$$C^* = \frac{z_1}{z_2} = z, \quad C^{**} = \frac{A}{B}.$$

Since z^* and z^{**} are essentially given by the projections $(C^*, C^{**}) \mapsto C^*$ and $(C^*, C^{**}) \mapsto C^{**}$, we thus have

$$z^*(t_1, \ldots, t_5) = z \quad \text{and} \quad z^{**}(t_1, \ldots, t_5) = \frac{A}{B}.$$

The first equation is of paramount importance. Writing the canonical resolvent in the short form

$$P^*(X) = \prod_{j=1}^{5} (X - t_j) = X^5 + 5a(u, v, w) X^2 + 5b(u, v, w) X + c(u, v, w),$$

we have

$$q(z^*(t_1, \ldots, t_5)) = w^*(a(u, v, w), b(u, v, w), c(u, v, w), \sqrt{\delta(u, v, w)})$$

$$= q(z) = w.$$

Here $\delta(u, v, w)$ is the discriminant of the canonical quintic resolvent. (It has the same expression as δ in Subsection C with a, b, c replaced by $a(u, v, w), b(u, v, w), c(u, v, w)$.)

The system

$$a(u, v, w) = a,$$

$$b(u, v, w) = b,$$

$$c(u, v, w) = c,$$

can be inverted, provided that we also match the square roots of the discriminants $\sqrt{\delta(u, v, w)} = \sqrt{\delta}$:

$$u = u(a, b, c, \sqrt{\delta}),$$

$$v = v(a, b, c, \sqrt{\delta}),$$

$$w = w(a, b, c, \sqrt{\delta}).$$

We will do this inversion explicitly at the end of this subsection. By solving this system we attain that the roots of the canonical quintic P and the roots of the quintic resolvent P^* coincide *as sets*. On the other hand, we saw that there exists an $\mathcal{A}_5$-equivariant correspondence between these sets of roots. Imposing this we obtain

$$z_j = t_j, \quad j = 1, \ldots, 5.$$

With this we can now describe how to solve a given irreducible quintic. First we use the Tschirhaus transformation to reduce the quintic to a canonical form $P(z) = z^5 + 5az^2 + 5bz + c$. This amounts to solving a quadratic equation. We also compute the discriminant δ from the explicit form given in Subsection C. Then we substitute the coefficients a, b, c into the right-hand sides of the equations in the inverted system above and obtain u, v, and w. We now solve the icosahedral equation for this particular value of w to obtain z as a ratio of hypergeometric functions as in Subsection B. By working out the forms Ω_j and Ξ_j using the particular values of u and v and z we obtain

$$t_j = t_j(u, v, z), \quad j = 1, \ldots, 5.$$

Since $t_j = z_j$, these are the five roots of our quintic. Tracing our steps back, we see that $z_1, \ldots, z_5$ depend rationally on z^*. In particular, when $K = k(z_1, \ldots, z_5)$ is the splitting field of the canonical quintic, we obtain that z^* generates K over k. This was the missing piece in the proof of Theorem 19.

Our final task is to invert the system above. Using the explicit forms of the coefficients of the canonical quintic in Subsection D, we write the system as

$$w \cdot a = 8u^3 + 12u^2v + \frac{6uv^2 + v^3}{1 - w},$$

$$\frac{w \cdot b}{3} = -4u^4 + \frac{6u^2v^2 + 4uv^3}{1 - w} + \frac{3v^4}{4(1 - w)^2},$$

$$\frac{w \cdot c}{3} = 48u^5 - \frac{40u^3v^2}{1 - w} + \frac{15uv^4 + 4v^5}{(1 - w)^2}.$$

A small miracle (due to our ad hoc approach) happens here. In trying to cancel the top terms in u and v, if we multiply the first equation by $-4v^2/(1 - w)$ and the second by $12u$, and then add all three then everything cancels on the right-hand sides! Rearranging the terms on the left-hand side, we obtain

$$\frac{v^2}{1 - w} = \frac{12ub + c}{12a}.$$

We call this the fundamental relation. Continuing, we can create cubic binomial terms as follows:

$$-uc + \frac{v^2}{1 - w}b = -\frac{9}{4w}\left(4u^2 - \frac{v^2}{1 - w}\right)^3$$

and

$$a^2 - \frac{4}{81}\frac{1 - w}{v^2}(3ua + 2b)^2 = \frac{1}{w}\left(4u^2 - \frac{v^2}{1 - w}\right)^3.$$

Combining these, we obtain

$$a^2 - \frac{4}{81}\frac{1 - w}{v^2}(3ua + 2b)^2 = \frac{4}{9}\left(uc - \frac{v^2}{1 - w}b\right).$$

Using the fundamental relation, this becomes

$$a^2 - \frac{16}{27}\frac{a}{12ub + c}(3ua + 2b)^2 = \frac{4}{9}\left(uc - \frac{12ub + c}{12a}b\right).$$

This is a quadratic equation in u! The quadratic formula gives

$$u = u(a, b, c, \sqrt{\delta}) = \frac{(11a^3b + 2b^2c - ac^2) \pm a\sqrt{\delta}/(25\sqrt{5})}{24(a^4 - b^3 + abc)},$$

where the sign in front of $\sqrt{\delta}$ needs to be determined.

Since u is now known in terms of $a, b, c,$ and $\sqrt{\delta}$, we can use the fundamental relation in the first equation containing a cubic binomial term and solve for w. We obtain

$$w = w(a, b, c, \sqrt{\delta}) = \frac{(48u^2a - 12ub - c)^2}{64a^2(12u(ac - b^2) - bc)}.$$

Finally, the first equation of our system can be rewritten as

$$\left(12u^2 + \frac{v^2}{1 - w}\right)v = wa - 8u^3 - 6u\frac{v^2}{1 - w}.$$

Using once again the fundamental relation, we finally arrive at

$$v = v(a, b, c, \sqrt{\delta}) = -\frac{96u^3a + 72u^2b + 6uc - 12a^2w}{144u^2a + 12ub + c}.$$

To determine the sign of $\sqrt{\delta}$, we set $\sqrt{\delta} = \sqrt{\delta(u, v, w)}$, where

$$\delta(u, v, w) = \prod_{1 \le j < l \le 5} (t_j - t_l)^2$$

is the discriminant of the canonical resolvent. It is enough to work out $\sqrt{\delta(u, v, w)}$ for $u = 1$ and $v = 0$. For these values of the parameters our system reduces to

$$a = \frac{8}{w}, \quad b = -\frac{12}{w}, \quad c = \frac{144}{w}.$$

Substituting these and $u = 1$ into the formula for u with the sign ambiguity, we obtain

$$\pm\sqrt{\delta} = 12^4 \cdot 25\sqrt{5}\frac{1 - w}{w^3}.$$

Remark.

Note that as a good check of our computations we can also substitute the values of a, b, c above into the discriminant formula in

Subsection C and get

$$\delta = 12^8 \cdot 5^5 \frac{(1-w)^2}{w^6}.$$

On the other hand, for $u = 1$ and $v = 0$, t_j reduces to $s_j = 12 \, (\mathcal{I}/\mathcal{H}) \, \Xi_j$, and hence,

$$\sqrt{\delta(1, 0, w)} = \prod_{1 \leq j < l \leq 5} (s_j - s_l) = 12^{10} \frac{\mathcal{I}^{10}}{\mathcal{H}^{10}} \prod_{1 \leq j < l \leq 5} (\Xi_j - \Xi_l).$$

The last product is a degree-80 icosahedral invariant, so that it must be a linear combination of $\mathcal{J}^2\mathcal{H}$ and $\mathcal{H}^4$. Comparing coefficients, we obtain

$$\prod_{1 \leq j < l \leq 5} (\Xi_j - \Xi_l) = -25\sqrt{5}\mathcal{J}^2\mathcal{H}.$$

Using $w : w - 1 : 1 = \mathcal{H}^3 : -\mathcal{J}^2 : 1728\,\mathcal{I}^5$, we finally end up with the following:

$$\sqrt{\delta(1, 0, w)} = -12^{10} \cdot 25\sqrt{5}\left(\frac{\mathcal{I}^5}{\mathcal{H}^3}\right)^2\left(\frac{\mathcal{J}^2}{\mathcal{H}^3}\right) = -12^4 \cdot 25\sqrt{5}\,\frac{1-w}{w^3}.$$

For $\sqrt{\delta} = \sqrt{\delta(1, 0, w)}$, we thus need to choose the *negative sign* in front of $\sqrt{\delta}$.

Problems

1. Verify the following identities:

$$Hess\,(\Phi) = -48\sqrt{3}i\Psi,$$
$$Hess\,(\Psi) = 48\sqrt{3}i\Phi,$$
$$Jac\,(\Phi, \Psi) = 32\sqrt{3}i\Omega,$$
$$Hess\,(\Omega) = -25\Phi\Psi,$$
$$Jac\,(\Omega, \Phi) = -4\Psi^2,$$
$$Jac\,(\Omega, \Psi) = -4\Phi^2.$$

2. Interpret the zeros of the degree-8 forms $\Xi_j, j = 1, \ldots, 5$, geometrically.

3. Locate the axes of the half-turns $S^j U S^{-j}$, $S^j V S^{-j}$, $S^j W S^{-j}$, $j = 1, \ldots, 5$, in our model of the icosahedron.

4. (a) Work out the resolvent of the five octahedral forms using the following steps. First consider the product

$$\prod_{j=1}^{5}(\mathcal{X} - \Omega_j) = \mathcal{X}^5 + a_1\mathcal{X}^4 + a_2\mathcal{X}^3 + a_3\mathcal{X}^2 + a_4\mathcal{X} + a_5,$$

where $\mathcal{X}$ is used as a variable. Using the description of the invariants of the icosahedral group, conclude that this quintic resolvent is

$$\mathcal{X}^5 - 10\mathcal{I}\,\mathcal{X}^3 + 45\mathcal{I}^2\,\mathcal{X} - \mathcal{J}.$$

Introduce the new variable

$$X = 12\frac{\mathcal{I}^2}{\mathcal{J}}\mathcal{X}$$

(depending only on $z = z_1/z_2$), and use the second table in Section 24 to arrive at the quintic icosahedral resolvent polynomial

$$P^*(X) = \prod_{j=1}^{5}(X - r_j^*) = X^5 - \frac{5}{6(1-w)}X^3 + \frac{5}{16(1-w)^2}X - \frac{1}{12(1-w)^2}.$$

(b) Describe the geometry of the "solution set" $(r_1^*(w), \ldots, r_5^*(w)) \in \mathbf{C}^5$ as w varies in $\hat{\mathbf{C}}$ and $q(z) = w$ as follows. Introduce the sums of various powers

$$\sigma_l = \sum_{j=1}^{5}(r_j^*)^l, \quad l = 1, \ldots, 4,$$

and verify

$$\sigma_1 = 0, \quad \sigma_2 = \frac{5}{3(1-w)}, \quad \sigma_3 = 0, \quad \sigma_4 = \frac{5}{36(1-w)^2}.$$

Notice that $\sigma_2^2 = 20\sigma_4$. Define

$$\mathcal{D} = \left\{[r_1^* : \ldots : r_5^*] \in \mathbf{CP}^4 \mid \sum_{j=1}^{5}r_j^* = \sum_{j=1}^{5}(r_j^*)^3 = 0\right\}$$

and

$$\mathcal{F} = \left\{[r_1^* : \cdots : r_5^*] \in \mathbf{CP}^4 \mid \sum_{j=1}^{5}r_j^* = 0, \left(\sum_{j=1}^{5}(r_j^*)^2\right)^2 = 20\sum_{j=1}^{5}(r_j^*)^4\right\},$$

and prove that $\mathcal{D}$ and $\mathcal{F}$ are smooth algebraic surfaces in $\mathbf{CP}^3$ of degree 3 and 4, respectively, ($\mathbf{CP}^3$ is identified with the linear slice $\mathbf{CP}_0^3$ of $\mathbf{CP}^4$ defined by $\sum_{j=1}^{5}r_j^* = 0$). Show that the projective points that correspond to the solutions of the quintic icosahedral resolvent above fill the smooth algebraic curve $\mathcal{D} \cap \mathcal{F} \subset \mathbf{CP}^3$ of degree 12. (This problem is rich in history. The quintic resolvent for $\mathcal{X}$ is called the *Brioschi quintic*. The surface $\mathcal{D}$ was named by Clebsch the *diagonal surface*, since this surface contains 15 diagonals of the coordinate pentahedron. The intersection of the diagonal surface $\mathcal{D}$ with the canonical surface $\mathcal{Q}$ is a (smooth) curve called the *Bring–Jerrard curve*.[6] A generating line through the point $[z_1 : \cdots : z_5] \in \mathcal{Q}$, corresponding to the roots of a canonical quintic, intersects the Bring–Jerrard curve in 3 points, and these intersection points can be obtained by extracting suitable square and cube roots. There is a beautiful geometric interpretation of the Bring–Jerrard curve in terms of stellated dodecahedra[7].)

5. (a) Derive an icosahedral resolvent polynomial of degree 6 based on the 6 quartic forms that vanish on the 6 pairs of antipodal vertices of the icosahedron as follows. Set $\phi_\infty = 5z_1^2 z_2^2$, and apply the icosahedral substitutions S^j to ϕ_∞ to obtain

$$\phi_j(z_1, z_2) = (\omega^j z_1^2 + z_1 z_2 - \omega^{4j} z_2^2)^2, \quad j = 0, \dots, 4.$$

Comparing coefficients, show that the resolvent form satisfies

$$\phi^6 - 10\mathcal{I}\phi^3 + \mathcal{H}\phi + 5\mathcal{I}^2 = 0.$$

(b) Work out the action of the icosahedral generators S and W on ϕ_∞ and $\phi_j, j = 0, \dots, 4$, and verify

$$S : \phi_\infty \mapsto \phi_\infty, \quad \phi_j \mapsto \phi_{j+1(\bmod 5)}, \quad j = 0, \dots, 4;$$

$$W : \phi_\infty \leftrightarrow \phi_0, \quad \phi_1 \leftrightarrow \phi_4, \quad \phi_2 \leftrightarrow \phi_3.$$

Interpret these transformation rules as congruences

$$S : j' \equiv j + 1 (\bmod 5),$$

$$W : j' \equiv -\frac{1}{j} (\bmod 5).$$

Show that the icosahedral group (the Galois group of this resolvent) is isomorphic to $SL(2, \mathbf{Z}_5)/\{\pm I\}$:

$$j' \equiv -\frac{aj + b}{cj + d} (\bmod 5), \quad ad - bc \equiv 1 (\bmod 5), \quad a, b, c, d \in \mathbf{Z}.$$

---

[6] For a very detailed exposition, see R. Fricke, *Lehrbuch der Algebra*, Vol. II, Braunschweig, 1926. For a recent text, see J. Shurman, *Geometry of the Quintic*, John Wiley and Sons, 1997.

[7] See P. Slodowy, *Das Ikosaeder und die Gleichungen fünften Grades*, in Mathematische Miniatüren, Band 3, Birkhäuser-Basel, 1986.

26

SECTION

The Fourth Dimension

♣ In Section 22, we encountered a number of objects in the 4-dimensional Euclidean space, such as S^3. Although it sits in $\mathbf{R}^4$, S^3 is essentially a 3-dimensional object. In fact, the generalized stereographic projection gives rise to a diffeomorphism $h_N : S^3 - \{N\} \to \mathbf{R}^3$ where $N = (0, 0, 0, 1)$ is the North Pole and $\mathbf{R}^3$ is the linear subspace of $\mathbf{R}^4$ spanned by the first three coordinate axes. (In coordinates, the inclusion $\mathbf{R}^3 \subset \mathbf{R}^4$ is given by assigning to $(a, b, c) \in \mathbf{R}^3$ the point $(a, b, c, 0) \in \mathbf{R}^4$.) The definition of h_N is the same as its 2-dimensional brother; $h_N(p)$, $N \neq p \in S^3$, is the unique intersection point of $\mathbf{R}^3$ and the line through N and p. (Notice that N, O, p and $h_N(p)$ all lie in a 2-dimensional plane which cuts a great circle from $S^3 \subset \mathbf{R}^4$.) The stereographic projection h_N enables us to visualize objects that lie in S^3 as objects in $\mathbf{R}^3$; in fact, this was used to create Color Plate 10 (Clifford tori).

Remark.
◇ Quaternions can be used to give a simple and explicit form of the stereographic projection. If we let $\mathbf{R}^3$ correspond to the linear space $\mathbf{H}_0$ of pure quaternions and $1 \in \mathbf{H}$ correspond to the North

380

Pole N, then $h_N : S^3 \to \mathbf{H}_0$ is given by

$$h_N(q) = \frac{q - \Re(q)}{1 - \Re(q)}, \quad 1 \neq q \in S^3.$$

Indeed, since 1, q, and $h_N(q)$ are on the same line, we have

$$t(1 - q) = 1 - h_N(q)$$

for some $t > 0$. Taking the real part of both sides and using the fact that $h_N(q)$ is pure, we obtain

$$t = \frac{1}{1 - \Re(q)}.$$

Substituting this back, the formula follows.

♣ The purpose of this section is to give a few examples that will enable us to get a feel for objects whose natural environment is $\mathbf{R}^4$. We begin with a cube[1] in $\mathbf{R}^4$. The best way to understand how a 4-dimensional cube and an ordinary 3-dimensional cube are related is to observe how a 3-dimensional cube is derived from a 2-dimensional cube, i.e., a square. As for the latter, given a flat square in space, choose a vector perpendicular to the plane of the square (Figure 26.1). We assume that the magnitude of the vector is given by the side length of the square. Performing spatial translation with this vector, we see that the square sweeps a 3-dimensional cube. During this process the number of vertices doubles, the number of edges doubles plus the old vertices sweep four new vertices, and finally, the number of faces doubles plus the old edges sweep four new faces.

Obtaining a 4-dimensional cube is now remarkably simple; just repeat this process! Take a 3-dimensional cube in $\mathbf{R}^4$, choose a vector perpendicular to the 3-dimensional linear subspace spanned by the cube and watch the object that is swept by translating the cube along this vector (Figure 26.2).

In particular the same counting principles hold, and we see that a 4-dimensional cube has 2×8 vertices, $(2 \times 12) + 8$ edges, and $(2 \times 6) + 12$ faces!

---

[1]The first pioneers of 4-dimensional geometric vision were Möbius (c. 1827), Schläfli, Cayley, and Grassmann (c. 1853).

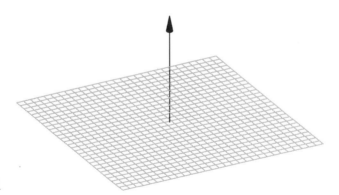

Figure 26.1

The skeptical reader may now say, "This is all humbug; the 4-dimensional cube is drawn on a 2-dimensional sheet of paper, so there is no way to get a feel for this object!" It is true, the picture you see is the projection of a 4-dimensional cube to a 2-dimensional plane, but remember that we use the same trickery every day on the chalkboard to visualize 3-dimensional objects!

Remark.
Observing how the 3-dimensional tetrahedron is derived from the 2-dimensional triangle, it is easy to obtain a model of the 4-dimensional tetrahedron. We already encountered this object in Section 20; in fact, the pentagonal graph in Figure 20.4 gives a 2-dimensional projection of the regular tetrahedron in $\mathbf{R}^4$. (Can you identify the five tetrahedral faces?)

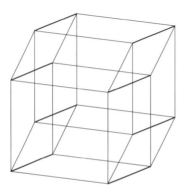

Figure 26.2

All 4-dimensional *regular polytopes*[2] were discovered and classified by Schläfli around 1853. Bypassing some computational details, we now sketch[3] this classification.

Note first that the relevant concepts (flag, regularity, vertex figure, reciprocality, etc.) introduced in Section 17 for regular polyhedra can be extended to convex polytopes in 4 dimensions. In particular, a (4-dimensional) polytope has vertices, edges, faces, and 3-dimensional faces, called *cells*. For example, the 4-dimensional tetrahedron, or *pentatope*, has $1 + 4$ tetrahedral cells, and the 4-dimensional cube, or *hypercube*, has $2 + 6$ cubical cells.

The Schläfli symbol of a regular polytope is a triple $\{a, b, c\}$, where $\{a, b\}$ is the Schläfli symbol of a typical cell, and each edge of the polytope is shared by c cells. A closer look at Figure 20.4 and Figure 26.2 shows that the Schläfli symbol of the pentatope is $\{3, 3, 3\}$, and the Schläfli symbol of the hypercube is $\{4, 3, 3\}$.

By a bit of a stretch of Euclid's argument in Section 17 we can list all *possible* Schläfli symbols. Take c congruent cells (regular polyhedra) of Schläfli symbol $\{a, b\}$, and cluster them around a single common edge (with consecutive cells pasted together along a single commmon face) in a fanlike pattern. This can be done in 3 dimensions. (For analogy, paste three, four, or five congruent equilateral triangles together to form a fanlike pattern sharing a single common vertex in the plane.) There are exactly two faces (one for the initial cell, and one for the terminal cell) in the fan sharing the common edges that have not been pasted together. Now place the entire configuration in 4 dimensions, and paste these remaining two faces together. Notice that as you move the two faces together, the entire configuration pops out in the fourth dimension! (Continuing the analogy, the fans of the equilateral triangles above pop out in 3 dimensions to form the sides of triangular, square, and pentagonal pyramids, the basic ingredients to make the tetrahedron, octahedron, and icosahedron!)

For the existence of the fan of c cells, it is clearly necessary that

$$c\delta < 2\pi,$$

---

[2] In 4-dimensions we use the word *polytope* rather than *polyhedron*.

[3] For a detailed account, see H.S.M. Coxeter, *Regular Polytopes*, Pitman, 1947.

where δ is the dihedral angle of the cell with Schläfli sybol $\{a, b\}$. As we will see below, this constraint will give all possible Schläfli symbols $\{a, b, c\}$ of a regular polytope. Thus, we need to work out the dihedral angle for each of the Platonic solids.

We begin with the tetrahedron. In Figure 17.16, the grey triangle (cut out from the tetrahedron by a symmetry plane) is isosceles, and its larger angle at the midpoint of the edge is the dihedral angle δ. We have $\sin(\delta/2) = 1/\sqrt{3}$, or $\delta \approx 70.5287793°$. The constraint above gives $3, 4, 5$ as the possible values of c. We obtain that the possible Schläfli symbols for a regular polytope with tetrahedral cells are $\{3, 3, 3\}$, $\{3, 3, 4\}$, and $\{3, 3, 5\}$.

The dihedral angle of a cube is $\delta = \pi/2$, so that $c = 3$ and the only possible Schläfli symbol is $\{4, 3, 3\}$.

A symmetry plane of the octahedron containing two opposite vertices and two midpoints of sides cuts out a rhombus from the octahedron, and δ is the larger angle of the rhombus at the midpoints. We have $\tan(\delta/2) = \sqrt{2}$, or $\delta \approx 109.4712206°$. Thus, $c = 3$, and the only possible Schläfli symbol is $\{3, 4, 3\}$.

A good look at Figure 17.30 shows that the dihedral angle of the icosahedron satisfies $\tan(\delta/2) = \tau/(\tau - 1)$, where τ is the golden section. A bit of computation gives $\delta \approx 138.1896851°$. We see that there cannot be a regular polytope with icosahedral cells.

Finally, the dihedral angle of the dodecahedron can be obtained from the roof-proof (Problem 20 of Section 17). The perpendicular bisector of the ridge of a roof is a symmetry plane that cuts out an isosceles triangle from the roof, and the larger angle (at the midpoint of the ridge) of this triangle is δ. From the metric properties of the roof, we have $\tan(\delta/2) = \tau$, or $\delta \approx 116.5650512°$. We obtain $c = 3$ and the Schläfli symbol $\{5, 3, 3\}$.

Summarizing, we see that the possible Schläfli symbols of a regular polytope are

$$\{3, 3, 3\}, \quad \{4, 3, 3\}, \quad \{3, 3, 4\}, \quad \{3, 4, 3\}, \quad \{5, 3, 3\}, \quad \{3, 3, 5\}.$$

The first two give the Schläfli symbols of the pentatope and the hypercube. Amazingly, the rest are also realized as Schläfli symbols of regular polytopes! To complete the classification, we now give brief geometric descriptions for each of these new pentatopes. (For pictures, see H.S.M. Coxeter, *Regular Polytopes*, Pitman, 1947, or

D. Hilbert and S. Cohn-Vossen, *Geometry and Imagination*, Chelsea, New York, 1952.)

Recall that 4 alternate vertices of the cube are the vertices of an inscribed tetrahedron. Each omitted vertex corresponds to a face of the tetrahedron. (For an omitted vertex, the vertex figure of the cube is parallel to the face.) Analogously, 8 alternate vertices of a hypercube are the vertices of a regular polytope. This polytope has 16 tetrahedral cells, and for this reason it is called the *16-cell*. Eight tetrahedra correspond to each of the omitted vertices, and 8 other tetrahedra are inscribed in each of the 8 cubic cells of the hypercube. The 16-cell has 24 edges that are the diagonals of the 24 square faces of the hypercube. (Since we are selecting alternate vertices of the hypercube, we get only one diagonal for each square face.) It follows that each edge must be surrounded by 4 tetrahedra, so that the Schläfli symbol is $\{3, 3, 4\}$. The 16-cell has $16 \times 4/2 = 32$ triangular faces.

A plane projection of the 16-cell can be obtained by starting with the vertices of a regular octagon, and connecting all but opposite pairs of vertices by line segments.

The regular polytope with Schläfli symbol $\{3, 4, 3\}$, the so-called *24-cell*, can be obtained from the 16-cell as follows. The vertices of the 24-cell are the midpoints of the 24 edges of the 16-cell. It has 24 octahedral cells. Eight of these cells are the vertex figures of the 16-cell, and 16 are inscribed in the 16 tetrahedra of the 16-cell. (For analogy, in Figure 17.20 of the truncated tetrahedron, 4 of the 8 faces of the inscribed octahedron are the vertex figures of the tetrahedron, and the 4 remaining faces are inscribed in the faces of the tetrahedron.) Each edge of the 24-cell is surrounded by 3 octahedra. The 24-cell has $24 \times 12/3 = 96$ edges and $24 \times 8/2 = 96$ triangular faces.

It is easier to visualize the remaining two regular polytopes projected to the 3-sphere S^3 from the origin. For the spherical regular polytope with Schläfli symbol $\{5, 3, 3\}$, we first construct a typical cell, a spherical dodecahedron in S^3. Recall from above that the dihedral angle of the Euclidean dodecahedron is $\delta \approx 116.5650512°$. If this dodecahedron is small, its metric properties are close to those of the spherical dodecahedron of the same size. At the other extreme, if a small spherical dodecahedron is inflated, at one stage it will cover an entire hemisphere in S^3, and hence the dihedral angle will become 180°. By continuity, there must be a spherical dodecahedron whose dihedral angle is 120°. As expected, this puffy (and tiny) dodecahedron is almost indistinguishable from its Euclidean brother. Due to the fact that the dihedral angle is $120° = 2\pi/3$, exactly 3 of these dodecahedra fit together to share a common edge. Hence, if we are able to tessellate S^3 with spherical dodecahedra of this size, the regular polytope obtained must have Schläfli symbol $\{5, 3, 3\}$.

It turns out that this construction is possible, and we need 120 dodecahedra for the tessellation. For this reason, this polytope is called the *120-cell*.

The explicit construction of the 120-cell is technical. There is an easy way, however, to see how these 120 dodecahedra fit together in S^3, and it is based on the fact that, up to adjustment by an isometry, the centroids of the dodecahedral cells can be considered as the 120 elements of the binary icosahedral group $\mathbf{I}^*$ in S^3 discussed in Section 23. Recall that in terms of the Clifford decomposition of S^3, $\mathbf{I}^*$ is made up of the vertices of two regular decagons inscribed in the orthogonal circles $C_{\pm 1}$, and the rest appear (in two groups of 50) in the Clifford tori $C_{\pm 1/\sqrt{5}}$ (Figure 23.5). In view of this, the 120-cell can be constructed as follows. First make a "necklace" of 10 (spherical) dodecahedra such that the centroids of the dodecahedra are the 10 elements of $\mathbf{I}^*$ on C_1. It turns out that these dodecahedra have dihedral angle 120° as above. A pair of consecutive dodecahedra in the necklace are pasted together at a common pentagonal face. Each of the 5 edges of this common face is the shared edge of two other pentagonal faces, one from each of the consecutive dodecahedra. These two faces meet at a dihedral angle of 120°, so that another dodecahedron can be pasted in. Since we have five edges (of the common pentagonal face), we can paste in 5 extra dodecahedra around the two consecutive dodecahedra in the necklace. This cluster of 5 dodecahedra makes a "bulge" in the necklace. Since the necklace has 10 places (of consecutive dodecahedra) for this construction, we can add 10 bulges to the necklace, a total of 10×5 dodecahedra. These, along with the original 10 dodecahedra, use up 60 dodecahedra, and give a "bumpy" polyhedral Clifford torus in S^3. As computation shows, the centroids of the dodecahedra in the bulges make up the 50 elements of $\mathbf{I}^*$ in $C_{1/\sqrt{5}}$.

Finally, the entire construction can be repeated for C_{-1} and $C_{-1/\sqrt{5}}$, and the two bumpy Clifford tori fit together to form the 120-cell. (This visualization of the 120-cell also reveals that the faces of each dodecahedral cell are contained in the perpendicular bisectors of the line segments connecting the centroid (in $\mathbf{I}^*$) of the cell and the $2+5+5$ nearby elements in $\mathbf{I}^*$.)

The 120-cell has $120 \times 12/2 = 720$ faces (by double counting the faces of the 120 dodecahedra), $120 \times 30/3 = 1200$ edges (since 3 dodecahedra share a common edge), and $120 \times 20/4 = 600$ vertices (since 4 dodecahedra meet at a common vertex).

The reciprocal of the 120-cell is the *600-cell*, a regular polytope with 600 tetrahedral cells and Schläfli symbol {3, 3, 5}. A typical cell is a puffy spherical tetrahedron with dihedral angle 72°. As the Schläfli symbol suggests, exactly 5 share a common edge. At each vertex exactly 20 tetrahedra meet. (Indeed, consider the configuration of rays tangent to the edges at a vertex. This configuration intersected with a sphere around the vertex (within the tangent space of S^3 at the vertex) gives the vertices of a regular polyhedron. This polyhedron must have the largest number of vertices; that is, it must be a dodecahedron.) Since each of the 600 tetrahedra contributes 4 vertices to the tesselation, and this way the total number of vertices is overcounted 20 times, we see that the number of vertices of the tesselation is $600 \times 4/20 = 120$. Once again, it turns out that up to an adjustment of the tessellation in S^3 by an isometry, these 120 vertices make up the binary icosahedral group $\mathbf{I}^* \subset S^3$. Finally, note that the 600-cell has $600 \times 4/2 = 1200$ faces, and $600 \times 6/5 = 720$ edges.

We summarize our classification of regular polytopes in Table 26.1.

To search for new objects, we now go back to Section 16. Recall that we had difficulty realizing certain nonorientable surfaces in $\mathbf{R}^3$ because of self-intersections. We will now clarify this.

First some examples: Recall that the infinite Möbius band is obtained by continuously sliding and tilting a straight line along a

Table 26.1

Polytope	Schläfli symbol	V	E	F	C
Pentatope	{3, 3, 3}	5	10	10	5
Hypercube	{4, 3, 3}	16	32	24	8
16-cell	{3, 3, 4}	8	24	32	16
24-cell	{3, 4, 3}	24	96	96	24
120-cell	{5, 3, 3}	600	1200	720	120
600-cell	{3, 3, 5}	120	720	1200	600

circle so that in a full round the line completes a half-turn. The motion of the intersection point of the line with the unit circle S^1 takes place in the Euclidean plane $\mathbf{R}^2$ and is conveniently parameterized by $\theta \mapsto (\cos\theta, \sin\theta)$, $\theta \in \mathbf{R}$ (Figure 26.3).

The motion of the line takes place in another copy of $\mathbf{R}^2$ perpendicular to this, as shown in Figure 26.4.

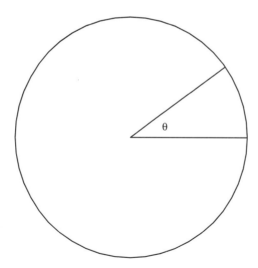

Figure 26.3

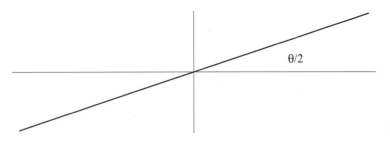

Figure 26.4

At $(\cos\theta, \sin\theta) \in S^1$, this line must have slope $\tan(\theta/2)$ because of the half-turn matching condition. It can be parameterized by

$$t \mapsto (t\cos(\theta/2), t\sin(\theta/2)), \quad t \in \mathbf{R}.$$

We obtain that the infinite Möbius band can be parameterized by

$$(\theta, t) \mapsto \left(\cos\theta, \sin\theta, t\cos\left(\frac{\theta}{2}\right), t\sin\left(\frac{\theta}{2}\right)\right) \in \mathbf{R}^2 \times \mathbf{R}^2 = \mathbf{R}^4.$$

Encouraged by being able to *imbed* the infinite Möbius band into $\mathbf{R}^4$, we now take a look at two other nonorientable surfaces: the Klein bottle K^2 and the real projective plane $\mathbf{R}P^2$. In Section 16 we realized K^2 by rotating a lemniscate in the same way as we rotated a line to obtain the Möbius band. Another (the plumber's) way of depicting the Klein bottle in $\mathbf{R}^3$ is given in Color Plate 11. Apart from the trunk, this picture of K^2 is obtained by cutting and pasting pieces of tori and cylinders. These pieces can actually be given by analytical formulas! Slicing K^2 along its symmetry plane (with, say, a hacksaw), we obtain two halves of the Klein bottle, one of which is shown in Color Plate 12. Taking a closer look, we see that it is a Möbius band.[4] So is the second half, and by pasting we indeed get K^2!

---

[4]Here is a limerick for the Klein bottle:

A mathematician named Klein
Thought the Möbius band was divine.
Said he, "If you glue
The edges of two
You'll get a weird bottle like mine."

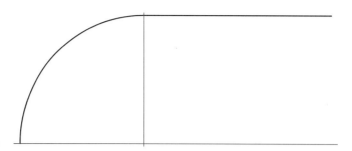

Figure 26.5

Remark.

◇ One fine point is worth noting. At the welding circles, the surface is only once differentiable. This is because if a straight line and a circle meet with the same tangent, as in Figure 26.5, then the joined curve is continuously differentiable only up to first order. (Indeed, the graph above is given by $f : [-1, \infty) \to \mathbf{R}$, $f(x) = \sqrt{1 - x^2}$, $-1 \leq x \leq 0$, and $f(x) = 1$, $x \geq 0$. Now differentiate at $x = 0$.) This lack of sufficient smoothness is a minor technical problem. As shown in differential topology, there exists an infinitely many times differentiable surface arbitrarily close to the one given.

♣ This representation of K^2 in $\mathbf{R}^3$ self-intersects in a circle, where the slim 3/4 torus penetrates into the trunk. To get rid of this self-intersection, we now imbed $K^2(\subset \mathbf{R}^3)$ into $\mathbf{R}^4$ using the inclusion $\mathbf{R}^3 \subset \mathbf{R}^4$, and then modify the surface locally around this circle so that the modified surface will not self-intersect in $\mathbf{R}^4$. For a good lower dimensional analogy, consider the intersection of the coordinate axes in $\mathbf{R}^2$; imbed $\mathbf{R}^2$ into $\mathbf{R}^3$ and "lift" one axis near the origin. The "lifting" is obtained by introducing the so-called *bump-function* $f : \mathbf{R}^2 \to \mathbf{R}$, whose graph is depicted in Figure 26.6.

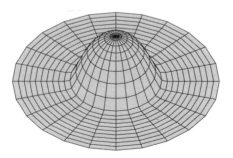

Figure 26.6

Analytically a bump is given by

$$f(p) = \begin{cases} \exp\left(\dfrac{1}{|p|^2 - 1}\right), & |p| < 1 \\ 0, & |p| \geq 1. \end{cases}$$

◇ Notice that f is infinitely many times differentiable even along the unit circle! (To show this, differentiate the exponential function at points $|p| \neq 1$, let $|p| \to 1$, and use the fact from calculus that exponential growth is faster than any polynomial growth, that is, $\lim_{x \to \infty} P(x)/e^x = 0$ for any polynomial P.)

♣ Bump functions can be "localized." The function $f_r : \mathbf{R}^2 \to \mathbf{R}$, $r > 0$, defined by $f_r(p) = f(rp)/r$, $p \in \mathbf{R}^2$, becomes "supported" on the disk $D_{1/r} = \{p \in \mathbf{R}^2 \mid |p| < 1/r\}$ (that is, $D_{1/r}$ is the locus of points on which f_r is nonzero).

It is now obvious how to steer clear the two axes of $\mathbf{R}^2$ in $\mathbf{R}^3$. Just take a bump function f and project one axis to the graph of f.

This process not only gives us a good clue but tells us what to do to avoid self-intersections of K^2 in $\mathbf{R}^4$. Just take the model of K^2 in $\mathbf{R}^3$ with a self-intersection circle assumed to be centered at the origin. Take a bump function $f : \mathbf{R}^3 \to \mathbf{R}$ (and notice that we defined f so that it immediately generalizes to any dimensions), localize it on a sufficiently small open ball (so as to avoid interaction with other parts of K^2), and project the pipe that penetrates to the trunk to the graph of f_r. We obtain K^2 in $\mathbf{R}^4$ without self-intersections! This process can be generalized to any compact surface. A fancy way of saying this is that any compact surface can be smoothly imbedded into $\mathbf{R}^4$. (Imbedding refers to no self-intersections.)

Remark.

Taking this a little further, some apparently strange phenomena emerge when dealing with objects in $\mathbf{R}^4$. For example, it is clear what we mean by two circles being "linked" in $\mathbf{R}^3$. (Recall from Section 22 that this happens to any two orbits of S^1 acting on S^3 defining the Hopf map.) Contrary to the evidence demonstrated by some magicians, it is quite obvious that these two circles cannot be unlinked by a continuous motion (homotopy) without breaking one of the circles. The exact proof of this is quite easy once we are acquainted with some elements of homotopy theory. On the other hand, if we place the circles (linked in $\mathbf{R}^3$) in $\mathbf{R}^4$, they can be unlinked by moving one circle away from the coordinate space $\mathbf{R}^3$ in the fourth direction. The popularly stated conclusion is that one cannot be chained in $\mathbf{R}^4$!

The imbedding result above applies to the real projective plane as well. Instead of repeating the previous argument, we will directly realize $\mathbf{R}P^2$—well, not quite in $\mathbf{R}^4$, but in S^4. Leaving a point out, we can use stereographic projection for S^4 to get back to $\mathbf{R}^4$. This imbedding will be defined by a map named after Veronese. We first define the *Veronese map* $f : S^2 \to S^4$ by

$$f(x, y, s) = \left(\frac{1}{\sqrt{6}}(2x^2 - y^2 - s^2), \frac{1}{\sqrt{6}}(2y^2 - x^2 - s^2), \right.$$

$$\left. \frac{1}{\sqrt{6}}(2s^2 - x^2 - y^2), \sqrt{3}xy, \sqrt{3}xs, \sqrt{3}ys \right),$$

$$(x, y, s) \in S^2 \subset \mathbf{R}^3.$$

This is actually a map from $\mathbf{R}^3$ to $\mathbf{R}^6$, but we claim that it sends the unit sphere $S^2 \subset \mathbf{R}^2$ given by $x^2 + y^2 + s^2 = 1$ to the unit sphere $S^5 \subset \mathbf{R}^6$. Indeed, we work out the sum of squares of the coordinates:

$$|f(x, y, s)|^2 = (1/6)((2x^2 - y^2 - s^2)^2 + (2y^2 - x^2 - s^2)^2$$

$$+ (2s^2 - x^2 - y^2)^2) + 3x^2y^2 + 3x^2s^2 + 3y^2s^2$$

$$= (x^2 + y^2 + s^2)^2.$$

Restricting to spheres, we thus land in S^5. Taking a closer look at the first three components of f, we see that they add up to zero. Thus the image of f lies in the 5-dimensional linear subspace of $\mathbf{R}^6$ given by the normal vector $(1, 1, 1, 0, 0, 0) \in \mathbf{R}^6$. Think of this linear subspace as a copy of $\mathbf{R}^5$. This copy of $\mathbf{R}^5$ cuts from $S^5 \subset \mathbf{R}^6$ a copy of the 4-dimensional unit sphere S^4! Thus, abusing the notation a bit, we think of the Veronese map being given by $f : S^2 \to S^4$. The components of f are homogeneous quadratic polynomials in the variables x, y, and s. In particular, $f(-x, -y, -s) = f(x, y, s)$ for all $(x, y, s) \in S^2$. This means that f takes the same value on each antipodal pair of points in S^2. It thus defines a map from the quotient $\mathbf{R}P^2 = S^2/\{\pm I\}$, so we end up with a map from $\mathbf{R}P^2$ to S^4. It is now a technical matter to check that this is an imbedding—in particular, it is one-to-one, smooth, and its inverse (from the image to $\mathbf{R}P^2$) is also smooth. As noted above, we can now apply stereographic projection to obtain $\mathbf{R}P^2$ imbedded into $\mathbf{R}^4$.

The Roman surface discussed in Section 19 can be obtained from the Veronese map as follows: Project $\mathbf{R}^6$ to the linear subspace $\mathbf{R}^3$ spanned by the last three coordinate axes. The Veronese map followed by this projection gives a map $g : S^2 \to \mathbf{R}^3$ defined by $g(x, y, s) = (xy, xs, ys)$, $(x, y, s) \in S^2 \subset \mathbf{R}^3$, where we also deleted the coefficient $\sqrt{3}$ by performing a central dilatation. Since $x^2 + y^2 + s^2 = 1$, the components $a = xy, b = xs, c = ys$ satisfy the equation

$$a^2 b^2 + b^2 c^2 + c^2 a^2 = abc$$

defining the Roman surface!

Remark.

$\diamond$ A more subtle property of the Veronese imbedding is that $\mathbf{R}P^2$ sits in S^4 as a "soap bubble," a property that we term *minimal*. This intuitively means that if we take a small piece of $\mathbf{R}P^2$ bounded by a wireframe, the soap film in S^4 stretched over the frame is given by the Veronese map. Omitting the details, we only hint that minimality is closely connected to the fact that the polynomial components of the Veronese map are harmonic.

♠ Our ambition to understand surfaces such as the Klein bottle and the real projective plane drove us right into four dimensions. The inevitable conclusion is that if we want to do decent mathematics, we have to be able to handle any dimensions. Continuing this multidimensional thought eventually leads us to twenty-first-century mathematics. Since we began these notes with the dawn of mathematical thinking, it is perhaps appropriate to finish it with a note from the nineties. Remember that in Section 17 we classified all finite subgroups of direct spatial isometries. Among these, a prominent role was played by the tetrehedral, octahedral, and icosahedral groups. They, or more appropriately, their binary double covers, can be realized as finite subgroups of quaternions in $S^3 \subset \mathbf{H}$. Based on analogy with $\mathbf{R}P^2 = S^2/\{\pm I\}$, in Section 23 we arrived at the tetrahedral $S^3/\mathbf{T}^*$, octahedral $S^3/\mathbf{O}^*$, and icosahedral $S^3/\mathbf{I}^*$ "manifolds." Can we view these in some perhaps high-dimensional spheres in the same way as we viewed $\mathbf{R}P^2$ in S^4? The answer is yes. In fact, DeTurck and Ziller[5] showed in 1992 that all these polyhedral manifolds can be imbedded minimally (as spatial soap bubbles) into spheres of large dimension.

♣ We are now getting dangerously close to the research interests of the author[6] of these Glimpses. It is time to say goodbye. I hope that you found enough motivation in the sublime beauty of the objects discussed here to carry on with your studies!

Problems

1. Show that the 4-dimensional cube ($[0, 1]^4 \subset \mathbf{R}^4$) can be sliced by 3-dimensional hyperplanes to obtain a 3-dimensional cube, a regular tetrahedron, and a regular octahedron.

2. (a) Show that each edge of the 4-dimensional cube is surrounded by three 3-dimensional cubes. Interpret the Schläfli symbol $\{4, 3, 3\}$ of the 4-dimensional cube.

   (b) Work out the number of vertices, edges, and faces of an n-dimensional cube.

3. (a) Draw several 2-dimensional projections of a 4-dimensional tetrahedron.

   (b) Show that each edge of the 4-dimensional tetrahedron is surrounded by three 3-dimensional tetrahedra. Interpret the Schläfli symbol $\{3, 3, 3\}$ of the 4-dimensional tetrahedron.

4. ♠ Let $\mathcal{H}$ denote the linear space of harmonic homogeneous quadratic polynomials in three real variables. Show that the components of the Veronese map span $\mathcal{H}$ so that $\mathcal{H}$ is 5-dimensional. (With respect to the L^2-scalar product on $\mathcal{H}$

---

[5]See "Minimal Isometric Immersions of Spherical Space Forms in Spheres," *Comment. Math. Helvetici* 67 (1992) 428–458.

[6]The author determined the modular structure of the "space" of spherical soap bubbles in any dimensions in "Eigenmaps and the Space of Minimal Immersions Between Spheres," *Indiana U. Math. J.* 46 (1997) 637–658. See also the author's new monograph *Finite Möbius Groups, Minimal Immersions of Spheres, and Moduli*, Springer, 2002.

defined by integration on S^2, the first three components of the Veronese map are the vertices of an equilateral triangle; the last three components form an orthonormal basis in the orthogonal complement of the triangle.)

5. Consider a regular polytope with Schläfli symbol $\{a, b, c\}$. Show that the vertex figures have Schläfli symbol $\{b, c\}$.

Film

T. Banchoff: *The Hypercube*: *Projections and Slicing*, Brown University; International Film Bureau Inc. (322 S. Michigan Ave., Chicago, IL 60604-4382).

No epilogue, I pray you—for your play needs no excuse.

—W. Shakespeare,
A Midsummer's Night's Dream

A Sets

In native set theory, the concept of a set is undefined. You may try to say that a set is a family or collection of objects, but these words are just synonyms. We usually denote sets by uppercase Latin letters: A, B, C, etc. If a is an element of A, we write $a \in A$. Otherwise $a \notin A$. A set A is *contained in a set* B, or A is a *subset of* B, written as $A \subset B$ if $c \in A$ implies $c \in B$. Two sets A and B are *equal*, $A = B$, if $A \subset B$ and $B \subset A$. The *Cartesian product* of two sets A and B is the set $A \times B$ consisting of pairs (a, b) with $a \in A$ and $b \in B$. The *union* and *intersection of* two sets A and B are defined by

$$A \cup B = \{c \mid c \in A \text{ or } c \in B\}$$

and

$$A \cap B = \{c \mid c \in A \text{ and } c \in B\}.$$

A *map* f from a set A to a set B, denoted by $f : A \to B$, is a rule that assigns to each element $a \in A$ an element $b \in B$. If b is assigned to a, then we write $b = f(a)$ and say that b is the value of f on a. The set A is the *domain* of f, the set B is the *range* of f, and the *image* of f is the subset $f(A) = \{b \in B \mid b = f(a) \text{ for some } a \in A\}$ of B.

The *graph* of f is the subset of $A \times B$ defined by $\{(a, f(a)) \mid a \in A\}$. The map $f : A \to B$ is *one-to-one* if $a_1 \neq a_2$ implies $f(a_1) \neq f(a_2)$. The map f is *onto* if $f(A) = B$. Finally, $f : A \to B$ is a *one-to-one correspondence* if f is one-to-one and onto. In this case the *inverse* $f^{-1} : B \to A$ of f exists, with $f(a) = b$ if $f^{-1}(b) = a$, $a \in A$, $b \in B$.

Two sets have the same *cardinality* if there is a one-to-one correspondence between them. If A is finite, $|A|$ denotes the number of elements in A. A set is *countable* if it has the same cardinality as the set of positive integers. The set of all real numbers is not countable.

If $f : A \to B$ and $g : B \to C$ are maps, then the composition $g \circ f : A \to C$ is the map given by $(g \circ f)(a) = g(f(a))$, $a \in A$. The inverse of f, if it exists, can be defined by saying that $f^{-1} \circ f = I_A$ and $f \circ f^{-1} = I_B$. Here $I_A : A \to A$ is the *identity map* on A—that is, $I_A(a) = a$, $a \in A$,—and similarly, I_B is the identity map on B.

B APPENDIX

Groups

An operation on a set A is a *map* $f : A \times A \to A$. It associates to a pair $(a_1, a_2) \in A \times A$ the element $f(a_1, a_2)$ of A. We usually write $f(a_1, a_2) = a_1 * a_2$, $a_1, a_2 \in A$, and call $*$ an *operation* on A. The operation $*$ on A is *associative* if

$$(a_1 * a_2) * a_3 = a_1 * (a_2 * a_3), \quad a_1, a_2, a_3 \in A,$$

and *commutative* if

$$a_1 * a_2 = a_2 * a_1, \quad a_1, a_2 \in A.$$

A *group* G is a set on which an associative operation $* : G \times G \to G$ is given such that

1. There exists a unique element $e \in G$, called the *identity*, such that $e * g = g * e = g$ for all $g \in G$;
2. Each element $g \in G$ possesses a unique *inverse* $g^{-1} \in G$ such that $g * g^{-1} = g^{-1} * g = e$.

A group G is called *abelian* if it is commutative. In specific groups the operation $*$ is written as addition or multiplication. Additive terminology always assumes that the group is Abelian.

If a group G is finite, then the number of elements $|G|$ in G is called the *order* of G.

$H \subset G$ is a *subgroup* if the group operation $*$ on G restricts to H, and defines a group structure on H.

The intersection of two subgroups is always a subgroup. Given two groups G and H the Cartesian product $G \times H$ carries a group structure. In fact (with obvious notations) we define $(g_1, h_1) * (g_2, h_2) = (g_1 * g_2, h_1 * h_2)$, $g_1, g_2 \in G$, $h_1, h_2 \in H$.

A *homomorphism* $f : G \to H$ between groups is a map of the underlying sets such that

$$f(g_1 * g_2) = f(g_1) * f(g_2), \quad g_1, g_2 \in G.$$

The image of a homomorphism $f : G \to H$ is always a subgroup of H. The *kernel* of $f : G \to H$ is the subgroup defined by $\ker(f) = \{g \in G \mid f(g) = e\}$. A homomorphism is one-to-one iff its kernel is trivial (that is, consists of the identity element alone). A homomorphism that is a one-to-one correspondence between the underlying sets is called an *isomorphism*. Two groups G and H are called *isomorphic*, written as $G \cong H$, if there is an isomorphism $f : G \to H$. The composition of homomorphisms is a homomorphism. The inverse of an isomorphism is an isomorphism.

Given a group G and a subset $\Gamma \subset G$, the smallest subgroup of G that contains Γ is called the *subgroup generated by* Γ; it is denoted by $\langle \Gamma \rangle$. In particular, if $\langle \Gamma \rangle = G$, then G is generated by Γ; if Γ is finite, we say that G is *finitely generated*.

If $g \in G$, the subgroup $\langle g \rangle$ consists of all integral powers of g, that is, $\langle g \rangle = \{e, g^{\pm 1}, g^{\pm 2}, \ldots\}$ ($g^0 = e$). This is called the *cyclic subgroup generated by g*. If all integral powers of g are distinct, then $\langle g \rangle$ is called *infinite cyclic*. If two powers of g with distinct exponents coincide, then there is a least positive integer n such that $g^n = e$. In this case g is said to have order n, and $\langle g \rangle$ consists of n elements: $\langle g \rangle = \{e, g, \ldots, g^{n-1}\}$. We say that $\langle g \rangle$ is *cyclic of order n*. Two cyclic groups are isomorphic iff their underlying sets have the same cardinality. Infinite cyclic groups are denoted by the symbol C_∞, and finite cyclic groups of order n are denoted by C_n.

If G is an abelian group, then the elements of finite order in G form a subgroup called the *torsion subgroup* G_{tor}. If, in addition, G

is finitely generated, then G_{tor} is finite, and we have

$$G \cong G_{\text{tor}} \times C_\infty^r,$$

where r is called the *rank* of G.

A *permutation* on a set $A = \{a_1, \ldots, a_n\}$ is a one-to-one correspondence $f : A \rightarrow A$. A permutation is usually given as

$$\begin{pmatrix} a_1 & \ldots & a_n \\ f(a_1) & \ldots & f(a_n) \end{pmatrix}$$

We usually assume that $A = \{1, 2, \ldots, n\}$. The set of all permutations form a group $\mathcal{S}_n$, called the *symmetric group on n letters*. We have $|\mathcal{S}_n| = n!$ A subgroup of $\mathcal{S}_n$ is called a *permutation group*. Every finite group is isomorphic to a permutation group. A permutation in $\mathcal{S}_n$ is called a *transposition* if it switches two elements in $\{1, 2, \ldots, n\}$ and leaves the other elements fixed. Every permutation can be written as the composition of finitely many transpositions. This decomposition is not unique, but the parity (even or odd) of the number of transpositions occuring in the composition is. We say that a permutation is *even* if it can be written as a composition of an even number of transpositions. Otherwise the permutation is odd. The alternating group $\mathcal{A}_n$, $n \geq 2$, is the subgroup of $\mathcal{S}_n$ that consists of all even permutations. We have $|\mathcal{A}_n| = n!/2$.

Given a group G and a subgroup H, the *left-coset* of H by $g \in G$ is the subset $gH = \{gh \mid h \in H\}$. The set of left-cosets gives a partition of the underlying subset of G. The quotient G/H of G by H is the set of all left-cosets:

$$G/H = \{gH \mid g \in G\}.$$

The number of distinct left-cosets is the *index of H in G* denoted by $[G : H] = |G/H|$. (This notation usually assumes that G/H is finite.) G/H is a group under the multiplication $(g_1H)(g_2H) = (g_1g_2)H$, $g_1, g_2 \in G$, iff H is a normal subgroup in G; that is, iff $gHg^{-1} = H$ for all $g \in G$. The kernel of a homomorphism $f : G \rightarrow H$ is always normal, and the quotient group $G/\ker(f)$ is isomorphic to the image of f. A subgroup of index 2 (such as $\mathcal{A}_n$ in $\mathcal{S}_n$) is always normal, and the quotient is C_2.

A group is *simple* if it has no normal subgroups other than itself and the trivial subgroup (consisting of the identity element alone). The alternating group $\mathcal{A}_n$ is simple for $n \geq 5$.

A group G is said to *act on a set A* if, to each element $g \in G$, there is associated a one-to-one correspondence $f_g : A \rightarrow A$. In addition, we require that $f_e = I_A$, $f_{g^{-1}} = (f_g)^{-1}$, $g \in G$, and $f_{g_1 g_2} = f_{g_1} \circ f_{g_2}$, $g_1, g_2 \in G$. Given an action of G on A, we usually write g instead of f_g. If G acts on A, then the *orbit* through $a \in A$ is the subset

$$G(a) = \{g(a) \mid g \in G\} \subset A.$$

The *isotropy* at $a \in A$ is the subgroup

$$G_a = \{g \in G \mid g(a) = a\}.$$

The map that sends the left-coset gG_a, $g \in G$, to $g(a)$ establishes a one-to-one correspondence between the quotient G/G_a and the orbit $G(a)$. G acts *transitively* on $G(a)$ in the sense that any two points in the orbit $G(a)$ can be carried into each other by suitable elements of G.

C APPENDIX

Topology

Given a set X, a topology on X is a family $\mathcal{T} = \{U_\alpha \subset X \mid \alpha \in A\}$ of subsets of X indexed by a set A with the following properties:

1. $\cup_{\alpha \in B} U_\alpha \in \mathcal{T}$ for any $B \subset A$;
2. $U_\alpha \cap U_\beta \in \mathcal{T}$ for any $\alpha, \beta \in A$;
3. $\emptyset, X \in \mathcal{T}$.

The elements of a topology on X are called *open subsets* of X. If a topology $\mathcal{T}$ is given on X, then X is said to be a *topological space* (with $\mathcal{T}$ suppressed). A topological space X is *Hausdorff* if for any two distinct points $x_1, x_2 \in X$ there exist disjoint open sets U_1 and U_2 such that $x_1 \in U_1$ and $x_2 \in U_2$. (The Hausdorff property is sometimes included in the definition of topology.)

Given a set X, a *distance function* on X is a map $d : X \times X \to \mathbf{R}$ such that

1. $d(x_1, x_2) \geq 0$, $x_1, x_2 \in X$, and equality holds iff $x_1 = x_2$;
2. $d(x_1, x_2) = d(x_2, x_1)$, $x_1, x_2 \in X$;
3. $d(x_1, x_2) + d(x_2, x_3) \geq d(x_1, x_3)$, $x_1, x_2, x_3 \in X$ (triangle inequality).

A set X with a distance function d is called a *metric space*. Given a metric space X with a distance function d, the metric ball with center at $x_0 \in X$ and radius $r > 0$ is defined as

$$B_r(x_0) = \{x \in X \mid d(x, x_0) < r\}.$$

A subset $U \subset X$ is called *open* if for every $x_0 \in U$ there exists $\varepsilon > 0$ such that $B_\varepsilon(x_0) \subset U$. In other words, the open sets are unions of metric balls. The triangle inequality ensures that metric balls are open. The open sets form a topology $\mathcal{T}_d$ on X called the metric topology. The *metric topology* is always Hausdorff.

In the set $\mathbf{R}$ of all real numbers, the Euclidean distance between two real numbers $x_1, x_2 \in \mathbf{R}$ is $d(x_1, x_2) = |x_1 - x_2|$. This defines the Euclidean (metric) topology on $\mathbf{R}$. An open set in $\mathbf{R}$ is nothing but the union of (countably many) open intervals in $\mathbf{R}$. Similarly, in the n-dimensional number space $\mathbf{R}^n$ the Euclidean distance defines the Euclidean topology in $\mathbf{R}^n$.

Given a topological space X with topology $\mathcal{T}$, a subset $Y \subset X$ inherits a topology from $\mathcal{T}$ by declaring that the open sets in Y are intersections of the elements in $\mathcal{T}$ and Y. We say that Y *carries the subspace topology*.

Given two topological spaces X and Y, the *Cartesian product $X \times Y$* carries a topology by declaring that the open sets in $X \times Y$ are unions of sets $U \times V$, where U is open in X and V is open in Y.

A subset $Y \subset X$ of a topological space X with topology $\mathcal{T}$ is called *closed* if its complement $X - Y$ is open, that is, $X - Y \in \mathcal{T}$.

Given a subset $Y \subset X$ of a topological space X, the *closure $\bar{Y}$* of Y is the intersection of all closed sets in X that contain Y. The closure $\bar{Y}$ of Y is the smallest closed set that contains Y. In the Euclidean topology, the closure of an open interval $(a, b) \subset \mathbf{R}$ is the closed interval $[a, b] \subset \mathbf{R}$.

A map $f : X \to Y$ between topological spaces is *continuous* at $x_0 \in X$ if for every open set V in Y that contains $f(x_0)$ there exists an open set U in X containing x_0 such that $f(U) \subset V$. This generalizes the usual Cauchy definition of continuity of real functions. A function $f : X \to Y$ is continuous if it is continuous at every point of X. $f : X \to Y$ is continuous iff the inverse image $f^{-1}(V)$ of any open set V in Y is open in X. A one-to-one correspondence $f : X \to Y$ between topological spaces such that f and f^{-1} are continuous is called a *homeomorphism*. Two topological spaces are homeomorphic if there is a homeomorphism between them.

Given a topological space X and a partition $\mathcal{P}$ of X into mutually disjoint subsets $\{X_\alpha \mid \alpha \in A\}$, the family $X/\mathcal{P}$ of these disjoint subsets can be made a topological space by requiring that a subset V of $X/\mathcal{P}$ be open if the union U of X_α's that participate in V is open in X. This definition is used to define "pasting" topological spaces as follows: Given two topological spaces X and Y and a homeomorphism $f : X_1 \to Y_1$ between subspaces $X_1 \subset X$ and $Y_1 \subset Y$, pasting X and Y along f is the quotient $X \cup_f Y = (X \cup Y)/\mathcal{P}_f$, where $\mathcal{P}_f$ is a partition of $X \cup Y$ defined as follows: Any point in $(X - X_1) \cup (Y - Y_1)$ is a one-element subset of the partition $\mathcal{P}_f$ and, for $x \in X_1$, $\{x, f(x)\}$ is a two-element subset in $\mathcal{P}_f$. (In $(X \cup Y)/\mathcal{P}_f$, the points that are "pasted together" are x and $f(x)$, $x \in X_1$.)

A topological space X is *compact* if from any covering of X by open subsets, finitely many members can be extracted that still cover X. A closed subspace of a compact Hausdorff space is compact. The continuous image of a compact topological space is compact. $\mathbf{R}$ with respect to the Euclidean topology is noncompact. A subspace $X \subset \mathbf{R}^n$ is compact iff X is closed and contained in a metric ball.

D

Smooth Maps

A real-valued function $f : U \to \mathbf{R}$ of an open set U in $\mathbf{R}^n$ is said to be *smooth* if f possesses continuous partial derivatives up to any order on U. A map $f : U \to \mathbf{R}^m$ is smooth if each component of f is smooth. Given any subsets $X \subset \mathbf{R}^n$ and $Y \subset \mathbf{R}^m$, a map $f : X \to Y$ is smooth if there exists an open set $U \subset \mathbf{R}^n$ covering X and a smooth extension of $\tilde{f} : U \to \mathbf{R}^m$ of f, that is $\tilde{f}|X = f$.

A smooth map $f : X \to Y$ between subsets $X \subset \mathbf{R}^n$ and $Y \subset \mathbf{R}^m$ is a *diffeomorphism* if f is a one-to-one correspondence and $f^{-1} : Y \to X$ is also smooth. $X \subset \mathbf{R}^n$ and $Y \subset \mathbf{R}^m$ are diffeomorphic if there is a diffeomorphism $f : X \to Y$. In this case $m = n$.

Given a smooth map $f : U \to V$ between open sets $U \subset \mathbf{R}^n$ and $V \subset \mathbf{R}^m$, the *Jacobi matrix* of f at x_0 is the $m \times n$ matrix given by $\mathrm{Jac}\,(f)_{x_0} = (\partial f^i / \partial x^j)_{i=1, j=1}^{m,n}$, where the partial derivatives are evaluated at x_0. Multiplication by $\mathrm{Jac}\,(f)_{x_0}$ gives a linear map $(df)_{x_0} : \mathbf{R}^n \to \mathbf{R}^m$ called the *differential of f at x_0*. If $\mathrm{Jac}\,(f)_{x_0}$ is non-singular, then $m = n$ and f is a local diffeomorphism at x_0. This means that there exists an open set $U_0 \subset U$ covering x_0 and an open set $V_0 \subset V$ covering $f(x_0)$ such that the restriction $f|U_0$ is a diffeomorphism of U_0 to V_0. The determinant of the Jacobi matrix of f at x_0 is called the *Jacobian of f at x_0*. If it is positive, then we say that f is *orientation preserving* at x_0. If it is negative, we say that f is *orientation reversing* at x_0. If Jac (f) is nonsingular on a connected open set U, then either f is orientation preserving at every point of U or f is orientation reversing at every point of U. In this case we simply say that f is orientation preserving or orientation reversing.

In particular, if $f : U \to \mathbf{C}$ is a complex function defined on an open set $U \subset \mathbf{C}$, and at $z_0 \in U$ the complex derivative $f'(z_0)$ is nonzero, then Jac $(f)_{z_0}$ is nonsingular and f is a local diffeomorphism at z_0. Moreover, the Jacobian of f at z_0 is $|f'(z_0)|^2$, so f is orientation preserving at z_0.

Composition of smooth maps corresponds to composition of their differentials, which in turn corresponds to the product of their Jacobi matrices. By the product theorem for determinants, composition follows the usual parity rule; for example, the composition of two orientation-reversing diffeomorphisms is orientation preserving.

The Hypergeometric Differential Equation and the Schwarzian

APPENDIX E

We summarize[1] here some facts on the solutions of *second-order homogeneous linear differential equations* of the form

$$z'' = p(w)z' + q(w),$$

where p and q are *rational functions* of the complex variable $w \in \mathbf{C}$. A point w_0 is said to be an *ordinary point* of the differential equation if p and q have removable singularities at w_0, i.e., if w_0 is not a pole for p and q. This concept is also extended to $w_0 = \infty$ as follows. The substitution $\tilde{w} = 1/w$ interchanges 0 and ∞, and since

$$\frac{dz}{dw} = -\tilde{w}^2 \frac{dz}{d\tilde{w}},$$

$$\frac{d^2 z}{dw^2} = 2\tilde{w}^3 \frac{dz}{d\tilde{w}} + \tilde{w}^4 \frac{d^2 z}{d\tilde{w}^2},$$

the transformed differential equation has the form

$$\frac{d^2 z}{d\tilde{w}^2} = -\left(\frac{2}{\tilde{w}} + \frac{1}{\tilde{w}^2} p\left(\frac{1}{\tilde{w}} \right) \right) \frac{dz}{d\tilde{w}} + \frac{1}{\tilde{w}^4} q\left(\frac{1}{\tilde{w}} \right) z.$$

---

[1] For more information, see L. Ahlfors, *Complex Analysis*, McGraw-Hill, Inc., 1979.

We say that $w_0 = \infty$ is an ordinary point of the original equation if $\tilde{w}_0 = 0$ is an ordinary point for the transformed equation. This is the case when $-(2w + w^2 p(w))$ and $w^4 q(w)$ have removable singularities at $w_0 = \infty$.

If w_0 is an ordinary point of our differential equation, then p and q are analytic at w_0, and hence there exists a local solution z (defined on an open neighborhood of w_0) with arbitrarily prescribed values $z(w_0) = z_0$ and $z'(w_0) = z_0'$. The local solution z is analytic at w_0, and its germ at w_0 is uniquely determined. (Indeed, a simple induction in the use of the Cauchy inequality shows that a formal power series expansion $z = \sum_{n=0}^{\infty} a_n (w - w_0)^n$ has positive radius of convergence.) Analytic continuation along curves that avoid the poles of p and q extends any local solution to a global solution. The global solution is, in general, multiple-valued.

We now begin to study the behavior of the solutions of our differential equation near a singular point w_0. We first assume that w_0 is the simplest possible singularity in the sense that both p and q have only simple poles at w_0. For simplicity, we assume that $w_0 = 0$. (This can always be attained by replacing $w - w_0$ with w.) We write the Laurent expansions of p and q at $w_0 = 0$ as follows:

$$p(w) = \frac{p_{-1}}{w} + p_0 + p_1 w + \cdots,$$

$$q(w) = \frac{q_{-1}}{w} + q_0 + q_1 w + \cdots.$$

Once again, if we substitute the formal power series $z = \sum_{n=0}^{\infty} a_n w^n$ into the differential equation, we find that only a_0 can be chosen arbitrarily, and the recurrence formula for the coefficients a_n works only under the condition that $p_{-1} \notin \{0, 1, 2, \ldots\}$. With this restriction there is always an analytic solution in a neighborhood of w_0.

The most general case for which this process gives an explicit solution is that in which w_0 is a *regular singular point* of the differential equation, i.e., when p has at most a simple pole at w_0, and q has at most a double pole at w_0. Assume that w_0 is a regular singular point. We seek a solution of the form $z = (w - w_0)^\alpha g(w)$, where g is analytic and nonzero at w_0. Once again, setting $w_0 = 0$

and substituting, we find that g satisfies the differential equation

$$g'' = \left(p(w) - \frac{2\alpha}{w}\right)g' + \left(q(w) + \frac{\alpha p(w)}{w} - \frac{\alpha(\alpha - 1)}{w^2}\right)g.$$

If we choose α to satisfy the *indicial equation*

$$\alpha(\alpha - 1) = p_{-1}\alpha + q_{-2},$$

where

$$q(w) = \frac{q_{-2}}{w^2} + \frac{q_{-1}}{w} + q_0 + q_1 w + \cdots,$$

then in the differential equation for g, the coefficients have at most simple poles at $w_0 = 0$, so that the previous discussion applies. We obtain that the original differential equation has a solution of the form $z = w^\alpha g(w)$, where g is analytic and nonzero at $w_0 = 0$, provided that $p_{-1} - 2\alpha \notin \{0, 1, \ldots\}$. Here α is one of the two *indicial roots* α_1, α_2, solutions of the indicial equation. Since $\alpha_1 + \alpha_2 = p_{-1} + 1$, the condition for the existence of a solution $z = w^\alpha g(z)$ as above for $\alpha = \alpha_1$ is equivalent to $\alpha_2 - \alpha_1 \notin \{1, 2, \ldots\}$. Switching the roles of α_1 and α_2, we see that our differential equation has two linearly independent solutions of the form $z_1 = w^{\alpha_1}g_1(w)$ and $z_2 = w^{\alpha_2}g_2(w)$, provided that $\alpha_1 - \alpha_2$ is not an integer.

Theorem.

If w_0 is a regular singular point of the differential equation

$$z'' = p(w)w' + q(w),$$

then there exist linearly independent solutions of the form

$$(w - w_0)^{\alpha_1}g_1(w) \quad \text{and} \quad (w - w_0)^{\alpha_2}g_2(w),$$

where α_1, α_2 are the solutions of the indicial equation

$$\alpha(\alpha - 1) = p_{-1}\alpha + q_{-2},$$

and g_1, g_2 are analytic and nonzero at w_0, provided that $\alpha_1 - \alpha_2$ is not an integer. If $\alpha_2 - \alpha_1$ is a nonnegative integer, then there is always one solution of the form $(w - w_0)^{\alpha_1}g_1(w)$, where g_1 is analytic and nonzero at w_0.

A simple analysis shows that our differential equation with one or two regular singular points (and all other points ordinary) can be explicitly solved. To get something new, we thus assume that we have three regular singular points. Since a linear fractional transformation does not change the structure of the differential equation (including the character of the singularities), we may assume that the three regular singular points are 0, 1, and ∞ (in fact, by the results of Section 12, any three points in the extended complex plane $\hat{\mathbf{C}}$ can be mapped to 0, 1, and ∞ by a suitable linear fractional transformation). Since 0 and 1 are regular singular points (and all other points are ordinary), we must have

$$p(w) = \frac{A}{w} + \frac{B}{w-1} + P(w),$$

$$q(w) = \frac{C}{w^2} + \frac{D}{w} + \frac{E}{(w-1)^2} + \frac{F}{w-1} + Q(w),$$

where P and Q are polynomials. Since the singularity at ∞ is also regular, $2w + w^2 p(w)$ must have at most a simple pole at ∞, and $w^4 q(w)$ must have at most a double pole at ∞. The former condition amounts to the vanishing of P, the latter to the vanishing of Q and $D + F$. We obtain

$$p(w) = \frac{A}{w} + \frac{B}{w-1},$$

$$q(w) = \frac{C}{w^2} - \frac{D}{w(w-1)} + \frac{E}{(w-1)^2}.$$

We now bring in the indicial equations. At $w_0 = 0$ the indicial equation is

$$\alpha(\alpha - 1) = A\alpha + C.$$

Letting α_1, α_2 denote the indicial roots, we have $A = \alpha_1 + \alpha_2 - 1$ and $C = -\alpha_1\alpha_2$. Similarly, at $w_0 = 1$, the indicial equation is

$$\beta(\beta - 1) = B\beta + E.$$

If β_1, β_2 are the roots, then $B = \beta_1 + \beta_2 - 1$ and $E = -\beta_1\beta_2$. Finally, since at $w_0 = \infty$ the leading coefficients of $-2w - w^2 p(w)$ and $w^4 q(w)$ are $-(2 + A + B)$ and $C - D + E$, the indicial equation at $w_0 = \infty$ is

$$\gamma(\gamma - 1) = -(2 + A + B)\gamma + C - D + E.$$

If γ_1, γ_2 are the roots, then we have $\gamma_1 + \gamma_2 = -A - B - 1$ and $\gamma_1 \gamma_2 = -C + D - E$. Expressing A, B, C, D, E in terms of the six roots, our differential equation takes the form

$$z'' + \left(\frac{1 - \alpha_1 - \alpha_2}{w} + \frac{1 - \beta_1 - \beta_2}{w - 1} \right) z'$$

$$+ \left(\frac{\alpha_1 \alpha_2}{w^2} - \frac{\alpha_1 \alpha_2 + \beta_1 \beta_2 - \gamma_1 \gamma_2}{w(w - 1)} + \frac{\beta_1 \beta_2}{(w - 1)^2} \right) z = 0,$$

and

$$\alpha_1 + \alpha_2 + \beta_1 + \beta_2 + \gamma_1 + \gamma_2 = 1.$$

This is called the *hypergeometric differential equation*. The solutions of the hypergeometric differential equation are called *hypergeometric functions*. By construction, the hypergeometric differential equation is the canonical form of a homogeneous second-order differential equation with exactly three regular singular points (and ordinary points elsewhere). The hypergeometric differential equation is often written in the equivalent form

$$z'' - \frac{z'}{w(w - 1)} \left((1 - \alpha_1 - \alpha_2) - (1 + \gamma_1 + \gamma_2)w \right)$$

$$+ \frac{z}{w^2(w - 1)^2} \left(\alpha_1 \alpha_2 - (\alpha_1 \alpha_2 - \beta_1 \beta_2 + \gamma_1 \gamma_2)w + \gamma_1 \gamma_2 w^2 \right) = 0.$$

From now on, we assume that none of the differences $\alpha_1 - \alpha_2$, $\beta_1 - \beta_2$, $\gamma_1 - \gamma_2$ is an integer. To simplify and at the same time solve the hypergeometric differential equation, we recall that the substitution $w^\alpha g(w)$ determines for g a similar differential equation. As a simple computation shows, the indicial roots of the differential equation for g are $\alpha_1 - \alpha$ and $\alpha_2 - \alpha$, and the indicial roots at ∞ are $\gamma_1 + \alpha$ and $\gamma_2 + \alpha$. Introducing (somewhat prematurely) Riemann's notation, we express this as

$$P\left\{\begin{matrix} 0 & 1 & \infty & \\ \alpha_1 & \beta_1 & \gamma_1 \,, & w \\ \alpha_2 & \beta_2 & \gamma_2 & \end{matrix}\right\} = z^\alpha P\left\{\begin{matrix} 0 & 1 & \infty & \\ \alpha_1 - \alpha & \beta_1 & \gamma_1 + \alpha \,, & w \\ \alpha_2 - \alpha & \beta_2 & \gamma_2 + \alpha & \end{matrix}\right\}.$$

Here the Riemann P-function on the left-hand side stands for *all* locally defined solutions of our differential equation, and the equality means that any local solution is equal to a solution of the differential equation for g multiplied by z^α, and the indicial roots undergo the changes as indicated. (We are not going into the proper interpretation of relations like this; in particular, we will not discuss unicity in any detail. We find, however, this notation undeniably suitable for tabulating the indicial roots.) We can also separate the factor $(w - 1)^\beta$ from g and obtain

$$P\left\{\begin{matrix} 0 & 1 & \infty & \\ \alpha_1 & \beta_1 & \gamma_1 \,, & w \\ \alpha_2 & \beta_2 & \gamma_2 & \end{matrix}\right\}$$

$$= w^\alpha (w - 1)^\beta P\left\{\begin{matrix} 0 & 1 & \infty & \\ \alpha_1 - \alpha & \beta_1 - \beta & \gamma_1 + \alpha + \beta \,, & w \\ \alpha_2 - \alpha & \beta_2 - \beta & \gamma_2 + \alpha + \beta & \end{matrix}\right\}.$$

Up to this point, α and β did not have any preassigned values. We now set $\alpha = \alpha_1$ and $\beta = \beta_1$. After separating the factor $w^{\alpha_1}(w-1)^{\beta_1}$, the remaining functions

$$P\left\{\begin{matrix} 0 & 1 & \infty & \\ 0 & 0 & \gamma_1 + \alpha_1 + \beta_1 \,, & w \\ \alpha_2 - \alpha_1 & \beta_2 - \beta_1 & \gamma_2 + \alpha_1 + \beta_1 & \end{matrix}\right\}$$

satisfy the hypergeometric differential equation

$$z'' + \left(\frac{c}{w} + \frac{1 - c + a + b}{w - 1}\right) z' + \frac{ab}{w(w - 1)} z = 0.$$

Here we used classical notation

$$a = \alpha_1 + \beta_1 + \gamma_1, \qquad b = \alpha_1 + \beta_1 + \gamma_2, \qquad c = 1 + \alpha_1 - \alpha_2,$$

noting also that

$$c - a - b = \beta_2 - \beta_1.$$

Multiplying out, we can write this differential equation as

$$w(1 - w)z'' + (c - (a + b + 1)w)z' - abz = 0.$$

This equation has a solution in the form $z = \sum_{n=0}^{\infty} A_n w^n$. In fact, substituting we find an embarrassingly simple recurrence relation that, up to a constant multiple, gives the solution

$$F(a, b, c, w) = 1 + \frac{a \cdot b}{1 \cdot c} w + \frac{a(a + 1) \cdot b(b + 1)}{1 \cdot 2 \cdot c(c + 1)} w^2$$

$$+ \frac{a(a + 1)(a + 2) \cdot b(b + 1)(b + 2)}{1 \cdot 2 \cdot 3 \cdot c(c + 1)(c + 2)} w^3 + \cdots.$$

In a similar vein, setting $\alpha = \alpha_2$ and $\beta = \beta_1$, after separating the factor $w^{\alpha_2}(w - 1)^{\beta_1}$, the remaining functions are

$$P \left\{ \begin{array}{ccc} 0 & 1 & \infty \\ \alpha_1 - \alpha_2 & 0 & \gamma_1 + \alpha_2 + \beta_1, \\ 0 & \beta_2 - \beta_1 & \gamma_2 + \alpha_2 + \beta_1 \end{array} \quad w \right\}.$$

Rewriting this in terms of the parameters a, b, c, we find that $w^{1-c} F(1 + a - c, 1 + b - c, 2 - c, w)$ is another solution linearly independent of $F(a, b, c, w)$. (The factor $w^{1-c} = w^{\alpha_2 - \alpha_1}$ makes up the quotient $w^{\alpha_2}(w - 1)^{\beta_1} / w^{\alpha_1}(w - 1)^{\beta_1}$.)

Linearly independent solutions near 1 can be found using the identity

$$P \left\{ \begin{array}{ccc} 0 & 1 & \infty \\ \alpha_1 & \beta_1 & \gamma_1, \\ \alpha_2 & \beta_2 & \gamma_2 \end{array} \quad w \right\} = P \left\{ \begin{array}{ccc} 0 & 1 & \infty \\ \beta_1 & \alpha_1 & \gamma_1, \\ \beta_2 & \alpha_2 & \gamma_2 \end{array} \quad 1 - w \right\}.$$

Solutions near ∞ can be written down in an analogous way.

Summarizing, we find that up to various multiplicative powers of the independent variable (depending on the indicial roots at 0, 1, and ∞), all solutions of the hypergeometric differential equation can be written in terms of the hypergeometric functions F.

We now return to the general setting and consider again our differential equation

$$z'' = p(w)z' + q(w)z,$$

where p and q are rational functions. Given two linearly independent but possibly multiple-valued solutions z_1 and z_2, any solution of the differential equation can be written as a linear combination of z_1 and z_2. Consider the ratio $z = z_1/z_2$. (We are overusing the symbol z a little here. Whether z stands for the ratio z_1/z_2 or for the general variable in our differential equation will be clear in the text.) If instead of z_1 and z_2 we choose a different pair of linearly independent solutions, then the corresponding ratio will be a linear fractional transformation applied to z. Indeed, if $\tilde{z}_1 = az_1 + bz_2$ and $\tilde{z}_2 = cz_1 + dz_2$ ($a, b, c, d \in \mathbf{C}$, $ad - bc = 1$), then $\tilde{z} = \tilde{z}_1/\tilde{z}_2 = (az + b)/(cz + d)$.

Remark.

Considering closed paths based at a fixed point (and away from the poles of p and q), analytic continuation of solutions along these paths give rise to the concept of *monodromy*, a homomorphism of the fundamental group of $\hat{\mathbf{C}}$ minus the poles of p and q into $SL(2, \mathbf{C})$. In 1873 Schwarz classified all hypergeometric differential equations with finite monodromy.

This motivates us to seek a differential expression $\mathcal{S}$ that remains invariant under all linear fractional transformations. Among the four quantities a, b, c, d there is a relation $ad - bc = 1$, so that $\mathcal{S}$ must be a third-order differential operator. An explicit expression for $\mathcal{S}$ can be derived in complete generality as follows. We let z be an analytic function and $\tilde{z} = (az + b)/(cz + d)$. Suppressing the variable $w \in \mathbf{C}$, we write this as

$$cz\tilde{z} + d\tilde{z} - az - b = 0.$$

Differentiating, we obtain

$$c(z'\tilde{z} + z\tilde{z}') + d\tilde{z}' - az' = 0,$$

$$c(z''\tilde{z} + 2z'\tilde{z}' + z\tilde{z}'') + d\tilde{z}'' - az'' = 0,$$

$$c(z'''\tilde{z} + 3z''\tilde{z}' + 3z'\tilde{z}'' + z\tilde{z}''') + d\tilde{z}''' - az''' = 0.$$

We view these three equations as a linear system for a, c, and d. Its determinant, after canceling the highest-order derivative mixed terms by column operations, is

$$-\begin{vmatrix} 0 & \tilde{z}' & z' \\ 2z'\tilde{z}' & \tilde{z}'' & z'' \\ 3z''\tilde{z}' + 3z'\tilde{z}'' & \tilde{z}''' & z''' \end{vmatrix}.$$

This determinant must vanish, since $ad - bc = 1$. Expanding and grouping similar terms together, we obtain

$$\left(\frac{\tilde{z}''}{\tilde{z}'}\right)' - \frac{1}{2}\left(\frac{\tilde{z}''}{\tilde{z}'}\right)^2 = \left(\frac{z''}{z'}\right)' - \frac{1}{2}\left(\frac{z''}{z'}\right)^2.$$

This tells us that $\mathcal{S}$ should be defined as

$$\mathcal{S}(z) = \left(\frac{z''}{z'}\right)' - \frac{1}{2}\left(\frac{z''}{z'}\right)^2.$$

We call $\mathcal{S}(z)$ the *Schwarzian* of z. By our computations, the Schwarzian is invariant under linear fractional transformations. Conversely, if the Schwarzians of two analytic functions coincide, then the two functions differ by a linear fractional transformation. This follows from unicity of solutions of third-order differential equations.

We now return to the main line. If z_1 and z_2 are linearly independent solutions of our differential equation, then the Schwarzian $\mathcal{S}(z)$ of the ratio $z = z_1/z_2$ will not depend on the particular choice of z_1 and z_2, but only on the coefficients p and q. To work out this dependence explicitly, we fix a choice for z_1 and z_2. Since they both satisfy our differential equation, we have

$$z_1''z_2 - z_1z_2'' = p(w)(z_1'z_2 - z_1z_2').$$

Using this, we work out the ingredients of the Schwarzian derivative of z. We have

$$z' = \frac{z_1'z_2 - z_1z_2'}{z_2^2}.$$

Logarithmic differentiation gives

$$\frac{z''}{z'} = (\ln z')' = (\ln(z_1'z_2 - z_1z_2'))' - 2(\ln z_2)' = \frac{z_1''z_2 - z_1z_2''}{z_1'z_2 - z_1z_2'} - 2\frac{z_2'}{z_2}.$$

Combining this with the above, we obtain

$$\frac{z''}{z'} = p - 2\frac{z_2'}{z_2}.$$

Finally, we compute

$$\mathcal{S}(z) = \left(\frac{z''}{z'}\right)' - \frac{1}{2}\left(\frac{z''}{z'}\right)^2 = -\left(-p + 2\frac{z_2'}{z_2}\right)' - \frac{1}{2}\left(-p + 2\frac{z_2'}{z_2}\right)^2$$

$$= p' - \frac{1}{2}p^2 - 2\frac{z_2'' - pz_2'}{z_2} = p' - \frac{1}{2}p^2 - 2q.$$

Summarizing, we obtain that if z_1 and z_2 are linearly independent solutions of our differential equation, then $z = z_1/z_2$ has Schwarzian derivative

$$\mathcal{S}(z) = p' - \frac{1}{2}p^2 - 2q.$$

We see that to every third-order differential equation of the form

$$\mathcal{S}(z) = s,$$

with s a rational function, there corresponds a homogeneous second-order linear differential equation via

$$s = p' - \frac{1}{2}p^2 - 2q.$$

The rational function s obviously does not determine the coefficients p and q uniquely.

F

Galois Theory

We review here some basic ingredients of Galois theory.[1] A *field K* is a set equipped with two operations: $+$ (addition) and $\times$ (multiplication), such that K is an abelian group with respect to the addition, $K - \{0\}$ is an abelian group with respect to the multiplication, and the distributive law holds in K. The *prime subfield* of a field K is the intersection of all subfields of K. The prime subfield is either isomorphic to the field $\mathbf{Q}$ of rational numbers, or the field $\mathbf{Z}_p$, for some prime p. In the former case, we say that K has *zero characteristic*, in the latter that K has *characteristic p*. All fields considered here will be of characteristic zero.

Given a *field extension* $k \subset K$ (k is a subfield of K), written as K/k, the *degree* $[K : k]$ of K/k is the dimension of K as a linear space over k. If $k \subset L \subset K$ is a chain of field extensions, then $[K : k] = [K : L][L : k]$. We say that the field extension K/k is *finite* if $[K : k]$ is finite. For example, a field extension K/k is *quadratic* if $[K : k] = 2$. A quadratic field extension K/k can be obtained by adjoining a square root $\sqrt{\delta}$ to k, where $\delta \in k$ but $\sqrt{\delta} \notin k$; we write the extension as $K = k(\sqrt{\delta})$. A *k-automorphism* of K is an automorphism of K that

---

[1] For more details, see M. Artin, *Algebra*, Prentice Hall, 1991, or I. Stewart, *Galois Theory*, Chapman and Hall, 1973.

fixes k. The *Galois group* $G = G(K/k)$ of the extension K/k is the group of k-automorphisms of K. If K/k is a finite field extension, then the Galois group G is finite. For any finite field extension K/k, the order $|G|$ of G divides $[K : k]$. We call K/k a *Galois extension* if $|G| = [K : k]$. If K/k is Galois then the fixed field K^G is equal to k. Conversely, if G is a finite group of automorphisms of a field K and $k = K^G$ denotes its fixed field, then K/k is a Galois extension with Galois group G. Given a Galois extension K/k with Galois group G, there is a one-to-one correspondence between the subgroups H of G and the intermediate fields $k \subset L \subset K$. The correspondence is given by $L = K^H$ and $H = G(K/L)$. We also have $[K : L] = |H|$ and $[L : k] = [G : H]$. In addition, L/k is Galois iff H is normal in G, and in this case, $G(L/k) = G/H$.

Let $k[z]$ denote the ring of polynomials with coefficients in k in the variable z. Given a nonconstant monic polynomial $P \in k[z]$, a *splitting field* for P over k is an extension field K such that P factors into linear factors in K, and K is the smallest extension of k with this property. A splitting field K is generated over k by the roots of P. A splitting field always exists and is unique up to a k-isomorphism, an isomorphism between the splitting fields that fixes the common ground field k. A splitting field over k is always a Galois extension of k. Conversely, any Galois extension of k is the splitting field of some polynomial $P \in k[z]$. If $P \in k[z]$ is irreducible, then the Galois group G of the splitting field of P is uniquely determined by P. For this reason, we say that G is the *Galois group of P*.

Given a splitting field K of a polynomial $P \in k[z]$ of degree n over k with roots $z_1, \ldots, z_n$, the Galois group G acts on the roots faithfully, so that G can be thought of as a subgroup of the symmetric group $\mathcal{S}_n$ on n letters. The action on the roots is transitive iff P is irreducible over k. The *discriminant* δ of P is

$$\delta = \prod_{1 \le j < l \le n} (z_j - z_l)^2.$$

The discriminant is nonzero iff P has distinct roots. In particular, this is the case if P is irreducible (since otherwise, P and its derivative P' would have a common divisor). Being fixed by $\mathcal{S}_n$, δ is contained in k. In addition, $\sqrt{\delta} \in k$ iff G is a subgroup of the alternating group $\mathcal{A}_n \subset \mathcal{S}_n$. In general, we always have

$G(K/k(\sqrt{\delta})) \subset \mathcal{A}_n$, so that by adjoining $\sqrt{\delta}$ to k (at the expense of a quadratic extension) we can make G a subgroup of $\mathcal{A}_n$.

Let K/k be a Galois extension with Galois group G. Given $z^* \in K - k$, consider the orbit

$$G(z^*) = \{z_1^*, \ldots, z_{n^*}^*\}$$

and the polynomial

$$P^*(z) = \prod_{j=1}^{n^*}(z - z_j^*).$$

Then $P^* \in k[z]$ is irreducible and $[k(z^*) : k] = n^*$. In particular, n^* divides $|G|$. If K is the splitting field of a polynomial $P \in k[z]$ with roots $z_1, \ldots, z_n$, then z^* is a rational function of the roots $z_1, \ldots, z_n$. In this case we call P^* a *resolvent polynomial* of P.

EXAMPLE
Let K be the splitting field of an irreducible quartic $P \in k[z]$ with roots z_1, z_2, z_3, z_4. Let

$$z_1^* = (z_1 + z_2)(z_3 + z_4),$$
$$z_2^* = (z_1 + z_3)(z_2 + z_4),$$
$$z_3^* = (z_1 + z_4)(z_2 + z_3).$$

These elements are distinct, since P is irreducible (cf. the argument at the end of Section 6). The symmetric group $\mathcal{S}_4$ acts on $\{z_1^*, z_2^*, z_3^*\}$ transitively. If the Galois group $G \subset \mathcal{S}_4$ is also transitive on $\{z_1^*, z_2^*, z_3^*\}$, then the resolvent cubic

$$P^*(z) = \prod_{j=1}^{3}(z - z_j^*)$$

of P is irreducible over k, and G is $\mathcal{A}_4$ or $\mathcal{S}_4$ according as $\sqrt{\delta} \in k$ or $\sqrt{\delta} \notin k$.

Let $w \in \mathbf{C}$, and consider the splitting field of the polynomial

$$P(z) = z^n - w,$$

where $n \geq 2$ is an integer. We choose k to be generated over $\mathbf{Q}$ by w and the primitive nth root of unity $\omega = e^{2\pi i/n}$. The complex roots of P are the nth roots of w, and if z^* denotes one of them, then the roots of P can be listed as

$$z^*, z^*\omega, z^*\omega^2, \ldots, z^*\omega^{n-1}.$$

It follows that the splitting field K of P is generated over k by a single root: $K = k(z^*)$. If w is not an nth power in k, then $[K : k] = n$, and the Galois group $G(K/k)$ is a cyclic group of order n.

Theorem.

Let $n \geq 2$ be an integer. Let k be a field and assume that k contains the primitive nth root of unity $\omega = e^{2\pi i/n}$. Assume that K is a Galois extension of k with Galois group G, a cyclic group of order n. Then there exists $w^* \in k$ such that K is the splitting field of the polynomial $z^n - w^*$, and K is generated by any of the roots. Moreover, for a given root $z^* \in K$, $K = k(z^*)$, there is an isomorphism

$$\varphi : G \to C_n$$

such that $\sigma^{-1}(z^*) = \varphi(\sigma)(z^*)$, $\sigma \in G$.

Remark.

We call a solution $z^* \in K$ of the equation $z^n = w^*$, $w^* \in k$, a *radical* over k. We also write $z^* = \sqrt[n]{w^*}$.

In general, let $P \in k[z]$ and let K be the splitting field of P with Galois group G. Consider a composition series of G:

$$G = G_0 \supset G_1 \supset \cdots \supset G_N = \{1\}.$$

Here, each G_{m+1} is a maximal normal subgroup in G_m, and thus, the factor G_m/G_{m+1} is simple, $m = 0, \ldots, N - 1$. The composition series of G is not unique, but by the Jordan–Hölder theorem, up to order, the factor groups are. Consider the corresponding chain of field extensions

$$k = k_0 \subset k_1 \subset \cdots \subset k_N = K,$$

where k_m is the fixed field of G_m, $m = 1, \ldots, N - 1$. Since G_{m+1} is normal in G_m, k_{m+1} is a Galois extension of k_m with Galois group G_m/G_{m+1}. Being a Galois extension, k_{m+1} can be thought of as the splitting field of a polynomial P_m over k_m. We see that finding the roots of P amounts to solving the polynomial equations $P_m(z) = 0$, for all $m = 0, \ldots, N - 1$.

Assume now that G is solvable, that is, each factor G_m/G_{m+1} is abelian, thereby cyclic of (prime) order $p_m = |G_m/G_{m+1}|$. By the theorem above, k_{m+1} is generated over k_m by a radical $\sqrt[p_m]{w_m}$, $w_m \in k_m$. Applying this procedure inductively, we obtain that every element of K (including the roots $z_1, \ldots, z_n$ of P) can be expressed by radicals over k (in the obvious sense). Since the converse is clear, we see that the roots of a polynomial $P \in k[z]$ are expressible by radicals over k iff the Galois group of the splitting field of P over k is solvable.

All subgroups of $\mathcal{S}_4$ are solvable. It follows that the roots of polynomials of degree 4 or less are expressible by radicals. We derived the explicit root formulas in Section 6. In contrast, an irreducible quintic with Galois group $\mathcal{A}_5$ has no radical formula, since $\mathcal{A}_5$ is simple.

Solutions for 100 Selected Problems

Section 1.

1. (a) Write the number in the form $3k$, $3k + 1$, or $3k + 2$, take the squares of these integers, and consider the remainders modulo 3. (b) Write the number in the form $7k$, $7k \pm 1$, $7k \pm 2$, or $7k \pm 3$, take the cubes of these, and consider the remainders modulo 7.

2. By Problem 1(b), we need to show that $7a - 1$ is not a square. This follows as in 1(b) by squaring the numbers $7k$, $7k \pm 1$, $7k \pm 2$, and $7k \pm 3$, and studying the remainders modulo 7.

4. Let $a = a_1 10^{n-1} + a_2 10^{n-2} + \cdots + a_n$, where $a_1, a_2, \ldots, a_n \in \{0, 1, \ldots, 9\}$, $a_1 \neq 0$. Write $10 = 9 + 1$, and use the binomial theorem to expand the powers of 10 in the expression for a.

5. Since $\mathbf{Z}_p$ is a field, the statement is equivalent to the vanishing of

$$\sum_{a=1}^{p-1}[1/a^2] = \sum_{a=1}^{p-1}[a]^{-2} = \sum_{a=1}^{p-1}[\bar{a}]^2,$$

where $[\bar{a}]$ is the inverse of $[a]$ in $\mathbf{Z}_p$. Now notice that the last sum is the sum of squares of the nonzero elements in $\mathbf{Z}_p$.

425

Section 2.

1. $\sqrt[n]{a}$ is a rational solution of the equation $x^n = a$.

2. Write $\cos 1 = \sum_{k=0}^{\infty}(-1)^k/(2k+1)!$ and repeat, with appropriate modifications, the proof of irrationality of e as in the remark following Theorem 1.

3. If a/b is a root of $P(x) = 8x^3 - 6x - 1$, then $a = \pm 1$ and $b|8$. Now check all possible combinations.

4. Using the geometric series formula, a Riemann sum can be worked out as

$$\sum_{k=1}^{m}(\alpha^k)^n(\alpha^k - \alpha^{k-1}) = \frac{\alpha - 1}{\alpha}\sum_{k=1}^{m}\alpha^{(n+1)k}$$

$$= (\alpha - 1)\alpha^n\frac{\alpha^{m(n+1)} - 1}{\alpha^{(n+1)} - 1}$$

$$= \frac{a^{n+1} - 1}{\alpha^n + \alpha^{n-1} + \cdots + \alpha + 1}.$$

For $m \to \infty$, we have $\alpha = \sqrt[m]{a} \to 1$, so that the Riemann sum converges to

$$\frac{a^{n+1} - 1}{n + 1}.$$

5. This is a primary example for induction. For another proof, differentiate the geometric series formula with respect to x, substitute $x = 1$ and obtain $1 + 2 + \cdots + n = \lim_{x \to 1}\frac{d}{dx}\left(\frac{1-x^{n+1}}{1-x}\right)$. To evaluate the limit, use L'Hospital's rule.

7. As in Problem 5, differentiate the geometric series formula with respect to x twice and substitute $x = 1$.

8. The pyramid staircase can be cut into a large square pyramid of volume $n^3/3$, and two sets each consisting of n triangular prisms of heights $1, 2, \ldots, n$, common base area $1/2$, and total volume $(1 + 2 + \cdots + n)/2$. Notice that the two sets of prisms overlap in n small square pyramids with total volume $n/3$, and this has to be subtracted from the volume. The volume of the pyramidal staircase is thus $n^3/3 + (1 + 2 + \ldots + n) - n/3$. Now use Problem 5.

10. Write $a = \sum_{n=0}^{N} c_n 2^n$ with $c_n \in \{0, 1\}$. Notice that $c_n = 1$ iff a_n is odd.

11. (c) $1/\,Tri\,(n) = 2(1/n - 1/(n+1))$.

Section 3.

2. Split the triangle into 3 subtriangles by drawing line segments from the center of the inscribed circle to the vertices. The sum of the areas of the subtriangles, $ar/2 + br/2 + cr/2$, must be equal to the area of the original triangle, $ab/2$. We obtain $r = ab/(a+b+c)$. Without loss of generality we may assume that a, b, c are relatively prime. Since (a, b, c) is Pythagorean, we have $a = 2st$, $b = t^2 - s^2$, and $c = t^2 + s^2$. Substitute these into the expression of r and simplify.

3. We may assume that a, b, c are relatively prime. Let $a = 2st$, $b = t^2 - s^2$, and $c = t^2 + s^2$ with $t > s$ relatively prime and of different parity. Since $2|a$ and $2|st$, we have $4|a$. Write $ab = 2st(t-s)(t+s)$ and notice that 3 must divide one of the numbers $s, t, t-s$, or $t+s$. (This follows by considering the remainders of t and s modulo 3). Thus $12|ab$. Similarly, 5 divides one of the numbers $s, t, t-s, t+s$, or $t^2 + s^2$. (This follows again by looking at the remainders of s and t modulo 5.)

4. For a, b, c consecutive, we have $a + c = 2b$. Substituting this into the Pythagorean equation, we obtain $4a = 3b$. Since a, b are relatively prime, $a = 3$ and $b = 4$ follow.

5. We claim that there is no solution a for $b \geq 2$. Notice first that a and $a + 2$ have the same parity, so that $a + 1$ must be even, say, $a + 1 = 2k$. Substituting, we have

$$(2k - 1)^{2b} + (2k)^{2b} = (2k + 1)^{2b}.$$

Expanding both sides by the binomial formula and grouping, conclude that k must divide b. On the other hand, dividing all terms by $(2k)^{2b}$, we have

$$\left(1 - \frac{1}{2k}\right)^{2b} + 1 = \left(1 + \frac{1}{2k}\right)^{2b};$$

in particular, the estimate

$$\left(1 + \frac{1}{2k}\right)^{2b} < 2$$

holds. Use the binomial formula again to conclude that $b/k <$ 1. This contradicts $k|b$.

6. If (a, b, c) is a Pythagorean triple with $a + b + c = ab/2$, then $(a - 4)(b - 4) = 8$.

7. For $n \geq 4$, let $a = n(n^2 - 3)$ and $b = 3n^2 - 1$.

8. (a) $(0, 0)$ is a double point; $y^2 = x^2(x + 2)$. Setting $y = mx$, $m \in \mathbf{Q}$, we obtain $x = m^2 - 2$ and $y = m(m^2 - 2)$. (b) $(-1, 0)$ is a double point; $y^2 = (x + 1)^2(x - 2)$. Setting $y = m(x + 1)$, $m \in \mathbf{Q}$, we obtain $x = m^2 + 2$ and $y = m(m^2 + 2)$.

9. Let (x_0, y_0), $y_0 \neq 0$, be a rational point on the Bachet curve given by $y^2 = x^3 + c$. Implicit differentiation gives $2yy' = 3x^2$, so that the slope m of the tangent line at (x_0, y_0) is $m = 3x_0^2/2y_0$. We need to couple the equation of the tangent line $y - y_0 = m(x - x_0)$ with Bachet's equation $y^2 = x^3 + c$ to obtain the coordinates of the intersection point. Since $y_0^2 = x_0^3 + c$, we write Bachet's equation as $y^2 - y_0^2 = x^3 - x_0^3$. After factoring, we have $(y - y_0)(y + y_0) = (x - x_0)(x^2 + xx_0 + x_0^2)$. Substituting $y - y_0 = m(x - x_0)$ and canceling the factor $x - x_0$, we obtain

$$m(y + y_0) = x^2 + xx_0 + x_0^2.$$

Substituting again $y - y_0 = m(x - x_0)$, we finally arrive at

$$m(2y_0 + m(x - x_0)) = x^2 + xx_0 + x_0^2.$$

This is a quadratic equation in x that has x_0 as a root, since $m = 3x_0^2/2y_0$. Factoring, we obtain

$$(x - x_0)\left(x - \frac{x_0^4 - 8cx_0}{4y_0^2}\right) = 0.$$

Bachet's formula follows.

10. $(1, 1)$ is a rational point. Write the equation of the circle as $(x^2 - 1) + (y^2 - 1) = 0$ and use $y - 1 = m(x - 1)$, $m \in \mathbf{Q}$, when factoring.

12. (a) $a^2 = 2b^2$ implies that $(2b - a)^2 = 2(a - b)^2$.

14. $78 = 2 \cdot 3 \cdot 13 = ab/2$ and $a^2 + b^2 = c^2$. In particular, a, b are relatively prime, since $6^2 + 26^2 = 712$ is not a square. Setting $a = 2st$, $b = t^2 - s^2$, and $c = t^2 + s^2$, we obtain $2 \cdot 3 \cdot 13 = st(t - s)(t + s)$ with $t > s$ relatively prime and of different parity. This is impossible.

Section 4.

1. Multiply the polynomial by the least common multiple of the denominators of the coefficients.

3. If c is a root of a polynomial P with integer coefficients, then $\sqrt{c}$ is a root of the polynomial Q, where $Q(x) = P(x^2)$.

4. Let P and Q be polynomials of degrees m and n with rational coefficients and leading coefficient one such that $P(c_1) = Q(c_2) = 0$. Let P and Q have roots $c_1 = \alpha_1, \ldots, \alpha_m$ and $c_2 = \beta_1, \ldots, \beta_n$. Consider the polynomial $R(x) = \prod_{i=1}^{m} \prod_{j=1}^{n} (x - (\alpha_i + \beta_j))$. The coefficients of R are symmetric in α_i and β_j, and therefore, by the fundamental theorem of symmetric polynomials, they can be expressed as polynomials (with integral coefficients) in the coefficients of P and Q. In particular, the coefficients of R are rational. Since $R(c_1 + c_2) = 0$, $c_1 + c_2$ is algebraic.

5. An isomorphism would associate to $\sqrt{3}$ an element $a + b\sqrt{2}$ and thus to 3 the element $(a + b\sqrt{2})^2$. Since $3 = 1 + 1 + 1$ in both fields, in $\mathbf{Q}(\sqrt{2})$ we have $3 = (a + b\sqrt{2})^2$.

7. Since $(x - (a + b\sqrt{p}))(x - (a - b\sqrt{p})) = x^2 - 2ax + a^2 + b^2 p$, the remainder is a polynomial of degree less than or equal to 1 with rational coefficients.

10. The existence and uniqueness of x_0 follow from the properties of the exponential function. Let m be rational. If x_0 is rational, then so is $e^{x_0} = mx_0 + 1$, and this contradicts the corollary to Theorem 1 in Section 2. Transcendentality of x_0 for algebraic m follows from Lindemann's theorem.

Section 5.

1. The equation of the circle is

$$\left(x - \frac{a - c}{2a}\right)^2 + \left(y + \frac{b}{2a}\right)^2 = \frac{(a - c)^2 + b^2}{4a^2}.$$

Setting $x = 1$ and solving for y, we obtain the quadratic formula for r_1 and r_2. A generalization of this construction to cubic polynomials would mean that the roots of a cubic polynomial with constructible coefficients would be constructible. This is false (cf. Problem 4 in Section 6).

2. $\sqrt[3]{-i} = \sqrt[3]{z(3\pi/2)} = z(\pi/2 + 2k\pi/3), \; k = 0, 1, 2.$ These give i and $\pm\sqrt{3}/2 - i/2$.

3. The two lines are parallel iff $(z_1 - z_2)/(z_3 - z_4)$ is real. Thus, we may assume that the two lines intersect. The condition $\Re\big((z_1 - z_2)/(z_3 - z_4)\big) = 0$ is invariant under translation (which amounts to adding a constant to each variable), so that we may assume that the lines intersect at the origin. Then z_1, z_2 and z_3, z_4 are real constant multiples of each other, and the condition reduces to $\Re(z_1/z_3) = 0$. Now write z_1 and z_3 in polar form and consider the argument of the ratio z_1/z_3.

5. The condition remains invariant when we translate, rotate, and scale each triangle individually. (For example, translation corresponds to the row operation of adding a constant multiple of the first row to the row representing the triangle.) We can thus reduce the problem to the case where $z_1 = w_1 = 0$ and $z_2 = w_2 = 1$. The condition now reduces to $z_3 = w_3$.

6. $2 \cdot 3 = (1 + \sqrt{5}i)(1 - \sqrt{5}i)$.

7. Letting $z_1 = a_1 + b_1 i$, and $z_2 = a_2 + b_2 i$, $a_1, a_2, b_1, b_2 \in \mathbf{N}$, we have $|z_1|^2|z_2|^2 = (a_1^2 + b_1^2)(a_2^2 + b_2^2) = |z_1 z_2|^2 = (a_1 a_2 - b_1 b_2)^2 + (a_1 b_2 + a_2 b_1)^2$. Any integer can be written as a power of 2 multiplied by numbers of the form $4k + 1$ and $4k + 3$. Now apply Fermat's theorem and the above.

Section 7.

1. (a) $N = (0, 0, 1), p = (a, b, c),$ and $h_N(p) = \big(a/(1 - c), b/(1 - c), 0\big)$ are collinear. (c) The stated expression for $h_N^{-1}(z)$ has unit length, and $N, z,$ and $h_N^{-1}(z)$ are collinear.

2. As in Problem 1, for $p = (a, b, c) \in S^2$, we have

$$h_S(p) = \frac{a + bi}{1 + c} \in \mathbf{C}.$$

Moreover,

$$h_S^{-1}(z) = \left(\frac{2z}{|z|^2 + 1}, -\frac{|z|^2 - 1}{|z|^2 + 1}\right) \in S^2, \; z \in \mathbf{C}.$$

Hence,

$$(h_N \circ h_S^{-1})(z) = \frac{2z/(|z|^2 + 1)}{1 + (|z|^2 - 1)/(|z|^2 + 1)} = \frac{z}{|z|^2}.$$

4. We have

$$|z - w|^2 = (z - w)(\bar{z} - \bar{w}) = |z|^2 - z\bar{w} - \bar{z}w + |w|^2$$

and

$$|1 - z\bar{w}|^2 = (1 - z\bar{w})(1 - \bar{z}w) = 1 - z\bar{w} - \bar{z}w + |z|^2|w|^2.$$

The stated inequality is thus equivalent to $0 < (1 - |z|^2)(1 - |w|^2)$.

Section 8.

4. (a) If P is an odd-degree polynomial with real coefficients, then $\lim_{x \to \pm\infty} P(x) = \pm \operatorname{sgn}(c_n) \infty$, where $\operatorname{sgn}(c_n)$ is the sign of the leading coefficient c_n of P. By continuity, P must have a real zero. (b) The polynomial Q_k, $k \in \mathbf{Z}$, has coefficients that are, up to sign, the elementary symmetric polynomials in the variables $\alpha_i + \alpha_j + k\alpha_i\alpha_j$, $1 \le i < j \le n$. But an elementary symmetric polynomial in these variables is symmetric in α_i, $i = 1, \ldots, n$, and hence it can be written as a real polynomial in the coefficients of P. Thus the coefficients of Q_k are real. The degree of Q_k is $n(n - 1)/2 = 2^{m-1}a(2^m a - 1)$, $m \ge 1$. Since $a(2^m a - 1)$ is odd, the induction hypothesis applies. Thus, for each k, there are indices $1 \le i < j \le n$ such that the root $\alpha_i + \alpha_j + k\alpha_i\alpha_j$ is a complex number. Since there are finitely many roots α_i but infinitely many choices of k, for some $1 \le i < j \le n$, $\alpha_i + \alpha_j$ and $\alpha_i\alpha_j$ are both complex numbers. The polynomial $z^2 - (\alpha_i + \alpha_j)z + \alpha_i\alpha_j = (z - \alpha_i)(z - \alpha_j)$ has complex coefficients, so that by the complex form of the quadratic formula, α_i and α_j are complex. (c) If P is a complex polynomial, then $P\bar{P}$ has real values; in particular, its coefficients must be real. By the above, $P\bar{P}$ has a complex root α. Since $|P|^2 = P\bar{P}$, α is also a complex root of P.

5. If z is a root of P with real coefficients, then $\overline{P(z)} = P(\bar{z}) = 0$, so that $\bar{z}$ is also a root of P. For another proof, divide P by the

quadratic polynomial $(x - z)(x - \bar{z}) = x^2 - 2\Re(z)x + |z|^2$ and study the remainder.

Section 9.

1. The regular hexagon is obtained by slicing the cube with a plane through the origin with normal vector $(1, 1, 1)$. The plane extensions of the three sides of the cube meeting at $(1, 1, 1)$ cut an equilateral triangle out of this plane, and the plane extensions of the other three sides of the cube meeting at $(-1, -1, -1)$ further truncate this triangle to a regular hexagon.

2. $T_v \circ R_\theta(p) \circ T_{-v}$ is a direct isometry, and it leaves q fixed. Thus it must be a rotation with center at q. The translations do not change the angle.

3. (a) The vertices of P_{2n} are those of P_n plus the midpoints of the circular arcs over the sides of P_n. Thus, half of a side of P_n and a side of P_{2n} meeting at a vertex of P_n are two sides of a right triangle whose third side is $1 - \sqrt{1 - (s_n/2)^2}$. We thus have

$$s_{2n}^2 = \left(\frac{s_n}{2}\right)^2 + \left(1 - \sqrt{1 - \left(\frac{s_n}{2}\right)^2}\right)^2.$$

This gives

$$s_{2n} = \sqrt{2 - \sqrt{4 - s_n^2}}.$$

In particular, since $s_4 = \sqrt{2}$, we have

$$s_{2^n} = \sqrt{2 - \sqrt{2 + \sqrt{2 + \cdots + \sqrt{2}}}},$$

with $n - 1$ nested square roots. (b) Let A_n denote the area of P_n. Since half of the sides of P_n serve as heights of the $2n$ isosceles triangles that make up P_{2n}, we have $A_{2n} = ns_n/2$. In particular, we have

$$A_{2^{n+1}} = 2^{n-1}s_{2^n} = 2^{n-1}\sqrt{2 - \sqrt{2 + \sqrt{2 + \cdots + \sqrt{2}}}},$$

with $n - 1$ nested square roots.

5. (a, b) is the normal vector of the reciprocal line. Thus, by symmetry, it is enough to consider the reciprocal of $z(0) = 1$. The radius of the circle inscribed in P_n is $r = |1 + z(2\pi/n)|/2$, and the line reciprocal to 1 is $x = r^2$. Now the statement follows from the identity $\Re(1 + z(\theta))/2 = |1 + z(\theta)|^2/4$.

Section 10.

2. Type 5.
3. Let $L = \{kv + lw \mid k, l \in \mathbf{Z}\}$ be a lattice. A half-turn about the midpoint of the lattice points $k_1v + l_1w$ and $k_2v + l_2w$ sends a point p to $(k_1 + k_2)v + (l_1 + l_2)w - p$.
4. Type 2.

Section 11.

1. Let $G \subset M\ddot{o}b\,(\hat{\mathbf{C}})$ be the group generated by isometries, dilatations with center at the origin, and one reflection in a circle. Since G contains all translations, G also contains all dilatations with arbitrary center, and all reflections in circles concentric to the given circle. Finally, using the translations again, G contains reflections in all circles with arbitrary center. Thus, $G = M\ddot{o}b\,(\hat{\mathbf{C}})$.

Section 12.

5. A linear fractional transformation that maps the unit disk D^2 to the upper half-plane H^2 is $z \mapsto i\frac{i+z}{i-z}$. Its inverse is $z \mapsto -i\frac{i-z}{i+z}$.
6. This transformation is a glide with axis the real axis, and the translation vector is 1.
7. Setting $z = |z|z(\theta)$, the image of a circle $|z| = r < 1$ is an ellipse with semimajor axis $r + 1/r$ and semiminor axis $r - 1/r$. The image of a half-line emanating from the origin is a hyperbola. The ellipses and the hyperbolas are confocal.

Section 13.

1. Problem 4 of Section 7 implies that the given linear fractional transformations are self maps of the unit disk D^2. Theorem 9 asserts that the group of linear fractional transformations of H^2

onto itself is 3-dimensional (with $a, b, c, d \in \mathbf{R}$ being the parameters subject to the constraint $ad - bc = 1$). Since D^2 and H^2 are equivalent through a linear fractional transformation (Problem 5 of Section 12), it follows that the group of linear fractional transformations that leave D^2 invariant is also 3-dimensional. On the other hand, the linear fractional transformations given in the problem form a group, and they depend on 3 parameters, $\Re(w)$, $\Im(w)$, and θ. Thus, these give all the linear fractional transformations preserving D^2.

4. We may assume that the triangle has vertices $i, ti, t > 1$, and $z(\theta), 0 < \theta < \frac{\pi}{2}$. Let α and β be the angles at $z(\theta)$ and at ti.
   (a) From the hyperbolic distance formula, we obtain

$$\cosh a = \frac{t^2 + 1}{2t}, \qquad \cosh b = \frac{1}{\sin \theta}, \qquad \cosh c = \frac{t^2 + 1}{2t \sin \theta}.$$

The Pythagorean theorem follows. (b) Let $r > 0$ be the radius and $c \in \mathbf{R}$ the center of the semicircle that represents the hyperbolic line through ti and $z(\theta)$. Let δ be the angle at c between the radial segment connecting c and $z(\theta)$, and the real axis. We have

$$\sin \beta = \frac{t}{r}, \qquad r \cos \delta = r \cos \beta + \cos \theta, \qquad r \sin \delta = \sin \theta.$$

The identity $\cos^2 \delta + \sin^2 \delta = 1$ gives

$$r^2 = r^2 \cos^2 \beta + 2r \cos \beta \cos \theta + 1,$$

so that we have

$$t^2 = r^2 \sin^2 \beta = 2r \cos \beta \cos \theta + 1.$$

Using this and $\alpha = \theta - \delta$, we compute

$$\cos \alpha = \cos(\theta - \delta) = \cos \theta \cos \beta + \frac{1}{r} = \frac{t}{r}\left(\frac{t^2 - 1}{2t} + \frac{1}{t}\right)$$

$$= \sin \beta \cosh a.$$

Switching the roles of a, b and α, β, we also have

$$\cos \beta = \sin \alpha \cosh b.$$

Now the identity

$$\sinh a \tan \beta = \tanh b$$

follows by eliminating α.

Section 14.

1. (a) A parabolic isometry g is the composition of two reflections in hyperbolic lines that meet at a common endpoint on the boundary of H^2. Conjugating g with an isometry that carries this endpoint to ∞, the two hyperbolic lines become vertical Euclidean lines, and the conjugated g has the form $z \mapsto z + a$, $a \in \mathbf{R}$. Now conjugate this with $z \mapsto \frac{1}{a}z$ to obtain $z \mapsto z + 1$.

2. Parabolic isometries have a unique fixed point on $\mathbf{R} \cup \{\infty\}$. Elliptic isometries have a unique fixed point in H^2. Hyperbolic isometries have two fixed points on $\mathbf{R} \cup \{\infty\}$.

3. Since a linear fractional transformation determines the corresponding matrix in $SL(2, \mathbf{C})$ up to sign, trace^2 is well-defined. Given an isometry g, the condition

$$g(z) = \frac{az + b}{cz + d} = z, \qquad a, b, c, d \in \mathbf{R},$$

for z to be a fixed point of g amounts to solving a quadratic equation with discriminant $(a - d)^2 - 4bc = (a + d)^2 - 4 = $ trace$^2(g) - 4$. If trace$^2(g) = 4$, then g has a unique fixed point on $\mathbf{R} \cup \{\infty\}$. Conjugating g by a suitable isometry, we may assume that this fixed point is ∞. This means that $c = 0$ and $b \neq 0$ (by unicity of the fixed point). As in Problem 1, we obtain that g is parabolic. If $0 \leq$ trace$^2(g) < 4$, then there are two fixed points of g, and they are conjugate complex numbers. Thus, there is a unique fixed point in H^2. The argument in the text shows that g is elliptic. Finally, if trace$^2(g) > 4$, then g has two fixed points on $\mathbf{R} \cup \{\infty\}$, which may be assumed to be 0 and ∞. We thus have $b = c = 0$, and g is hyperbolic.

5. Any positive integral power of a parabolic or hyperbolic isometry is of the same type.

6. If $z_0 \in \mathbf{R} \cup \{\infty\}$ is the unique fixed point of the parabolic g_1, then by the stated commutativity, $g_2(z_0)$ is left fixed by g_1, and we must have $g_2(z_0) = z_0$. Thus, g_2 is either parabolic or hyperbolic.

If $w_0 \in \mathbf{R} \cup \{\infty\}$ were another fixed point of g_2, then $g_1(w_0)$ would also be left fixed by g_2. This is a contradiction, since $g_1(w_0) \neq w_0, z_0$.

7. As shown in the text, the conformal group corresponding to $T = T_1$ is $SL(2, \mathbf{Z})$. It remains to consider the conformal group defined by T_2. The composition of reflections to the vertical sides of T_2 gives $z \mapsto z + 1$. The composition of $z \mapsto -\bar{z} + 1$ and $z \mapsto 1/\bar{z}$ gives $z \mapsto -1/z + 1$, so that $z \mapsto -1/z$ is also contained in this conformal group.

8. This follows from Example 8 by minor modifications.

Section 15.

2. Using the Euler formula for complex exponents, we have

$$\sin \theta = \frac{e^{i\theta} - e^{-i\theta}}{2i}.$$

The complex extension of sine is therefore defined by

$$\sin z = \frac{e^{iz} - e^{-iz}}{2i}.$$

Since the exponential function is periodic with period $2\pi i$, the sine function is periodic with period 2π. To show that $\sin z$ is one-to-one on any vertical strip $(k-1/2)\pi < \Re(z) < (k+1/2)\pi$, $k \in \mathbf{Z}$, write the difference of two sines as the product of a sine and a cosine, relate these to various complex exponentials, and recall the mapping properties of e^z on horizontal strips.

Section 16.

1. (b) Let $m_0 \in S^2$ be the midpoint of the shorter great circular arc connecting q_0 and $S(q_0)$. If $S^2(q_0) \notin C$, then consider the spherical triangle $\triangle m_0 S(q_0) S(m_0)$ and get a contradiction to the triangle inequality.

Section 17.

1. Label the vertices of the tetrahedron 1, 2, 3, 4. The three products of disjoint transpositions such as $(12)(34)$ correspond to the three half-turns around the midpoints of the three pairs

of opposite edges. In addition, we have four 60° rotations corresponding to products such as $(12)(13) = (123)$, and four 120° rotations corresponding to products such as $(13)(12) = (132)$.

2. (a) Label the vertices of the tetrahedron such that p corresponds to 1, e connects 1 and 2, and f has vertices 1, 2, 3. Then S_1 corresponds to the permutation (243), and S_2 corresponds to (132). The product is $(13)(24)$, and it corresponds to a half-turn. (c) Let S be the half-turn with axis connecting the midpoints of e_1 and e_2. The midpoints of the edges complementary to e_1 and e_2 form a quadrangle that has equal sides and parallel opposite edges. Since e_1 and e_2 are perpendicular, the quadrangle also has perpendicular adjacent edges, so it is a square. The quarter-turn around the axis of S followed by reflection in the plane of the square carries the two pieces into one another.

3. The symmetry groups of $\mathcal{T}$ and its reciprocal are the same.

4. The vertices of a pair of reciprocal tetrahedra are those of a cube. Take a good look at Color Plate 5a.

5. (a) Notice that the midpoint of an edge in a unit square has distance $\sqrt{5}/2$ from an opposite vertex.

7. (c) Use the fact that $Symm^+(\mathcal{I}) \cong \mathcal{A}_5$ is simple. For the image, compose ϕ with the even–odd homomorphism $\mathcal{S}_5 \to \{\pm 1\}$.

8. By the construction of the dodecahedron in the text, every vertex of the dodecahedron is the vertex of at most two cubes. The 5 inscribed cubes have the total of 40 vertices. Since the dodecahedron has 20 vertices, it follows that every vertex of the dodecahedron is the vertex of exactly two cubes. These two cubes have a common diagonal. There are exactly $20/2 = 10$ diagonals. The number of different pairs of cubes is also 10. Thus, the statement is true.

9. No symmetry carries the polar vertices to the equatorial vertices.

13. Since the vertex figures are regular, all the faces meeting at a vertex must have equal side lengths and face angles. It follows that all the faces are congruent (and regular). Reciprocally, all vertex figures must be congruent (and regular). Thus, all dihedral angles of adjacent faces must be equal.

15. (a) The sum is a vector left fixed by the entire symmetry group of the Platonic solid. Thus, it must be zero. (b) Any symmetry carries midpoints of edges to midpoints of edges. (c) Same as in (b). (d) The two Platonic solids are reciprocal. This follows by considering their Schläfli symbols.

16. Notice that at each vertex two hexagons and one pentagon meet.

17. (a) In Figure 17.28, the triangles $\triangle p_0 p_3 p_4$ and $\triangle p_1 p_3 p_4$ are similar (by the definition of the golden section). Thus all three angles at p_3 are congruent and thereby equal to $\pi/5$. It follows that the segment connecting p_0 and p_4 is the side of a decagon inscribed in a unit circle. (b) We may assume that the line segment from the midpoint of a side (of the inscribed triangle) to the circle has unit length. Insert two crucial right triangles and study their intersection.

18. The vertices of the octahedron are the midpoints of the faces of the cube circumscribed around the reciprocal pair of tetrahedra. The vertex figure at a vertex v of the octahedron is parallel to the face of the cube whose midpoint is v. Homothety of the two cubes follows. Since the edges of the reciprocal cube bisect those of the octahedron perpendicularly, the ratio of magnification is $\frac{1}{2}$.

19. The 20 vertices of the 5 tetrahedra are vertices of a regular polyhedron with symmetry group $\mathcal{A}_5$. Thus, it must be a dodecahedron. At a fixed vertex v of the colored icosahedron all 5 colors are represented. Consider the 5 triangular faces, one for each tetrahedron, that are extensions of the 5 icosahedral faces meeting at v. The 5 vertices closest to v in each of these triangular faces form a regular pentagon, since the order-5 rotation with axis through v permutes these vertices cyclically. The pentagon is perpendicular to this axis and hence parallel to the vertex figure of the icosahedron at v. Homothety of the two dodecahedra follows.

20. Take slices of the roof perpendicular to the base and use the Pythagorean theorem twice to conclude that the height of the roof is

$$\sqrt{1 - (\tau/2)^2 - ((\tau - 1)/2)^2} = 1/2.$$

Look for similar right triangles and write the defining equality of the golden section as

$$\frac{\tau/2}{1/2} = \frac{1/2}{(\tau - 1)/2}.$$

22. Inscribe a Euclidean dodecahedron in D^3 with vertices $v_1, \ldots, v_{20} \in S^2$. By the hyperbolic geometry of D^3, for $0 < t < 1$ there is a unique hyperbolic dodecahedron in D^3 with vertices $tv_1, \ldots, tv_{20}$. Let $\delta(t)$ be the common dihedral angle of this hyperbolic dodecahedron. The angle δ is a continuous decreasing function on $(0, 1)$. We have $\lim_{t \to 1} \delta(t) = \pi/3 < \pi/2$ (by symmetry), and the limit $\delta_0 = \lim_{t \to 0} \delta(t)$ is the dihedral angle of the Euclidean dodecahedron. We have $\tan \delta_0/2 = \tau$, where τ is the golden section. (Take a good look at Figure 17.30: Focus on the top horizontal edge of the icosahedron, and superimpose the reciprocal dodecahedron.) Thus, we have $\tan \delta_0 = 2\tau/(1 - \tau^2) = -2$, so that $\delta_0 = \arctan(-2) > \frac{\pi}{2}$. By continuity, there is a value $t_0 \in (0, 1)$ for which $\delta(t_0) = \frac{\pi}{2}$. The fact that hyperbolic dodecahedra with this dihedral angle tessellate D^3 follows, since at each vertex the three faces are mutually prependicular.

Section 19.

1. $\mathbf{R}P^2 \# \mathbf{R}P^2$ is homeomorphic to K^2. Cut a hole in $T^2 \# \mathbf{R}P^2$ so that it becomes the connected sum of a Möbius strip and a torus. Cut a hole in $\mathbf{R}P^2 \# \mathbf{R}P^2 \# \mathbf{R}P^2 = K^2 \# \mathbf{R}P^2$ so that it becomes the connected sum of a Möbius strip and a Klein bottle. Compare.

Section 20.

6. Use the Jordan curve theorem.
11. Find a state surrounded by an odd number of states.
12. If we assume that no more than three countries touch at one point (so that the corresponding graph is a triangulation), then a necessary and sufficient condition is that each country have an even number of neighbors.

13. Four colors are needed for a checkerboard, but no finite number will suffice for every map.

14. Six.

15. Take a complete graph on 27 vertices, and color each edge red or green. Then we can find a subgraph that is either (i) a triangle with all edges colored green, or (ii) a complete graph on eight vertices with all edges colored red.

16. Use a stereographic projection to go between maps on the sphere and maps on the plane. One point (the point at infinity) will be missing from the sphere, but that does not affect the number of colors, because countries that touch at only one point are not required to have different colors.

17. Establish customs stations at every frontier between neighboring countries, and build roads from each customs station to the capitals of both countries. If two roads should cross, rebuild them by changing the connections so that they do not cross. The capitals are the vertices and the roads are the edges of a planar graph. (The roads should not cross any frontiers except at the customs stations. This can be done if every country is pathwise connected.)

18. The chromatic number is n.

19. Consider the graph whose vertices are those of a regular pentagon plus the centroid, and whose edges are the sides of the pentagon plus the five radial segments from the centroid to the vertices of the pentagon.

20. There cannot be a vertex of degree 3; else we could remove that vertex, color the remaining graph with four colors, replace the vertex, and color it differently from its three neighbors. If there is a vertex of degree 4, and we use four colors to color every other vertex, then the four neighbors of the vertex of degree 4 must require four different colors, say red, green, yellow, and blue; otherwise, the missing color could be used to color the vertex of degree 4. We can now use the same argument as in the proof of the five color theorem, looking at red–green paths or yellow–blue paths, and swapping either red and green or yellow and blue for part of the graph.

21. The red–blue path might cross the red–green path in Figure 20.6.

Section 21.

1. Let $v \in \mathbf{R}^2$, $v \neq 0$. Since both $J_1 v$ and $J_2 v$ are perpendicular to v, we have $J_1 v = \pm J_2 v$. Moreover, $J_1^2 v = J_2^2 v = -v$. Since a linear map is uniquely determined by its values on a basis, we have $J_1 = \pm J_2$.
2. Define $P^{\alpha\beta}$ for $\alpha \neq \beta$, $\alpha, \beta = 1, \ldots, n$, using the relations in the text.

Section 22.

1. The map is a reflection in the plane $\mathbf{R} \times \mathbf{R} \cdot k$ in V.
5. This follows by adjusting the argument in the text to the quaternionic case. The action of $S^3 \subset \mathbf{H}$ on $\mathbf{H}^2$ is given by $g : (p, q) \mapsto (p \cdot g^{-1}, q \cdot g^{-1})$, $g \in S^3$.
6. (a) The parallel of latitude at r corresponds to rotations with fixed angle $\theta = 2 \cos^{-1}(r)$. (b) The group of spherical rotations and $S^3/\{\pm 1\}$ are isomorphic. $SU(2) = S^3$ is a subgroup of $SL(2, \mathbf{C})$, and $M\ddot{o}b\,(\hat{\mathbf{C}}) = SL(2, \mathbf{C})/\{\pm 1\}$.

Section 23.

2. A rotation generating the orbit of the 5 tetrahedra circumscribed around the colored icosahedron has angle $2\pi/5$. Its axis lies in the coordinate plane orthogonal to the first axis, and the slope of the axis is the golden section τ (see Figure 17.30). This is because at the common vertex $(0, 1, \tau)$ of the icosahedron and a golden rectangle all 5 colors are represented. This axis goes through

$$\left(0, \ \frac{1}{\sqrt{\tau^2 + 1}}, \ \frac{\tau}{\sqrt{\tau^2 + 1}} \right).$$

Using the notation in Problem 6 of Section 22, we have $\theta = 2\pi/5$ and

$$a = \cos\left(\frac{\pi}{5}\right) = \frac{\tau}{2},$$

$$b = \frac{\tau}{\sqrt{\tau^2 + 1}} \sin\left(\frac{\pi}{5}\right) = \frac{1}{2},$$

$$c = \frac{1}{\sqrt{\tau^2 + 1}} \sin\left(\frac{\pi}{5}\right) = \frac{1}{2\tau},$$
$$d = 0.$$

Hence, the pair of quaternions that corresponds to this rotation is

$$\pm \frac{1}{2}\left(\tau + i + \frac{j}{\tau}\right).$$

Section 26.

1. A vertex figure of the 4-dimensional cube is a regular 3-dimensional polyhedron with four vertices, since exactly four edges meet at a vertex. It is thus a regular tetrahedron. The 3-dimensional hyperplane through the origin with normal vector $(1, 1, 1, 1)$ gives an octahedral slice of the cube; the four 3-dimensional cubic faces that meet at $(1, 1, 1, 1)$ intersect this hyperplane in a regular tetrahedron, while the other four cubic faces meeting at $(-1, -1, -1, -1)$ further truncate this tetrahedron to yield the octahedron. For an analogy, consider Problem 1 in Section 9.

2. (a) Take a good look at Figure 26.2. The Schläfli symbol $\{4, 3, 3\}$ means that the 3-dimensional faces of the 4-dimensional cube are ordinary 3-dimensional cubes with Schläfli symbol $\{4, 3\}$, and each edge is surrounded by 3 of these. (b) Let V_n, E_n, and F_n denote the number of vertices, edges, and faces of the n-dimensional cube. We have $V_n = 2^n$, $E_{n+1} = 2E_n + V_n$, $E_1 = 1$, and $F_{n+1} = 2F_n + E_n$, $F_1 = 0$. These recurrences can be solved easily, and they give $E_n = n2^{n-1}$ and $F_n = \frac{n(n-1)}{2}2^{n-2}$.

3. (a) A 2-dimensional projection of the 4-dimensional tetrahedron is the pentagram star (top of Figure 20.4). (b) Each edge is surrounded by 3 ordinary tetrahedral faces; thus the Schläfli symbol is $\{3, 3, 3\}$.

4. Consider the general form of a quadratic polynomial in 3 variables, and impose harmonicity.

Index

Undergraduate Texts in Mathematics

Abbott: Understanding Analysis.

Anglin: Mathematics: A Concise History and Philosophy.
Readings in Mathematics.

Anglin/Lambek: The Heritage of Thales.
Readings in Mathematics.

Apostol: Introduction to Analytic Number Theory. Second edition.

Armstrong: Basic Topology.

Armstrong: Groups and Symmetry.

Axler: Linear Algebra Done Right. Second edition.

Beardon: Limits: A New Approach to Real Analysis.

Bak/Newman: Complex Analysis. Second edition.

Banchoff/Wermer: Linear Algebra Through Geometry. Second edition.

Berberian: A First Course in Real Analysis.

Bix: Conics and Cubics: A Concrete Introduction to Algebraic Curves.

Brémaud: An Introduction to Probabilistic Modeling.

Bressoud: Factorization and Primality Testing.

Bressoud: Second Year Calculus.
Readings in Mathematics.

Brickman: Mathematical Introduction to Linear Programming and Game Theory.

Browder: Mathematical Analysis: An Introduction.

Buchmann: Introduction to Cryptography.

Buskes/van Rooij: Topological Spaces: From Distance to Neighborhood.

Callahan: The Geometry of Spacetime: An Introduction to Special and General Relativity.

Carter/van Brunt: The Lebesgue–Stieltjes Integral: A Practical Introduction.

Cederberg: A Course in Modern Geometries. Second edition.

Childs: A Concrete Introduction to Higher Algebra. Second edition.

Chung: Elementary Probability Theory with Stochastic Processes. Third edition.

Cox/Little/O'Shea: Ideals, Varieties, and Algorithms. Second edition.

Croom: Basic Concepts of Algebraic Topology.

Curtis: Linear Algebra: An Introductory Approach. Fourth edition.

Devlin: The Joy of Sets: Fundamentals of Contemporary Set Theory. Second edition.

Dixmier: General Topology.

Driver: Why Math?

Ebbinghaus/Flum/Thomas: Mathematical Logic. Second edition.

Edgar: Measure, Topology, and Fractal Geometry.

Elaydi: An Introduction to Difference Equations. Second edition.

Exner: An Accompaniment to Higher Mathematics.

Exner: Inside Calculus.

Fine/Rosenberger: The Fundamental Theory of Algebra.

Fischer: Intermediate Real Analysis.

Flanigan/Kazdan: Calculus Two: Linear and Nonlinear Functions. Second edition.

Fleming: Functions of Several Variables. Second edition.

Foulds: Combinatorial Optimization for Undergraduates.

Foulds: Optimization Techniques: An Introduction.

Franklin: Methods of Mathematical Economics.

Frazier: An Introduction to Wavelets Through Linear Algebra.

Gamelin: Complex Analysis.

Gordon: Discrete Probability.

Hairer/Wanner: Analysis by Its History.
Readings in Mathematics.

Halmos: Finite-Dimensional Vector Spaces. Second edition.

Undergraduate Texts in Mathematics

Undergraduate Texts in Mathematics